Fundamental Food Microbiology

Bibek Ray

CRC Press
Boca Raton Boston London New York Washington, D.C.

Acquiring Editor: Robert A. Stern
Project Editor: Les Kaplan
Marketing Manager: Becky McEldowney
Cover design: Dawn Boyd
PrePress: Greg Cuciak
Manufacturing: Sheri Schwartz

Library of Congress Cataloging-in-Publication Data

Ray, Bibek
Fundamental food microbiology / Bibek Ray
p. cm.
Includes bibliographical references and index.
ISBN 0-8493-9442-2
1. Food—Microbiology. I. Title.
QR115.R39 1996
576′.163—dc20 95-4889
CIP

Direct all inquiries to CRC Press LLC, 2000 Corporate Blvd., N.W., Boca Raton, Florida 33431.

International Standard Book Number 0-8493-9442-2
Library of Congress Card Number 95-4889
Printed in the United States of America 2 3 4 5 6 7 8 9 0
Printed on acid-free paper

Dedication

To my parents, Hem and Kiron, and my family.

The Author

Bibek Ray, Ph.D., is professor of food microbiology in the Department of Animal Science at the University of Wyoming, Laramie. Professor Ray obtained a B.S. and M.S. in Veterinary Science from the University of Calcutta and University of Madras, in India, respectively. He received his Ph.D. in Food Science from the University of Minnesota in 1970 and joined the faculty in the Department of Food Science, North Carolina State University, and then the Department of Biology at Shaw University, both at Raleigh, North Carolina. He joined the University of Wyoming, Laramie, in 1981. At the University of Wyoming, he expanded his research to intestinal beneficial bacteria and bacteriocins of lactic acid bacteria along with his previous research activities in the area of microbial sublethal injury. He also teaches courses in Food Microbiology, Food Fermentation, and Food Safety. His laboratory has been involved in extensive and thorough studies in both basic and applied areas of bacteriocin, pediocin AcH from *Pediococcus acidilactici* H. In addition, his group is studying various aspects of bacteriocins produced by *Lactococcus, Leuconostoc, Lactobacillus,* and *Pediococcus*, as well as *Bacillus* and *Staphylococcus* spp. He has received research funding from the National Science Foundation, American Public Health Association, National Live Stock and Meat Board, U.S. Department of Agriculture, North Atlantic Treaty Organization (with Turkey), and Binational Agriculture Research Development Agency (with Israel), Wyoming Development Fund, and industry funds. At present, he is studying the combined effect of bacteriocins, ultrahigh hydrostatic pressure and pulse field electricity, and sublethal injury for the destruction of microbial cells and spores and its application in food preservation. This project is funded by the U.S. Army. In addition, Dr. Ray has been engaged in collaborative research programs with research institutes/universities in Turkey, Israel, India, and France.

Professor Ray has published over 100 research articles, reviews, book chapters, proceedings articles, and popular articles in the area of Food Microbiology. He has also edited two books: *Injured Index and Pathogenic Bacteria* (1989) and *Food Biopreservatives of Microbial Origin* (1992 with Dr. M. A. Daeschel), both published by CRC Press, Boca Raton, Florida. He is a member of the American Society for Microbiology, Institute of Food Technologists, and International Association of Milk Food and Environmental Sanitation. He is a Fellow of the American Academy of Microbiology and on the Editorial Board of the *Journal of Food Protection*. In 1994, Professor Ray was awarded the University of Wyoming Presidential Achievement Award in recognition of his excellence in academic performance.

The Authors

Preface

Between the time I first studied food microbiology as an undergraduate student and now the discipline has undergone a radical change. This change is well expressed by Dr. David Mossel of the Netherlands in his letter published in *ASM News* (59(10), 1993): from "no challenge in plate count and coliform scouting" to "linkage of molecular biology to food safety (also food bioprocessing and food stability) strategies — proclaim a new era in Food Microbiology." This transition was necessary to meet the changes that occurred in the food industry, especially in the United States and other developed countries. The necessary knowledge, techniques, and expertise for this transition were all available. This book reflects this transition from the traditional approach to an approach to meet the needs of those who are directly or indirectly interested in Food Microbiology.

Introductory Food Microbiology is a required course for undergraduates majoring in Food Science. In some form it is also taught in several other programs, namely Microbiology, Public Health, Nutrition and Dietetics, Veterinary Science, and others. For the majority of food scientists, except those majoring in Food Microbiology, this single course forms the basis of the study microorganisms and their interactions to food. Similarly, for the latter groups, Food Microbiology is probably the only course that provides with information on the interaction of food and microorganisms. This book was written with the major objective of relating interaction of microorganisms and food in relation to food bioprocessing, food spoilage, and foodborne diseases. Thus, it will be useful as a text in the introductory food microbiology courses taught under various programs and disciplines. In addition, it will be a valuable reference book for those who directly and indirectly are involved in food and microbiology, which includes individuals in academic institutions, research institutions, federal, state, and local government agencies, food industries, food consultants, and even food lobbyists.

The subject matter has been divided into seven sections. For undergraduate teaching, the first six sections can be taught as a semester or a quarter course; Section VII (Appendices) can be used as advanced information for an undergraduate course that contains materials that are either taught in other courses, such as advanced food microbiology or food safety courses and laboratory courses. The first section contains four chapters that describe the history of food microbiology, characteristics of microorganisms important in foods, their sources, and significance. The second section also has four chapters that deal with microbial growth and metabolism in food and the significance of microbial sublethal injury and sporulation in foods. In the third section, the different beneficial uses of microorganisms, which include bioprocessing, biopreservation, and probiotics, are explained in eight chapters. Section IV deals with spoilage of foods by microorganisms and their enzymes and methods used to determine food spoilage. In addition, a chapter is included to discuss some of the emerging spoilage bacteria. Section V has seven chapters and deals with foodborne pathogens associated with intoxications, infections, and toxico-infections, as well as those that are considered opportunistic pathogens and pathogenic parasites and algae. In addition, a chapter has been included on emerging pathogens and chapter on indicators of pathogens. In Section VI, different methods used to control undesirable microorganisms for the safety and stability of food are discussed. A chapter on new nonthermal methods and a chapter on hurdle concept in food preservation are included.

The materials in each chapter are arranged in logical, systematic and concise sequences. Tables, figures, and data have been used only when their presentation is necessary for better understanding. At the end of each chapter, a limited list of selected references and suggested questions have been included. To reduce confusion, especially for those who are not familiar with constant changes in microbial genera, three first letters have been used to identify genus of a species. The index has been prepared carefully so that the materials in the text can be easily found.

I thank Mrs. Deb Rogers for an excellent performance in typing the manuscript. Finally, I thank my students, who, over a period of the last 20 years, have suggested what they would like to have in a food microbiology course. Those suggestions have been followed in writing this book.

TABLE OF CONTENTS

SECTION I

INTRODUCTION TO MICROBES IN FOODS

Microorganisms are living entities of microscopic size and include bacteria, viruses, yeasts and molds (together designated as fungi), algae, and protozoa. While bacteria are classified as procaryotes (cells without definite nuclei), the fungi, algae, and protozoa are eucaryotes (cells with nuclei); viruses do not have regular cell structures and are classified separately. Microorganisms are present everywhere on earth, which includes humans, animals, plants and other living creatures, soil, water, and atmosphere, and they can multiply everywhere except in the atmosphere. Together, their numbers far exceed all other living cells on this planet. They were the first living cells to inhabit the earth over 3 billion years ago and since then have played important roles, many of which are beneficial to the other living systems.

Among the microorganisms, some molds, yeasts, bacteria, and viruses have both desirable and undesirable roles in our food. In this section, their importance in food, predominant microorganisms associated with food, and the sources from which they get in the food and their significance are presented in the following chapters:

Chapter 1: History and Development of Food Microbiology
Chapter 2: Characteristics of Predominant Microorganisms in Food
Chapter 3: Sources of Microorganisms in Food
Chapter 4: Significance of Microorganisms in Food

CHAPTER 1

History and Development of Food Microbiology

CONTENTS

I. INTRODUCTION

Except for a few sterile foods, all foods harbor one or more types of microorganism. Some of them have desirable roles in food, such as in the production of fermented food, while others cause food spoilage and foodborne diseases. To study the role of microorganisms in food and to control them when necessary, it is important to isolate them in pure culture and study their characteristics. Some of the most simple techniques in use today for these studies were developed over the last 300 years; a brief description is included here.

II. DISCOVERY OF MICROORGANISMS

The discovery of microorganisms[1-4] ran parallel with the invention and improvement of the microscope. Around 1658, Athanasius Kircher reported that, using a microscope, he had seen minute living worms in putrid meat and milk. The magnification power of his microscope was so low that he could not have seen bacteria. In 1664, Robert Hooke described the structure of molds. However, probably the first person to see different types of microorganisms, especially bacteria, was the Dutch businessman turned naturalist Anton van Leeuwenhoek, using a microscope that probably had not above 300× magnification power. He observed bacteria in saliva, rainwater, vinegar, and other materials, sketched the three morphological groups (spheroids or cocci, cylindrical rods or bacilli, and spiral or spirilla), and also described some to be motile. He called them animalcules and in 1675 reported his observations to the newly formed leading scientific organization, The Royal Society of London, where his observations were read with fascination. As fairly good microscopes were not easily available at the time, during the course of the next 100 years, other interested individuals and scientists only confirmed Leeuwenhoek's observations. In the 19th century, as a result of the Industrial Revolution, improved microscopes became more easily available, and that stimulated many inquisitive minds to see and describe creatures they discovered under a microscope. By 1838, Ehrenberg (who introduced the term *bacteria*) had proposed at least 16 species in four genera and by 1875 Cohn had developed the preliminary classification system of bacteria. Cohn also was the first to discover that some bacteria produced spores. Although, like bacteria, the existence of submicroscopic viruses was recognized in the mid-19th century, they were observed only after the invention of the electron microscope in the 1940s.

III. WHERE ARE THEY COMING FROM?

Following Leeuwenhoek's discovery, although there were no bursts of activity, some scientific minds did have the curiosity to determine where the animalcules, found to be present in many different objects, were coming from.[1-4] Society had just emerged from the Renaissance period and science, known as experimental philosophy, was in its infancy. The theory of spontaneous generation, i.e., the generation of some form of life from nonliving objects, had many strong followers among the educated and elite class. Since the time of the Greeks, the emergence of maggots from dead bodies and spoiled flesh was thought to be due to spontaneous generation. But around 1665, Redi disproved that theory by showing that the maggots in spoiled meat and fish could only appear if flies were allowed to contaminate them. The advocates of the spontaneous generation theory argued that the animalcules could not regenerate by themselves (biogenesis), but that they were present in different things only through abiogenesis (spontaneous generation). In 1749, Needham showed that boiled meat and meat broth, following storage in covered flasks, showed the presence of animalcules within a short time. This was used to prove the appearance of these animalcules by spontaneous generation. Spallanzani (1765) showed that boiling meat infusion in broth in a flask and sealing the flask immediately prevented the appearance of these microscopic organisms and thus disproved Needham's theory. This was the time when Antoine-Laurent Lavoisier and his co-workers showed the need of oxygen for life. The believers of abiogenesis rejected Spallanzani's observation, suggesting that there was not enough vital force (oxygen) present in the sealed flask for animalcules to appear through spontaneous generation. Later, Schulze (1830; by passing air through acid), Theodor Schwann (1838; by passing air through red hot tubes), and Schroeder (1854; by passing air through cotton) showed that bacteria failed to appear in boiled meat infusion even in the presence of air. Finally, in 1864, Louis Pasteur demonstrated that, in boiled infusion, bacteria could grow only if the infusions were contaminated with bacteria carried by dust particles in air. His careful and controlled studies proved that bacteria were able to reproduce (biogenesis) and life could not originate by spontaneous generation. John Tyndall, in 1870, showed that in a dust-free box, boiled infusion could be stored in dust-free air without microbial growth.

IV. WHAT ARE THEIR FUNCTIONS?

The involvement of invisible organisms in many diseases in humans was suspected as early as the 13th century by Roger Bacon. In the 16th century, Francostro of Verona suggested that many human diseases were transmitted by small creatures from person to person. This was also indicated by Kircher in 1658. In 1762, von Plenciz of Vienna suggested that different invisible organisms were responsible for different diseases. Schawnn (1837) and Hermann Helmholtz (1843) pointed out that putrefaction and fermentation were connected with the presence of the organisms derived from the air. Finally, Pasteur, in 1875, showed that wine fermentation from grapes and souring of wine were caused by microorganisms. He also proved that spoilage of meat and milk was associated with the growth of microorganisms. Later, he showed the association of microorganisms with several diseases in humans, cattle, and sheep, and later developed vaccines against several human and animal diseases, including the rabies virus. Robert Koch, in Germany (in the 1880s and 1890s), isolated bacteria in pure cultures responsible for anthrax, cholera, and tuberculosis. He also developed the famous Koch's postulate to associate a specific bacterium as a causative agent for a specific disease. Along with his associates, he also developed techniques of agar plating methods to isolate bacteria in pure cultures, the petri dish (by Petri in his laboratory), and staining methods for better microscopic observation of bacteria.

With time, the importance of microorganisms in human and animal diseases, soil fertility, plant diseases, fermentation, food spoilage and foodborne diseases, and other areas was recognized, and microbiology was developed as a separate discipline. Later, it was divided into several disciplines, such as medical microbiology, soil microbiology, plant pathology, and food microbiology.[1-4]

V. EARLY DEVELOPMENTS IN FOOD MICROBIOLOGY (PRIOR TO 1900 A.D.)

It is not known exactly when our ancestors recognized the importance of the invisible creatures, now designated as microorganisms, in food. But it had to be around 8000 B.C. in the Near East after they developed agriculture and animal husbandry.[3-6] They produced more foods than they could consume within the short growing season, and a portion of the produce was lost due to spoilage. They solved the problems and secured uniform food supplies throughout the year by developing different preservation techniques. Between 8000 and 2000 B.C., they used drying, cooking, smoking, salting, low temperature, baking, modified atmosphere, fermentation, spices, and honey to extend the storage life of different types of raw and processed foods. Although we are not sure if they had perceptions about the cause of foodborne diseases, they definitely associated food spoilage with some invisible factors and developed successful preventative measures.

From the time of the Greeks until the discovery of biogenesis, spoilage of foods, especially of meat and fish, was thought to be due to spontaneous generation, such as the development of maggots. When the presence of different types of bacteria in many foods was discovered, their appearance through spontaneous generation was explained to be the cause of food spoilage. Schawnn (1837) and Helmholtz (1843) associated the presence of microorganisms (bacteria) in food with both putrefactive and fermentative changes of foods. However, they did not believe in spontaneous generation, but they could not explain how microorganisms could bring about those changes. Finally, Pasteur resolved the mystery by explaining that contamination of foods with microorganisms from the environment and their subsequent metabolic activities and growth were the causes of fermentation of grapes, souring of milk, and putrefaction of meat.

Diseases caused by the consumption of certain foods (foodborne disease) was recognized at least during the Middle Ages. Ergot poisoning in Europe was related to the consumption of grains (infested with molds) in the 12th century. In 1857, consumption of raw milk was suspected to be the cause of typhoid fever. In 1870, Selmi related certain food poisoning with ptomaine (histamine). Gaertner was the first to isolate *Salmonella* from a meat implicated in a foodborne disease in 1888. Denys, in 1894, was able to establish *Staphylococcus aureus* with food poisoning and, in 1896, Ermengem isolated *Clostridium botulinum* from food. The association of many other pathogenic bacteria and viruses to foodborne diseases was established after 1900 A.D. This aspect is discussed in Section V.

Pasteur, in the 1860s, recognized the role of yeasts in alcohol fermentation. He also showed that souring of wine was due to growth of acetic acid–producing bacteria (*Acetobacter aceti*), and developed the pasteurization process (heating at 145°F for 30 min) to selectively eliminate these undesirable bacteria from wine. Like fermentation, cheese ripening was suggested by Martin in 1867 to be of microbial origin. John Lister, in 1873, was able to isolate milk-souring bacteria (*Lactococcus lactis*) by the serial dilution (dilution to extinction) procedure. Cienkowski, in 1878, isolated the bacteria (*Leuconostoc mesenteroides*) associated with slime formation in sugar (sucrose). In 1895, microbial enumeration of milk was developed by von Geuns. After 1900 A.D., the involvement of different microorganisms in food spoilage and food fermentation was demonstrated. These aspects are discussed in later chapters.[3-6]

VI. IMPORTANCE OF MICROORGANISMS IN FOODS

Since 1900 A.D., our understanding of the importance of microorganisms in food has increased greatly. Their role in food can be either desirable (food bioprocessing) or undesirable (foodborne diseases and food spoilage), which is briefly discussed here.[6]

A. Foodborne Diseases

Many pathogenic microorganisms (bacteria, molds, and viruses) can contaminate food during various stages of their handling between production and consumption. Consumption of these foods can cause foodborne diseases. Foodborne diseases not only can be fatal, but they can cause large economic losses. Foods of animal origin are associated more with foodborne diseases than foods of plant origin. Mass production of foods, introduction of new technologies in the processing and storage of foods, changes in food consumption patterns, and the increase in imports of food from other countries have increased the chances of large outbreaks as well as the introduction of new pathogens. Effective intervention technologies are being developed to ensure the safety of consumers against foodborne diseases. New methods are also being developed to effectively and rapidly identify the pathogens in contaminated foods.[6]

B. Food Spoilage

Except for sterile foods, all foods harbor microorganisms. Food spoilage stems from the growth of these microorganisms in food or is due to the action of microbial heat-stable enzymes. New marketing trends, the consumers' desire for foods that are not overly processed and preserved, extended shelf life, and chances of temperature abuse between production and consumption of foods have greatly increased the chances of food spoilage and, in some instances, with new types of microorganisms. The major concerns are the economic loss and wastage of food. New concepts are being studied to reduce contamination as well as control the growth of spoilage microbes in foods.

C. Food Bioprocessing

Many food-grade microorganisms are used to produce different types of fermented foods using raw materials from animal and plant sources. Consumption of these foods has increased greatly over the last 10 to 15 years and is expected to increase still more in the future. There have been great changes in the production and availability of these microorganisms (starter cultures) to meet the large demand. Also, novel and better strains are being developed using genetic engineering techniques.

Microbial enzymes are also being used to produce food and food additives. Genetic recombination techniques are being used to obtain better enzymes and from diverse sources. Many types of additives from microbial sources are being developed and used in food.

D. Food Biopreservation

Antimicrobial metabolites of desirable microorganisms are being used in foods in place of nonfood preservatives to control pathogenic and spoilage microorganisms in food. Economic production of these antimicrobial compounds and their effectiveness in food systems have generated wide interest.

E. Probiotics

Consumption of foods containing live cells of bacteria that have apparent health benefits has generated interest among consumers. The efficiency of these bacteria to produce these benefits is being investigated.

VII. FOOD MICROBIOLOGY AND FOOD MICROBIOLOGISTS

From the above discussion, it becomes apparent what, as a discipline, food microbiology has to offer. Prior to the 1970s, food microbiology was regarded as an applied science mainly involved in the microbiological quality control of food. Since then, the technology used in food production, processing, distribution and retailing, and food consumption patterns have changed dramatically. These changes have introduced new problems that no longer can be solved by just using applied knowledge. Thus, modern-day food microbiology needs to include a great deal of basic science to effectively solve the microbiological problems in food. The discipline not only includes the microbiological aspects of food spoilage and foodborne diseases and their effective control and bioprocessing of foods, it also includes basic information of microbial physiology, metabolism, and genetics. This information is helping to develop methods for rapid and effective detection of spoilage and pathogenic bacteria, to develop desirable microbial strains by recombinant DNA technology, to produce fermented foods of better quality, to develop thermostable enzymes in enzyme processing of food and food additives, to develop methods to remove bacteria from food and equipment surfaces, and to combine several control methods for effective control of spoilage and pathogenic microorganisms in food.

An individual who has completed courses in food microbiology (both lecture and laboratory) should gain knowledge in the following areas:

1. Determination of microbiological quality of foods and food ingredients using appropriate techniques

2. Determination of microbial type(s) involved in spoilage, health hazards, and the identification of the sources
3. Design corrective procedures to control the spoilage and pathogenic microorganisms in food
4. Identify how new technologies adapted in food processing can have specific microbiological problem and design methods to overcome the problem
5. Design effective sanitation procedures to control spoilage and pathogen problems in food processing facilities
6. Effective use of desirable microorganisms to produce fermented foods
7. Design methods to produce better starter cultures for use in fermented foods and probiotics
8. Food regulations (state, federal, international)

To be effective, in addition to the knowledge gained, one has to be able to communicate with different groups of people about the subject (food microbiology and its relation to food science). An individual with good common sense is always in a better position to sense a problem and correct it quickly.

REFERENCES

1. Bulloch, W., *The History of Bacteriology,* Oxford University Press, London, 1938.
2. De Kruif, P., *Microbe Hunters.,* Harcourt, New York, 1926.
3. Pelezar, M. J. and Chan, E. C. S., *Elements of Microbiology,* McGraw Hill, New York, 1981, 3.
4. Tortora, G. J., Funke, B. R., and Case, C. L., *Microbiology: An Introduction,* 4th ed., Benjamin/Cummings, New York, 1992, 6.
5. Jay, J. M., *Modern Food Microbiology,* 4th ed., Van Nostrand Reinhold, New York, 1991, 3.
6. Ray, B., The need for food biopreservation, in *Food Biopreservatives of Microbial Origin,* CRC Press, Boca Raton, FL, 1992, 1.

CHAPTER 1 QUESTIONS

1. Describe briefly the contributions of the following scientists in the development of microbiology and food microbiology:

a. Leeuwenhoek
b. Spallanzani
c. Louis Pasteur
d. Robert Koch

2. Define the “spontaneous generation” theory. Discuss how Redi succeeded in disproving this theory by his simple but foolproof theory.

3. Why did Needham’s experiments fail to disprove spontaneous generation for microbes, yet Pasteur succeeded in disproving that theory?

4. Discuss early observations and incidences that made scientists believe that microorganisms might be associated with food spoilage and foodborne diseases.
5. List and briefly describe the importance of microorganisms in foods.
6. List what a food microbiologist should know.

CHAPTER 2

Characteristics of Predominant Microorganisms in Food

CONTENTS

I. INTRODUCTION

The microbial groups important in foods contain several species and types of bacteria, yeasts, molds, and viruses. Although some algae and protozoa, as well as some worms (such as nematodes), are important in foods, they are not included among the microbial groups in this chapter. Some of the protozoa and worms associated with health hazards, and several algae associated with health hazards and bioprocessing (sources of vitamins, single cell proteins), are discussed in Chapters 15 and 26.

Bacteria, yeasts, molds, and viruses are important in food for three main reasons: namely, their ability to cause foodborne diseases and food spoilage and to produce food and food ingredients. Many bacterial species and some molds and viruses, but not yeasts, are involved in foodborne diseases. Most bacteria, molds, and yeasts, due to their ability to grow in foods (viruses cannot grow in foods), are potentially capable of causing food spoilage. Several species of bacteria, molds, and yeasts, recognized as safe and/or food grade, are used to produce fermented foods and food ingredients. Among the four major groups, bacteria constitute the largest group; due to their ubiquitous presence and rapid growth rate, even under conditions where yeasts and molds cannot grow, they are considered the most important in food spoilage, foodborne diseases, as well as in developing methods to control microorganisms in foods.

In this chapter, initially a brief discussion is included on the methods currently used in the classification and nomenclature of microorganisms. Later, important characteristics of microorganisms predominant in food are discussed.

II. CLASSIFICATION OF MICROORGANISMS

Living cellular organisms, on the basis of phylogenetic and evolutionary relationships, are grouped in five kingdoms, in which bacteria belong to procaryote (before nucleus) while the eucaryotic (with nucleus) molds and yeasts are grouped under fungi.[1-3] Viruses are not considered living cells and are not included in this classification system.

For the classification of yeasts, molds, and bacteria, several ranks are used after the kingdom. These are: divisions, classes, orders, families, genera (singular, genus), and species. The basic taxonomic group is the species. Several species with similar characteristics form a genus. Among eucaryotes, the species in the same genus are able to interbreed. This is not considered among procaryotes, although conjugal transfer of genetic materials exist among many bacteria. A family is made up of several genera and the same procedure is followed in the hierarchy. In food microbiology, ranks above species, genus, and family are seldom used. Among bacteria, a species is regarded as a collection of strains having many common features. A strain is the descendant of a single colony (single cell). Among the strains in a species, one is assigned as the type strain; it is used as a reference strain while comparing the characteristics of an unknown isolate.

To determine relatedness among bacteria, yeasts, and molds for taxonomic classification, several methods are used. In yeasts and molds, morphology, reproduction, biochemical nature of the macromolecules, and metabolic patterns are used along with other criteria. For bacteria morphology, Gram-stain characteristics, protein profiles, amino acid sequences of some specific proteins, base composition (mol % G + C), nucleic acid (DNA and RNA) hybridization, nucleotide base sequence, and computer-assisted numerical taxonomy are used. Protein profiles, amino acid sequence, base composition, DNA and RNA hybridization, and nucleotide base sequence are all directly or indirectly related to the genetic make-up of the organisms and thus provide a better chance of comparing the two organisms at the genetic level. In mol % G + C ratio, if two strains differ by 10% or more, they are most likely not related. Similarly, in the hybridization study, two strains are considered the same if their DNA have 90% or more homology. For the nucleotide base sequence, the sequences in 16S RNA among the strains are compared. The nucleotide sequence in 16S RNA is most conserved, so related strains should have high homology. In numerical taxonomy, many characteristics are compared. They can be morphological, physiological, biochemical, and otherwise. Each characteristic is given the same weight. Two strains in the same species should score 90% or higher.

Evolutionary relationships among viruses, if there are any, are not known. Their classification system is rather arbitrary and based on the types of disease they cause (such as hepatitis virus, causing inflammation of the liver), nucleic acid content (DNA or RNA, single stranded or double stranded), and morphological structures. In food, two groups of viruses are important: the bacterial viruses (bacteriophages) of starter culture bacteria and human pathogenic viruses associated with foodborne diseases.

III. NOMENCLATURE

The basic taxonomic group in bacteria, yeasts, and molds is the species, and each species is given a name.[1-3] The name has two parts (binomial name); the first part is the genus name and the second part is the specific epithet (adjective). Both parts are Latinized; when written, they are italicized (or underlined) with the first letter of the genus written in a capital letter (e.g., *Saccharomyces cerevisiae, Penicillium roquefortii,* and *Lactobacillus acidophilus*). A bacterial species can be divided into several subspecies (subsp. or ssp.) if they show minor but consistent differences in characteristics. Under such conditions, a trinominal epithet (subspecific epithet) is used (e.g., *Lactococcus lactis* ssp. *lactis* or *Lactococcus lactis* ssp. *cremoris*). In some instances, ranks below subspecies are used to differentiate strains recognized by specific characters (e.g., serovar, antigenic reaction; biovar, producing a specific metabolite; and phagovar, sensitive to a specific phage). Such ranks have no taxonomic importance but can have practical usefulness (e.g., *Lactococcus lactis* subsp. *lactis* biovar diacetilactis: a *Lactococcus lactis* subsp. *lactis* strain that produces diacetyl, an important flavor compound in some fermented dairy products). Each strain of a species should be identified with a specific strain number, which can be alphabetical letters or numerical or a mixture of both (e.g., *Pediococcus acidilactici* LB923). At the family level, bacterial names are used as plural adjectives in feminine gender and agree with the suffix "aceae" (e.g., *Enterobacteriaceae*). The species and strains in a genus can be represented collectively, using either spp. after genus (e.g., *Lactobacillus* spp.) or plural forms of the genus (e.g., lactobacilli).

The scientific names of bacteria are given according to the specifications of the International Code of Nomenclature of Bacteria. The International Committee on Systematic Bacteriology of the International Union of Microbiological Association examines the validity of each name and then publishes the Approved Lists of Bacterial Names from time to time. A new name (species, genus) must be published in the *International Journal of Systematic Bacteriology* before it is judged for inclusion in the approved list. Any change in name (genus, species) has to be approved by this Committee.

When writing the name of the same species more than once in an article, it is customary to use both genus and specific epithet for the first time and abbreviate the genus name subsequently. In *Bergey's Manual of Systematic Bacteriology,* only the first letter is used (i.e., *Listeria monocytogenes; L. monocytogenes*). The same system is used in most publications in the United States. However, it creates confusion when one article has several species with the same first letter in the genus (e.g., *Lactobacillus lactis, Leuconostoc lactis,* and *Lactococcus lactis* as *L. lactis*). In some European journals, more than one letter is used, but there is no definite system (e.g., *Lact. lactis, Leu. lactis, Lb. lactis,* and *List. monocytogenes*). In this book, to reduce confusion among students and others, many of whom may not be familiar with the rapid changes in bacterial nomenclature, a three-letter system is used (e.g., *Lis. monocytogenes, Leu. lactis; Sal. typhimurium* for *Salmonella; Shi. dysenteriae* for *Shigella; Sta. aureus* for *Staphylococcus*; and so on). In rare cases, a slight modification is used (e.g., *Lactococcus lactis* and *Lactobacillus lactis* are written as *Lac. lactis, Lab. lactis*, respectively).

The viruses, as indicated before, have not been given specific taxonomic names as given for bacteria. They are often identified with letters (alphabetical) and/or numerical designation (e.g., T4 or λ bacteriophages), the disease they produce (e.g., hepatitis A, causing liver inflammation), or by other methods (e.g., Norwalk viruses, causing a type of foodborne gastroenteritis in humans).

IV. MORPHOLOGY AND STRUCTURE OF MICROORGANISMS IN FOODS

A. Yeasts and Molds

Both yeasts and molds are eucaryotic, but yeasts are unicellular while molds are multicellular.[4] The eucaryotic cells are generally much larger (20 to 100 μm) than the procaryotic cells (1 to 10 μm). The cells have rigid cell walls and thin plasma membranes. The cell wall does not have mucopeptide, is rigid, and is composed of carbohydrates. The plasma membrane contains sterol. The cytoplasm is mobile (streaming) and contains organellae (mitochondria, vacuoles) that are membrane bound. Ribosomes are 80S type and attached to the endoplasmic reticulum. The DNA are linear (chromosomes), contain histone, and are enclosed in nuclear membrane. Cell division is by mitosis (i.e., asexual reproduction). Sexual reproduction, when it occurs, is by meiosis.

Molds are nonmotile, filamentous, and branched (Figure 2.1). The cell wall is composed of cellulose, chitin, or both. A mold (thallus) is composed of large numbers of filaments called hyphae. An aggregate of hyphae is called mycelium. A hypha can be nonseptate, septate-uninucleate, or septate-multinucleate. A hypha can be vegetative or reproductive. The reproductive hypha usually extend in the air and form exospores, either free (conidia) or in a sack (sporangium). Shape, size, and color of spores are used in taxonomic classification.

Yeasts are widely distributed in nature. The cells are oval, spherical, or elongated, about 5 to 30 × 2 to 10 μm in size (Figure 2.1). They are nonmotile. The cell wall contains polysaccharide (glycan), proteins, and lipids. The wall can have scars, indicating the site(s) of budding. The membrane is beneath the wall. The cytoplasm has a finely granular appearance for ribosomes and organellae. The nucleus is well defined with a nuclear membrane.

B. Bacterial Cells

Bacteria are unicellular, about 0.2 to 10 μm in size, and have three morphological forms: spherical (cocci), rod shaped (bacilli), and curved (coma) (Figure 2.1).[5,6] They can form associations such as clusters, chains (two or more), and tetrads. They can be motile or nonmotile. The cytoplasmic materials are enclosed in a rigid wall on the surface and a membrane beneath the wall. The nutrients in molecular form are transported from the environment through the membrane by several mechanisms. The membrane also contains energy-generating components. It also forms intrusions in the cytoplasm (mesosomes). The cytoplasmic material is immobile and does not

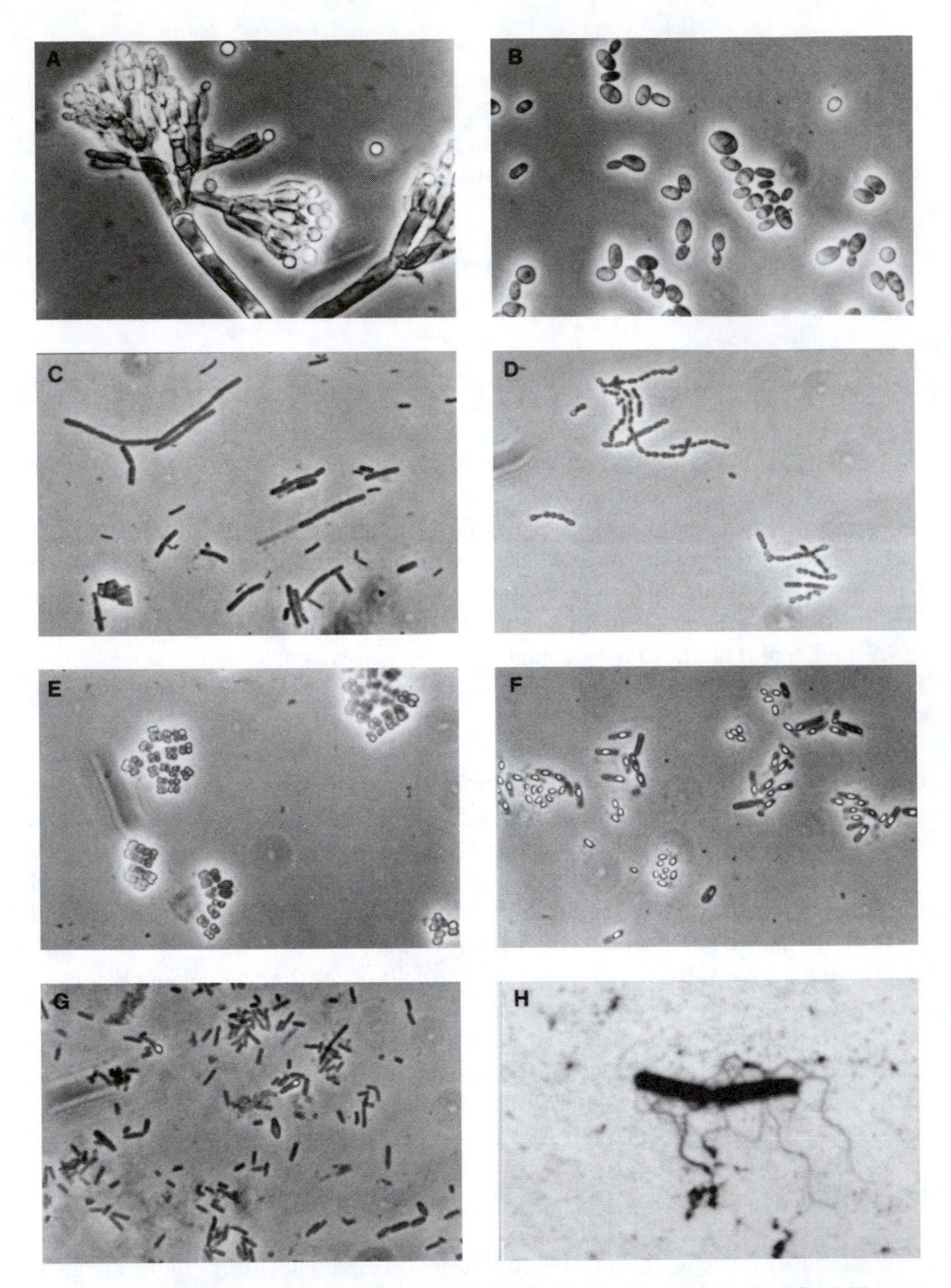

Figure 2.1 Photograph of microbial morphology. (a) Molds: Conidial head of *Penicillium* sp. showing conidiophore (stalk) and conidia; (b) Yeasts: *Saccharomyces cerevisiae*, some carrying buds; (c) Rod-shaped bacteria: *Bacillus* sp., single and chain; (d) Spherical-shaped bacteria: *Streptococcus* sp., chain; (e) Spherical-shaped bacteria: tetrads; (f) *Bacillus* cells carrying spores, center and off-center; (g) *Clostridium* cells, some carrying terminal spore (drumstick appearance); (h) Motile rod-shaped bacterium (*Clostridium* sp.) showing peretrichous flagella.

contain organellae enclosed in a separate membrane. The ribosomes are 70S type and are dispersed in the cytoplasm. The genetic materials (structural and plasmid DNA) are circular, not enclosed in nuclear membrane, and do not contain basic proteins such as histones. Both gene transfer and genetic recombination occur but do not involve gamete or zygote formation. The cell division is by binary fission. The procaryotic cells can also have flagella, capsules, surface layer proteins, and pili for specific functions. Some also form endospores (one per cell).

On the basis of Gram-stain behavior, the bacterial cells are grouped as Gram-negative and Gram-positive. The Gram-negative cells have a complex cell wall containing an outer membrane (OM) and a middle membrane (MM) (Figure 2.2). The OM is composed of lipopolysaccharides (LPS), lipoprotein (LP), and phospholipids. The phospholipid molecules are arranged in a bilayer with the hydrophobic part (fatty acids) inside and the hydrophilic part (glycerol and phosphate) on the outside. The LPS and LP molecules are embedded in the phospholipid layer. The OM has both limited transport and barrier functions. The resistance of Gram-negative bacteria to many enzymes (lysozyme, which hydrolyzes mucopeptide), hydrophobic molecules (SDS and bile salts), and antibiotics (penicillin) is due to the barrier property of the OM. LPS molecules also have antigenic properties. Beneath the OM is the MM, composed of a thin layer of peptidoglycan or mucopeptide embedded in the periplasmic materials that contain several types of proteins. Beneath the periplasmic materials is the plasma or inner membrane (IM) composed of a phospholipid bilayer in which many types of proteins are embedded.

The Gram-positive cells have a thick cell wall composed of several layers of mucopeptide (responsible for thick rigid structure) and two types of teichoic acids (Figure 2.2). Some species also have a layer over the cell surface, called surface layer protein (SLP). The wall teichoic acid molecules are linked to mucopeptide layers, and the lipoteichoic acid molecules are linked to both mucopeptide and the cytoplasmic membrane. The teichoic acids are negatively charged (due to phosphate groups) and may bind to or regulate the movement of cationic molecules in and out of the cell. Teichoic acids have antigenic properties and can be used to identify Gram-positive bacteria by serological means.

C. Viruses

Viruses are regarded as noncellular entities. The bacterial viruses (bacteriophages) important in food microbiology are widely distributed in nature.[7-9] They are composed of nucleic acids (DNA or RNA) and several proteins. The proteins form the head (surrounding the nucleic acid) and tail. A bacteriophage attaches itself to the surface of a host bacterial cell and inoculates its nucleic acid into the host cell. Subsequently, many phages form inside a host cell and are released outside following lysis of the cell. This is discussed in Chapter 12.

Several pathogenic viruses have been identified as causing foodborne diseases in humans. However, due to the difficulty of their detection in foods, the involvement of other pathogenic viruses in foodborne diseases is not currently known. The two most important viruses implicated in foodborne outbreaks are the hepatitis A and Norwalk viruses. Both are single-stranded RNA viruses. Hepatitis A is a small, naked

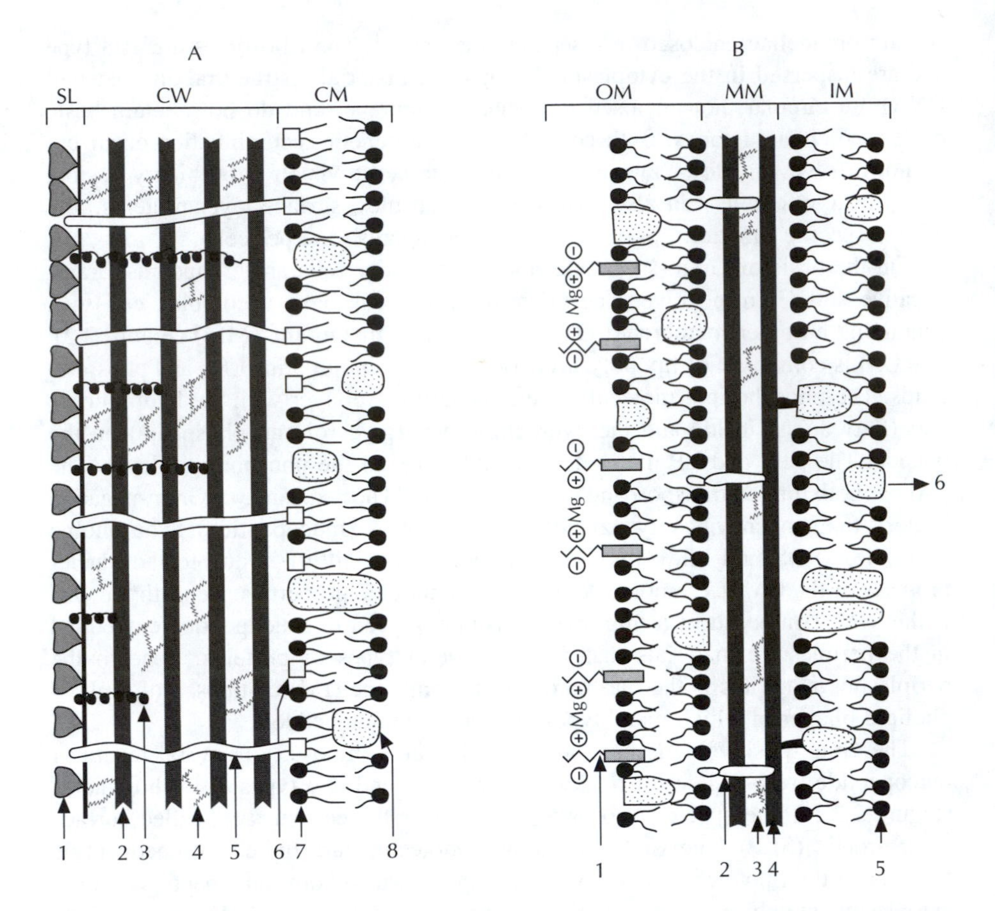

Figure 2.2 Schematic representations of cell envelopes of bacteria. (A) Gram-positive bacteria: SL: surface layer proteins with protein subunits (1); CW: cell wall showing thick mucopeptide backbone layers (2) covalently linked to peptides (4), wall teichoic acids (or teichouronic acid, 3), lipoteichoic acids (anchored to cytoplasmic membrane; 5); CM: cytoplasmic membrane with lipid bylayers containing phospholipids (7), glycolipids (6), and embedded proteins. (B) Gram-negative bacteria: OM: outer membrane containing lipopolysaccharide molecules stabilized by divalent cations (1), phospholipids (5), and proteins (6); MM, middle membrane containing thin mucopeptide layers (4) covalently linked to peptides (3) and lipoproteins (2); IM: inner membrane with phospholipid bilayers (5) and proteins (6).

polyhedral enteric virus of about 30 nm diameter. The RNA strand is enclosed in a capsid.

V. IMPORTANT MICROORGANISMS IN FOOD

A. Important Mold Genera

Molds are important in food because they can grow under conditions in which many bacteria cannot grow, such as low pH, low water activity (A_w), and high osmotic pressure.[4] They are important spoilage microorganisms. Many strains also produce mycotoxins and have been implicated in foodborne intoxication. Many are used in food bioprocessing. Finally, many are used to produce food additives and enzymes. Some of the most common genera of molds found in food are listed here.

Aspergillus: They are widely distributed and contain many species important in food. They have septate hyphae and produce asexual spores (black color) on conidia. Many are xerophilic (able to grow in low A_w) and can grow in grains, causing spoilage. They are also involved in spoilage of foods such as jams, cured ham, nuts, and fruits and vegetables (rot). Some species/strains produce mycotoxin (e.g., *Aspergillus flavus* produces aflatoxin). Many species/strains are also used in food and food additive processing. *Aspergillus oryzae* is used to hydrolyze starch by α-amylase in the production of sake. *Aspergillus niger* is used to process citric acid from sucrose and to produce enzymes like β-galactosidase.

Alternaria: They are also septate and form dark-colored spores on conidia. They cause rot in tomatoes and rancid flavor in dairy products. Species: *Alternaria tenuis*.

Geotrichum: They are septate and form rectangular arthrospores. They grow, forming a yeastlike cottony, creamy colony. They establish easily in equipment and often grow on dairy products (dairy mold). Species: *Geotrichum candidum*.

Mucor: They are widely distributed. They have nonseptate hyphae and produce sporangiophores. They produce cottony colonies. Some species are used in food fermentation and as a source of enzymes. They cause spoilage of vegetables. Species: *Mucor rouxii*.

Penicillium: They are widely distributed and contain many species. They have septate hyphae and form conidiophore on a blue-green, brushlike conidia head (Figure 2.1). Some species are used in food production, such as *Penicillium roquefortii* and *Penicillium camembertii* in cheese. Many species cause fungal rot in fruits and vegetables. They also cause spoilage of grains, breads, and meat. Some strains produce mycotoxins (e.g., Ochratoxin A).

Rhizopus: The hyphae are aseptate and form sporangiophores in sporangium. They are involved in the spoilage of many fruits and vegetables. *Rhizopus stolonifer* is the common black bread mold.

B. Important Yeast Genera

Yeasts are important in food due to their ability to cause spoilage. Many are also used in food bioprocessing. Some are used to produce food additives. Several important genera are briefly described below.[4]

Saccharomyces: Cells are round, oval, or elongated. It is the most important genus and contains heterogenous groups (Figure 2.1). *Saccharomyces cerevisiae* variants are used in baking for the leavening of bread and in alcoholic fermentation. They are also involved in spoilage of food with the production of alcohol and CO_2.

Pichia: They are oval to cylindrical cells and form pellicle in beer, wine, and brine to cause spoilage. Some are also used in oriental food fermentation. Species: *Pichia membranaefaciens*.

Rhodotorula: They are pigment-forming yeasts and can cause discoloration of foods, such as in meat, fish, and sauerkraut. Species: *Rhodotorula glutinis*.

Torulopsis: They have a spherical to oval structure. They cause spoilage of milk due to the ability to ferment lactose (*Torulopsis versatilis*). They also spoil fruit juice concentrates and acid foods.

Candida: Many spoil foods with high acid, salt, and sugar and form pellicle on the surface of liquids. Some can cause rancidity in butter and dairy products (*Candida lipolytica*).

C. Important Viruses

Viruses are important in food for three reasons.[7-9] Some are able to cause enteric disease and thus, if present in a food, can cause foodborne diseases. Hepatitis A and Norwalk viruses have been implicated in foodborne outbreaks. Several other enteric viruses, such as Poliovirus, Echovirus, and Coxsackievirus, have the potential of causing foodborne diseases. In some countries where the level of sanitation is not very high, they can contaminate foods and cause disease.

Some bacterial viruses (bacteriophages) are used in the identification of some pathogens (e.g., *Salmonella* spp.; *Staphylococcus aureus* strains) on the basis of the sensitivity of the cells to a series of bacteriophages at appropriate dilutions. Bacteriophages are used to transfer genetic traits in some bacterial species/strains by a process called transduction (e.g., in *Escherichia coli* and *Lactococcus lactis*).

Finally, some bacteriophages can be very important due to their ability to cause fermentation failure. Many lactic acid bacteria, used as starter cultures in food fermentation, are sensitive to different bacteriophages. They can infect and destroy starter culture bacteria, causing product failure. Among the lactic acid bacteria, bacteriophages have been isolated for many species in genera *Lactococcus, Streptococcus, Leuconostoc,* and *Lactobacillus*; no bacteriophage of *Pediococcus* is yet known. Methods are being studied to genetically engineer lactic starter cultures so that they become resistant to multiple bacteriophages (see Chapter 12).

D. Important Bacterial Genera

Bacterial classification is changing rapidly.[1-3] In *Bergey's Manual of Systematic Bacteriology*, published between 1984 and 1988, more than 420 bacterial genera are listed in 33 sections on the basis of their differences in characteristics. Since then, many other genera have been created, such as *Lactococcus* (former N-group or dairy *Streptococcus*) and *Carnobacterium* (some species previously included in *Lactobacillus*). In the 9th edition of *Bergey's Manual of Determinative Bacteriology* (1993), more than 560 genera are listed in 35 groups. Since then, many new genera have been proposed. Among these, a total of 48 genera, whose species are frequently associated with spoilage, health hazard, and bioprocessing of food, are included in Table 2.1. Species of other genera, besides these 48, can also be found in food, but their relative significance is not well established. Many species names in several genera are no longer valid and thus not included in the current *Bergey's Manual*. In this text, only those species and genera currently approved and listed in *Bergey's Manual* are used. Brief important characteristics of these genera and their importance in foods are described. Some descriptions are also presented in other chapters, such

Table 2.1 Genera of Bacteria Important in Foods

Section[a] (group[b]) and description		Family	Genera
2:	Gram-negative, aerobic/ microaerophilic, motile, helical/vibrioid	Not indicated	*Campylobacter*
4:	Gram-negative, aerobic, rods and cocci	*Pseudomonadaceae*	*Pseudomonas, Xanthomonas*
		Acetobacteraceae	*Acetobacter, Gluconobacter*
		Nisseriaceae	*Acinetobacter, Moraxella*
		Not indicated	*Alteromonas, Flavobacterium, Alcaligenes, Brucella, Psychrobacter*
5:	Gram-negative, facultative anaerobic, rods	*Enterobacteriaceae*	*Citrobacter, Escherichia, Enterobacter, Edwardsiella, Erwinia, Hafnia, Klebsiella, Morganella, Proteus, Salmonella, Shigella, Serratia, Yersinia*
		Vibrionaceae	*Vibrio, Aeromonas, Plesiomonas*
9:	Rickettsias	*Rickettsiaceae*	*Coxiella*
12: (17)	Gram-positive, cocci	*Micrococeaceae*	*Micrococcus, Staphylococcus*
		Not indicated	*Streptococcus, Enterococcus, Lactococcus, Leuconostoc, Pediococcus, Sarcina*
13: (18)	Gram-positive, Endospore-forming rods and cocci	Not indicated	*Bacillus, Sporolactobacillus, Clostridium, Desulfotomaculum*[c]
14: (19)	Gram-positive, nonsporing, regular rods	Not indicated	*Lactobacillus, Carnobacterium, Brochothrix, Listeria*
15:	Gram-positive,	Not indicated	*Corynebacterium,*

Table 2.1 Genera of Bacteria Important in Foods (continued)

Section[a] (group[b]) and description	Family	Genera
(20) nonsporing, irregular rods		*Brevibacterium, Propionibacterium, Bifidobacterium*

[a] Sections in *Bergey's Manual of Systematic Bacteriology.*
[b] Groups in *Bergey's Manual of Determinative Bacteriology.* Only those sections (or groups) containing bacteria important in food are listed in this table.
[c] *Disulfotomaculum* cells stain Gram negative.

as for pathogens in Chapters 23 to 25 and beneficial bacteria (bioprocessing) in Chapters 9 and 16.

1. Gram-Negative Aerobic Group

Campylobacter: Two species, *Campylobacter jejuni* and *Campylobacter coli,* are foodborne pathogens. Small (0.2 × 1 μm) microaerophilic, helical, motile cells found in the intestinal tract of humans, animals, and birds. Mesophiles.

Pseudomonas: Includes large numbers of species. Some important species in foods are *Pseudomonas fluorescens, Pseudomonas aeruginosa,* and *Pseudomonas putida*. Important spoilage bacteria. Capable of metabolizing a wide variety of carbohydrates, proteins, and lipids in foods. Found widely in the environment. Straight or curved (0.5 × 5 μm), aerobic, motile rods. Psychrotrophs.

Xanthomonas: Most characteristics are like *Pseudomonas*. Plant pathogens, and thus can cause spoilage of fruits and vegetables. *Xanthomonas campestris* strains are used to produce xanthan gum used as food stabilizers.

Acetobacter: Cells ellipsoidal to rod-shaped (0.6 × 4 μm), occur in single or short chains, motile or nonmotile, aerobic, and oxidize ethanol to acetic acid. Mesophiles, cause souring of alcoholic beverages and fruit juices, and used to produce vinegar (acetic acid). Can also spoil some fruits (rot). Widely distributed in plants and alcohol fermentation places. Important species: *Acetobacter aceti*.

Gluconobacter: Many characteristics are similar to *Acetobacter*. *Gluconobacter oxydans* causes spoilage of pineapples, apples, and pears (rot).

Acinetobacter: Cells are rods (1 × 2 μm), occur in pairs or small chains, show twitching motility due to presence of polar fimbriae. Strictly aerobic and grow between 20 to 35°C. Found in soil, water, and sewage. Important species: *Acinetobacter calcoaceticus*.

Moraxella: Very short rods frequently approaching coccoid shape (1 × 1.5 μm), occurring singly, in pairs, or short chains, may be capsulated. Twitching motility in some cells may be present. Optimum growth at 30 to 35°C. Found in animals and humans. Important species: *Moraxella lacunata*.

Alteromonas: Most currently assigned *Alteromonas* species are of marine origin and can be present in foods of marine origin. Need 100 m*M* NaCl for optimum growth (unlike *Pseudomonas*). *Alteromonas putrefaciens* (this species recently has been reclassified as *Shewanella putrefaciens*); due to many character similarities to

Pseudomonas, was previously designated as *Pseudomonas putrefaciens*. The strains are important in fish and meat spoilage. Psychrotrophs.

Flavobacterium: Cells are rods with parallel sides (0.5 × 3 μm), nonmotile, colonies are colored, some species can grow at low temperature (psychrotrophs) and cause spoilage of milk, meat, and other protein foods. Species: *Flavobacterium aquatile*.

Alcaligenes: Cells are rods or coccobacillary (0.5 × 1 μm), motile, present in water, soil, fecal material, associated with spoilage of protein-rich foods, generally mesophilic. Example of species: *Alcaligenes faecalis*.

Brucella: Coccobacilli (0.5 × 1.0 μm), mostly single, nonmotile. Different species cause disease in animals, including cattle, pigs, and sheep. They are human pathogens and have been implicated in foodborne brucellosis. *Brucella abortus* causes abortion in cows.

Psychrobacter: The genus was created in 1986 and contains one species: *Psychrobacter immobilis*. Cells are coccobacilli (1 × 1.5 μm), nonmotile, can grow at 5°C or below, optimum growth at 20°C, unable to grow at 35°C. Found in fish, meat, and poultry products.

2. Gram-Negative Facultative Anaerobes

Citrobacter: Straight rod (1 × 4 μm), single or pairs, usually motile, mesophiles, found in the intestinal contents of humans, animals, and birds and in the environment. Included in the coliform group as an indicator of sanitation. Important species: *Citrobacter freundii*.

Escherichia: Straight rods (1 × 4 μm), motile or nonmotile, mesophiles. Important species: *Escherichia coli*. Found in the intestinal contents of humans, warm-blooded animals, and birds. Many strains are nonpathogenic, but some strains are pathogenic to humans and animals. Involved in foodborne diseases. Used as an indicator of sanitation (theoretically nonpathogenic strains) in the coliform and fecal coliform groups.

Enterobacter: Straight rods (1 × 2 μm), motile, mesophiles, found in the intestinal contents of humans, animals, birds, and in the environment. Included in the coliform group as an indicator of sanitation. Important species: *Enterobacter aerogenes*.

Edwardsiella: Small rods (1 × 2 μm), motile, found in the intestines of cold-blooded animals and in fresh water. Can be pathogenic to humans, but involvement in foodborne disease has not been shown.

Erwinia: Small rods (1 × 2 μm), occur in pairs or short chains, motile, facultative anaerobes, optimum growth at 30°C. Many are plant pathogens and involved in the spoilage of plant products. Species: *Erwinia amylovora*.

Hafnia: Small rods (1 × 2 μm), motile, mesophiles, occur in intestinal contents of humans, animals, and birds and in the environment. Associated with food spoilage. Species: *Hafnia alvei*.

Klebsiella: Medium rods (1 × 4 μm), occur singly or in pairs, motile, and capsulated. Found in the intestinal contents of humans, animals, and birds, soil, water, and grains. Has been included in the coliform group as an indicator of sanitation. Mesophiles. Important species: *Klebsiella pneumoniae*.

Morganella: Small rods (0.5 × 1 μm), motile, mesophilic, found in the intestinal contents of humans and animals. Can be pathogenic but has not been implicated in foodborne disease. Species: *Morganella morganii*.

Proteus: Straight small rods (0.5 × 1.5 μm), highly motile, form swarm on media, some can grow at low temperature. Occur in the intestinal contents of humans and animals and the environment. Many are involved in food spoilage. Species: *Proteus vulgaris*.

Salmonella: Medium rods (1 × 4 μm), usually motile, mesophiles. There are over 2000 serovars and all are regarded as human pathogens. Found in the intestinal contents of humans, animals, birds, and insects. Major cause of foodborne diseases. Species: *Salmonella enteritidis* and *Salmonella typhimurium*.

Shigella: Medium rods, nonmotile, mesophiles. Found in the intestine of humans and primates. Associated with foodborne diseases. Species: *Shigella dysenteriae*.

Serratia: Small rods (0.5 × 1.5 μm), motile, colonies can be white, pink, or red. Some can grow at refrigerated temperature. Occur in the environment. Cause food spoilage. Species: *Serratia liquefaciens*.

Yersinia: Small rods (0.5 × 1 μm), motile or nonmotile, can grow at 1°C. Present in the intestinal contents of animals. One species, *Yersinia enterocolitica*, has been involved in foodborne disease outbreaks.

Vibrio: Curved rods (0.5 × 1.0 μm), motile, mesophiles. Found in freshwater and marine environments. Some species need NaCl for growth. Several species are pathogens and have been involved in foodborne disease (*Vibrio cholerae, Vib. parahaemolyticus,* and *Vib. vulnificus*), while some others can cause food spoilage (*Vib. alginolyticus*).

Aeromonas: Small rods (0.5 × 1.0 μm), occur singly or in pairs, motile, found in water environments. *Aeromonas hydrophila* has been suspected as a potential foodborne pathogen. Psychrotrophs.

Plesiomonas: Small rods (0.5 × 1.0 μm), motile, occur in fish and aquatic animals. *Plesiomonas shigelloides* has been suspected as a potential foodborne pathogen.

3. Rickettsias

Coxiella: Gram-negative, nonmotile, very small cells (0.2 × 0.5 μm), grows on host cells. Relatively resistant to high temperature (killed by pasteurization). *Coxiella burnetii* causes infection in cattle and has been implicated with Q fever in humans (especially from the consumption of unpasteurized milk).

4. Gram-Positive Cocci

Micrococcus: Spherical cells (0.2 to 2 μm), occurring in pairs, tetrads, or clusters, aerobic, nonmotile, some species produce yellow colonies, found in mammalian skin, can cause spoilage, mesophiles, resistant to low heat. Species: *Micrococcus luteus*.

Staphylococcus: Spherical cells (0.5 to 1 μm), occurring singly, in pairs, or clusters, nonmotile, mesophiles, facultative anaerobes, grow in 10% NaCl. *Staphy-*

lococcus aureus strains are frequently involved in foodborne diseases. *Sta. carnosus* is used in the processing of some fermented sausages. Main habitat is the skin of humans, animals, and bird.

Streptococcus: Cells spherical or ovoid (1 μm), occurring in pairs or chains, nonmotile, facultative anaerobes, mesophiles. *Streptococcus pyogenes* is pathogenic and has been implicated in foodborne diseases. Present as commensals in human respiratory tract. *Streptococcus thermophilus* is used in dairy fermentation. Can be present in raw milk. Can grow at 50°C.

Enterococcus: Spheroid cells (1 μm), occurring in pairs or chains, nonmotile, facultative anaerobes, some strains survive low heat (pasteurization), mesophiles. Normal habitat is the intestinal contents of humans, animals, and birds, and the environment. Can establish on equipment surfaces and used as an indicator of sanitation. They are important in food spoilage. Species: *Enterococcus faecalis*.

Lactococcus: Ovoid elongated cells (0.5 to 1.0 μm), occurring in pairs or short chains, nonmotile, facultative anaerobic, mesophiles but can grow at 10°C, produce lactic acid, used to produce many bioprocessed foods, especially fermented dairy foods. Species: *Lactococcus lactis* subsp. *lactis* and subsp. *cremoris*. Present in raw milk and plants. Several strains produce bacteriocins, some with relatively wide host range against Gram-positive bacteria and have potential as food biopreservatives.

Leuconostoc: Spherical or lenticular cells, occurring in pairs or chains, nonmotile, facultative anaerobes, heterolactic fermentators, mesophiles, but some species and strains can grow at 3°C or below. Some are used in food fermentation. Psychrotrophic strains are associated with spoilage (gas formation) of vacuum-packaged refrigerated foods. Found in plants, meat, and milk. Species: *Leuconostoc mesenteroides* subsp. *mesenteroides, Leu. lactis, Leu. carnosum. Leu. mesenteroides* subsp. *dextranicum* produces dextran while growing in sucrose. Several strains produce bacteriocins, some with wide antibacterial spectrum against Gram-positive bacteria that have potential as food biopreservatives.

Pediococcus: Spherical cells (1 μm), forms tetrads mostly present in pairs, nonmotile, facultative anaerobic, homolactic fermentators, mesophilic, but some can grow at 50°C. Some survive pasteurization. Some species and strains are used in food fermentation. Some can cause spoilage of alcoholic beverages. Found in vegetative materials and in some food products. Species: *Pediococcus acidilactici* and *Ped. pentosaceus*. Several strains produce bacteriocins, some with wide spectrum against Gram-positive bacteria and can be used as food biopreservatives.

Sarcina: Large spherical cells (1 to 2 μm), occurring in packets of eight or more, nonmotile, produce acid and gas from carbohydrates, facultative anaerobes, present in soil, plant products, and animal feces. Can be involved in spoilage of foods of plant origin. Species: *Sarcina maxima*.

5. Gram-Positive, Endospore-Forming Rods

Bacillus: Rod-shaped, straight cells, vary widely in size (small, medium, or large; 0.5 to 1 × 2 to 10 μm) and shape (thick or thin), single or in chains, motile or nonmotile, mesophilic or psychrotrophic, aerobic or facultative anaerobic, all form endospores that can be spherical or oval and large or small (one per cell). Spores are highly heat resistant. Includes many species, some of which are important in foods, due to their ability to cause foodborne disease (*Bacillus cereus*) and food spoilage, especially in canned products (*Bac. coagulans* and *Bac. stearothermophilus*). Enzymes of some species and strains are used in food bioprocessing (*Bacillus subtilis*). Present in soil, dust, plant products (especially spices). Many species and strains are capable of producing extracellular enzymes that hydrolyze carbohydrates, proteins, and lipids.

Sporolactobacillus: Slender medium-sized rods (1 × 4 μm), motile, microaerophilic, homolactic, form endospores (spore formation is rare in most media), but the spores are less heat resistant than *Bacillus* spores. Found in chicken feed and soil. Importance in food is not clearly known. Species: *Sporolactobacillus inulinus*.

Clostridium: Rod-shaped cells vary widely in size and shape, motile or nonmotile, anaerobic (some species are extremely sensitive to oxygen), mesophilic or psychrotrophic. Form endospores (oval or spherical) usually at one end of the cell, some species sporulate poorly. Spores are heat resistant. Found in soil, marine sediments, sewage, decaying vegetation, animal and plant products. Some are pathogens and important in food (*Clostridium botulinum, Clo. perfringens*), while others are important in food spoilage (*Clo. tyrobutyricum, Clo. saccharolyticum, Clo. laramie*). Some species are used as sources of enzymes to hydrolyze carbohydrates and proteins in food processing.

6. Gram-Negative, Endospore-Forming Rods

Desulfatomaculum: One species important in food is *Delsufatomaculum nigrificans*. The medium-sized cells are rod shaped, motile, thermophilic, strictly anaerobic, and produce H_2S. Endospores are oval and resistant to heat. Found in soil. Cause spoilage of canned food.

7. Gram-Positive, Nonsporulating Regular Rods

Lactobacillus: Rod-shaped cells vary widely in shape and size, some can be very long while some can be coccobacilli, appear in single or in small and large chains. Facultative anaerobic, most species are nonmotile, mesophilic (but some are psychrotrophs), can be homo- or heterolactic fermentors. Found in plant sources, milk, meat, and feces. Many are used in food bioprocessing (*Lactobacillus delbrueckii* subsp. *bulgaricus, Lab. helveticus, Lab. plantarum*) and some are used as probiotics (*Lab. acidophilus, Lab. reuteri, Lab. casei* subsp. *casei*). Some species can grow at low temperatures in products stored at refrigerated temperature (*Lab. sake, Lab. curvatus*). Several strains produce bacteriocins, some of which with wide spectrum can be used as food biopreservatives.

Carnobacterium: Similar in many characteristics to lactobacilli cells, found in meat and fish, facultative anaerobic, heterofermentative, nonmotile, can grow in foods, especially in meat products, stored at refrigerated temperature. Some strains produce bacteriocins. Species: *Carnobacterium piscicola*.

Brochothrix: Similar in many characteristics to lactobacilli, facultative anaerobic, homofermentative, nonmotile, found in meat. Capable of growing in refrigerated vacuum-packaged meat and meat products. Species: *Brochothrix thermosphacta*.

Listeria: Short rods (0.5 × 1 μm), occur singly or in short chains, motile, facultative anaerobic, can grow at 1°C. The species are widely distributed in the environment and have been isolated from different types of foods. The cells are killed by pasteurization. *Listeria monocytogenes* (some strains) are important foodborne pathogens.

8. Gram-Positive, Nonsporeforming Irregular Rods

Corynebacterium: Slightly curved rods, some cells stain unevenly, facultative anaerobes, nonmotile, mesophiles, found in the environment, plants, and animals. Some species have been found to cause food spoilage. *Corynebacterium glutamicum* is used to produce glutamic acid.

Brevibacterium: Cells can change from rod to coccoid shape, aerobic, nonmotile, mesophilic. Two species, *Brevibacterium linens* and *Brevibacterium casei,* have been implicated in the development of the aroma in several cheese varieties (surface ripened) due to the production of sulfur compounds (such as methanethiol). In other protein-rich products, they can cause spoilage (in fish). Found in different cheeses and raw milk.

Propionibacterium: Pleomorphic rods (0.5 × 2 μm), can be coccoid, bifid, or branched, present singly or in short chains, v and y configuration and in clumps with "Chinese character"-like arrangement. Nonmotile, anaerobic, mesophilic. Dairy propionibacteria are used in food fermentation (*Propionibacterium freudenreichii* in Swiss cheese). Produce proline and propionic acid. Found in raw milk, Swiss cheese, and silage.

Bifidobacterium: Cells are rods of various shapes, present as single or in chains, arranged in V or star-like shape, nonmotile, mesophilic, anaerobic. Metabolize carbohydrates to lactate and acetate. Found in the colons of humans, animals, and birds. Some species are used in probiotics (*Bifidobacterium bifidum, Bif. infantis, Bif. adolescentis*).

VI. IMPORTANT BACTERIAL GROUPS IN FOODS

Among the microorganisms found in foods, bacteria constitute major important groups.[10] This is not only because many different species can be present in foods, but is also due to their rapid growth rate, ability to utilize food nutrients, and their ability to grow under a wide range of temperatures, aerobiosis, pH, and water activity, as well as to better survive adverse situations, such as survival of spores at high temperature. For convenience, bacteria important in foods have been arbitrarily divided into several groups on the basis of similarities in certain characteristics. This grouping does not have any taxonomic significance. Some of these groups and their importance in foods are listed here.

A. Lactic Acid Bacteria

Those bacteria that produce relatively large quantities of lactic acid from carbohydrates. Include species mainly from genera *Lactococcus, Leuconostoc, Pediococcus, Lactobacillus*, and *Streptococcus thermophilus*.

B. Acetic Acid Bacteria

Those bacteria that produce acetic acid, such as *Acetobacter aceti*.

C. Propionic Acid Bacteria

Those bacteria that produce propionic acid and are used in dairy fermentation. Include species such as *Propionibacterium freudenreichii*.

D. Butyric Acid Bacteria

Those bacteria that produce butyric acid in relatively large amounts. Some *Clostridium* spp. such as *Clostridium butyricum*.

E. Proteolytic Bacteria

Those bacteria that are capable of hydrolyzing proteins, due to production of extracellular proteinases. Species in genera *Micrococcus, Staphylococcus, Bacillus, Clostridium, Pseudomonas, Alteromonas, Flavobacterium, Alcaligenes*, some in *Enterobacteriaceae*, and *Brevibacterium* are included in this group.

F. Lipolytic Bacteria

Able to hydrolyze triglycerides due to production of extracellular lipases. Species in genera *Micrococcus, Staphylococcus, Pseudomonas, Alteromonas,* and *Flavobacterium* are included in this group.

G. Saccharolytic Bacteria

Able to hydrolyze complex carbohydrates. Include some species in genera *Bacillus, Clostridium, Aeromonas, Pseudomonas,* and *Enterobacter.*

H. Thermophilic Bacteria

Able to grow at 50°C and above. Include some species from genera *Bacillus, Clostridium, Pediococcus, Streptococcus,* and *Lactobacillus*.

I. Psychrotrophic Bacteria

Able to grow at refrigerated temperature (≤5°C). Include some species from *Pseudomonas, Alteromonas, Alcaligenes, Flavobacterium, Serratia, Bacillus, Clostridium, Lactobacillus, Leuconostoc, Carnobacterium, Brochothrix, Listeria, Yersinia,* and *Aeromonas*.

J. Thermoduric Bacteria

Able to survive pasteurization temperature treatment. Include some species from *Micrococcus, Enterococcus, Lactobacillus, Pediococcus, Bacillus* (spores), and *Clostridium* (spores).

K. Halotolerant Bacteria

Able to survive high salt concentrations (≥10%). Include some species from *Bacillus, Micrococcus, Staphylococcus, Pediococcus, Vibrio,* and *Corynebacterium*.

L. Aciduric Bacteria

Able to survive at low pH (below 4.0). Include some species from *Lactobacillus, Pediococcus, Lactococcus, Enterococcus,* and *Streptococcus*.

M. Osmophilic Bacteria

Can grow in a relatively higher osmotic environment than other bacteria. Some species from genera *Staphylococcus, Leuconostoc,* and *Lactobacillus* are included in this group. They are much less osmophilic than yeasts and molds.

N. Gas-Producing Bacteria

Produce gas (CO_2, H_2, H_2S) during metabolism of nutrients. Include species from genera *Leuconostoc, Lactobacillus, Propionibacterium, Escherichia, Enterobacter, Clostridium,* and *Desulfotomaculum*.

O. Slime Producers

Produce slime due to synthesis of polysaccharides. Include some species or strains from *Xanthomonas, Leuconostoc, Alcaligenes, Enterobacter, Lactococcus,* and *Lactobacillus*.

P. Sporeformers

Ability to produce spores. Include *Bacillus, Clostridium,* and *Desulfotomaculum* spp. They are again divided into aerobic sporeformers, anaerobic sporeformers, flat sour sporeformers, thermophilic sporeformers, and sulfide-producing sporeformers.

Q. Coliforms

Include mainly species from *Escherichia, Enterobacter, Citrobacter,* and *Klebsiella* and used as index of sanitation.

R. Fecal Coliforms

Include mainly *Escherichia coli*. Also used as index of sanitation.

S. Enteric Pathogens

Includes pathogenic *Salmonella, Shigella, Campylobacter, Yersinia, Escherichia, Vibrio, Listeria,* hepatitis A, and others that can cause gastrointestinal infection.

Due to their importance in food, many laboratory methods are designed to detect a specific group instead of a specific genus or species. Similarly, control methods are sometimes designed to destroy or prevent growth of a specific group.

REFERENCES

1. Krieg, N. R., Ed., *Bergey's Manual of Systematic Bacteriology*, Vol. 1, Williams & Wilkins, Baltimore, 1984.
2. Sneath, P. H. A., Ed., *Bergey's Manual of Systematic Bacteriology,* Vol. II, Williams & Wilkins, Baltimore, 1986.
3. Holt, J. G., Ed., *Bergey's Manual of Determinative Bacteriology,* 9th Ed., Williams & Wilkins, Baltimore, 1993.
4. Beuchat, L. R., Ed., *Food and Beverage Mycology,* 2nd ed., Van Nostrand Reinhold, New York, 1987, 1.
5. Pelezar, M. J. and Chan, E. C. S., *Elements of Microbiology,* 2nd ed., McGraw Hill, New York, 1981, 69, 131, and 181.
6. Tortora, G. J., Funka, B. R., and Case, C. L., *Microbiology: An Introduction,* 4th ed., Benjamin/Cummings, Menlo Park, CA, 1992, 250.
7. Mata, M. and Ritzenhaler, P., Present state of lactic acid bacteria phage taxonomy, *Biochimie,* 70, 395, 1988.
8. Hill, C., Bacteriophages and bacteriophage resistance in lactic acid bacteria, *FEMS Microbiol. Rev.,* 12, 87, 1993.
9. Gerba, C. P., Viral diseases transmission by seafoods, *Food Technol.,* 42(3), 99, 1988.
10. Vanderzant, C. and Splittstoesser, D. F., Eds., *Compendium of Methods for the Microbiological Examination of Foods,* 3rd ed., American Public Health Assocciation, Washington, D.C., 1992.

CHAPTER 2 QUESTIONS

1. List five methods that are used in the classification of bacteria. Why are nucleotide sequences in 16S RNA used as an important technique in the classification?

2. Explain the term and give one example for each of the following in relation to bacterial nomenclature: (a) family, (b) genus, (c) species, (d) subspecies, and (e) biovar (use scientific method in the examples). Write the plural of *Lactobacillus, Staphylococcus, Enterococcus, Listeria,* and *Salmonella.*

3. List general differences in the morphology of yeasts, molds, bacteria, bacteriophages, and human viruses important in food.

4. List the differences in the chemistry and function of cell wall structures between Gram-positive and Gram-negative bacteria. How does this help determine Gram characteristics of an unknown bacterial isolate?

5. List four species of molds and two species of yeasts most important in food.

6. List the sections, as used in *Bergey's Manual of Systematic Bacteriology,* that contain important bacterial genera. In each section, list up to four genera.

7. List six important bacterial groups and include up to four genera in each group.

8. Explain how bacterial groups can be important to a food microbiologist.

CHAPTER 3

Sources of Microorganisms In Food

CONTENTS

I. INTRODUCTION

The internal tissues of healthy plants (fruits and vegetables) and animals (meat) are essentially sterile. Yet raw and processed (except sterile) foods contain different types of molds, yeasts, bacteria, and viruses. These microorganisms get into foods both from natural (including internal) sources and from external sources to which a food comes into contact from the time of production until the time of consumption. The natural sources for foods of plant origin include the surface of fruits, vegetables, and grains, and the pores in some tubers (e.g., radishes and onions). The natural sources for foods of animal origin include skin, hair, feathers, gastrointestinal tract, urogenital tract, respiratory tract, and milk ducts (teat canal) in udders of milk animals. Natural microflora exist in ecological balance with their hosts, and their types and levels vary greatly with the type of plants and animals as well as their geographical locations and the environmental conditions. Besides natural microorganisms, a food can be contaminated with different types of microorganisms coming from outside sources such as air, soil, sewage, water, humans, food ingredients, equipment, packages, and insects. Microbial types and their levels from these sources getting into foods vary widely and depend upon the degree of sanitation used during the handling of foods.

An understanding of the sources of microorganisms in food is important in order to develop methods to control access of some microorganisms in the food, to develop processing methods to kill them in food, and to determine the microbiological quality of food, as well as to set up microbiological standards and specifications of foods and food ingredients.

II. PREDOMINANT MICROORGANISMS IN DIFFERENT SOURCES

A. Plants (Fruits and Vegetables)

The inside tissue of foods from plant sources are essentially sterile, except for a few porous vegetables (e.g., radishes and onions) and leafy vegetables (e.g., cabbage and brussels sprouts). (See References 1 through 5). Some plants produce natural antimicrobial metabolites that can limit the presence of some microorganisms. Fruits and vegetables harbor microorganisms on the surface; their type and level vary with the soil condition, type of fertilizers and water used, and air quality. Molds, yeasts, lactic acid bacteria, and bacteria from genera *Pseudomonas, Alcaligenes, Micrococcus, Erwinia, Bacillus, Clostridium,* and *Enterobacter* can be expected from this source. Pathogens, especially enteric types, can be present if the soil is contaminated with untreated sewage. Diseases of the plants, damage of the surface before, during, and after harvest, long delays between harvesting and washing, and unfavorable storage and transport conditions after harvesting and before processing can greatly increase the microbial numbers as well as the predominant types.

Proper methods used during growing (such as use of treated sewage or other types of fertilizers), damage reduction during harvesting, quick washing with good quality water to remove soil and dirt, and storage at low temperature until processed can be used to reduce microbial load in foods of plant origin.

B. Animals, Birds, Fish, and Shellfish

Food animals and birds normally carry many types of indigenous microorganisms in the digestive, respiratory, and urogenital tracts, the teat canal in the udder, as well as in the skin, hooves, hair, and feathers. Their numbers, depending upon the specific organ, can be very high (large intestinal content can have as high as 10^{10} bacteria per gram). Many, as carriers, can carry pathogens such as *Salmonella* spp., pathogenic *Escherichia coli, Campylobacter jejuni, Yersinia enterocolitica,* and *Listeria monocytogenes* without showing symptoms. Laying birds have been suspected of asymptomatically carrying *Salmonella enteritidis* in the ovaries and contaminating the yolk during ovulation. Disease situations, such as mastitis in cows and intestinal, respiratory, and uterine infections as well as injury, can change the ecology of normal microflora. Similarly, poor husbandry resulting in fecal contamination on the body surface (skin, hair, feathers, and udder) and supplying contaminated water and feed (contaminated with salmonellae) can also change their normal microbial flora.

Fish and shellfish also carry normal microflora in the scales, skin, and digestive tracts. Water quality, feeding habits, and diseases can change the normal microbial types and level. Pathogens such as *Vibrio parahaemolyticus, Vibrio vulnificus,* and *Vibrio cholerae* are of major concern from these sources. Many spoilage and pathogenic microorganisms can get into foods of animal origin (milk, egg, meat, and fishery products) during production and processing. Milk can become contaminated with fecal materials on the udder surface, egg shells can be contaminated with fecal material, meat can be contaminated with the intestinal contents during slaughtering, and fish can become contaminated with intestinal contents during processing. Because of the specific nature, contamination of foods from animal sources with fecal materials is viewed with concern (possible presence of enteric pathogens).

In addition to enteric pathogens from fecal materials, meat from food animals and birds can be contaminated with several spoilage and pathogenic microorganisms from skin, hair, and feathers, namely *Staphylococcus aureus, Micrococcus* spp., *Propionibacterium* spp., *Corynebacterium* spp., and molds and yeasts.

Prevention of food contamination from these sources needs the use of effective husbandry of the live animals and birds. Also, testing animals and birds for pathogens and culling the carriers will be important in reducing their incidence in foods. During slaughter, washing, removing hair, defeathering, use of good quality water for washing carcasses (preferably with acceptable antimicrobial agents), care in removing digestive, urogenital, and respiratory organs without contaminating tissues, removing contaminated parts, and proper sanitation during the entire processing stage are necessary to keep the microbial quantity and quality at desirable levels. Proper cleaning of the udder prior to milking, cooling milk immediately after milking, temperature treatment as soon as possible, and sanitation at all stages are important in order to keep microbial levels low in milk. Eggs should be collected soon after laying and washed and stored using recommended procedures.

Fish and marine products should be harvested from unpolluted and recommended water. Proper sanitation should be used during processing. They should be stored properly to prevent further contamination and microbial growth.

C. Air

In the air, microorganisms are present in the dust. They do not grow in dust, but are transient and variable depending upon the environment. Their level is controlled by the degree of humidity, size and level of dust particles, temperature and air velocity, and resistance of microorganisms to drying. Generally, dry air with low dust content and higher temperature has a low microbial level. Spores of *Bacillus* spp., *Clostridium* spp., molds, and some Gram-positive bacteria (e.g., *Micrococcus* spp. and *Sarcina* spp.), as well as yeasts, can be predominantly present in air. If the surroundings contain a source of pathogens (e.g., animal farms or a sewage treatment plant), different types of bacteria, including pathogens and viruses (including bacteriophages), can be transmitted via the air.

Microbial contamination of food from the air can be reduced by removing the potential sources, controlling dust particles in the air (using filtered air), using positive air pressure, reducing the humidity level, and installing UV light.

D. Soil

Soil, especially the types used to grow agricultural produce and raise animals, contains several varieties of microorganisms. Because they can multiply in soil, their numbers can be very high (billions per gram). Many types of molds, yeasts, and bacterial genera (e.g., *Enterobacter, Pseudomonas, Proteus, Micrococcus, Enterococcus, Bacillus,* and *Clostridium*) can get into foods from the soil. Soil contaminated with fecal materials can be the source of enteric pathogenic bacteria and viruses in food. Sediments where fish and marine foods are harvested can also be a source of microorganisms in those foods. Removal of soil (and sediments) and avoiding soil contamination are used to reduce microorganisms in foods from this source.

E. Sewage

Sewage, especially when used as fertilizer in crops, can contaminate food with microorganisms. The most important types are different enteropathogenic bacteria and viruses. This can be of major concern with organically grown foods and many imported fruits and vegetables, where untreated sewage may be used as fertilizer.

To reduce the incidence of microbial contamination of foods from sewage it is better not to use sewage as fertilizer. If it is used, it should be efficiently treated to kill the pathogens. Also, washing foods following harvesting is important.

F. Water

Water is used to produce, process, and, under certain conditions, store foods. It is used for irrigation of crops, drinking by food animals, raising fishery and marine products, washing foods, processing (pasteurization, canning, and cooling of heated foods) and storage of foods (e.g., fish on ice), washing and sanitation of equipment, and processing and transportation facilities. Water is also used as an ingredient in many processed foods. Thus water quality can greatly influence the microbial quality of foods.

Wastewater can be recycled for irrigation. However, chlorine-treated potable water (drinking water) should be used in processing, washing, sanitation, and as an ingredient. Although potable water does not contain coliforms and pathogens (mainly enteric types), it can contain other bacteria capable of causing food spoilage; these include *Pseudomonas, Alcaligenes,* and *Flavobacterium*. Improperly treated water can contain pathogen and spoilage microorganisms. To overcome the problems, many food processors use water, especially as an ingredient, that has higher microbial quality than potable water.

G. Humans

Between production and consumption, foods come in contact with different people handling the foods. They include not only the people working in a food processing plant, but also those handling foods at restaurants, catering services, retail stores, and at home. They have been the source of pathogenic microorganisms in foods that later caused foodborne diseases, especially with ready-to-eat foods. Improperly cleaned hands, lack of aesthetic sense and personal hygiene, dirty clothes, and hair can be major sources of microbial contamination in foods. The presence of minor cuts and infection in hands and face and mild generalized diseases (e.g., flu, strep throat, and hepatitis A in an early stage) can amplify the situation. In addition to spoilage bacteria, pathogens such as *Staphylococcus aureus, Salmonella* spp., *Shigella* spp., pathogenic *Escherichia coli*, and hepatitis A can be introduced into foods from human sources.

Proper training of personnel in personal hygiene, regular checking of health, and maintaining efficient sanitary and aesthetic standards are necessary to reduce contamination from this source.

H. Food Ingredients

In prepared or fabricated foods, many ingredients or additives are included in different quantities. Many of these ingredients can be the source of both spoilage and pathogenic microorganisms. Various spices can possess very high populations of mold and bacterial spores. Starch, sugar, and flour can have spores of thermophilic bacteria. Pathogens have been isolated from dried coconut, egg, and chocolate.

The ingredients should be produced under sanitary conditions and given antimicrobial treatments. In addition, setting up acceptable microbial specifications for the ingredients will be important in reducing microorganisms in food from this source.

I. Equipment

A wide variety of equipment is used in harvesting, transportation, processing, and storage of foods. Many types of microorganisms from air, raw foods, water, and personnel can get into the equipment and contaminate foods. Depending upon the environment (moisture, nutrients, and temperature) and time, the microorganisms can multiply and, even from a low initial population, reach a high level and contaminate large volumes of foods. Also, when processing equipment is used continuously for a long period of time, microorganisms present initially can multiply and act as a continuous source of contamination in the product. In some equipment, small parts, inaccessible sections, and certain materials may not be efficiently cleaned and sanitized. They (dead spots) can serve as sources of both pathogenic and spoilage microorganisms in food. Similarly, small equipment, such as cutting boards, knives, spoons, and similar articles, due to improper cleaning, can be the source of cross-contamination. *Salmonella, Listeria, Escherichia, Enterococcus, Micrococcus, Pseudomonas, Lactobacillus, Leuconostoc, Listeria, Clostridium, Bacillus* spp., and yeasts and molds could get into the food from equipment.

Proper cleaning and sanitation of equipment at prescribed intervals are important in order to reduce microbial levels in food. In addition, developing means to prevent or reduce contamination from air, water, personnel, and insects will be important. Finally, in designing the equipment, possible microbiological problems need to be considered.

Many types of packaging materials are used in food. Since they are used in the products ready for consumption and in some cases without further heating, proper microbiological standards (or specifications) for packaging materials are necessary.

J. Miscellaneous

Foods may be contaminated with microorganisms from several other sources, namely wrapping materials, containers, flies, vermin, birds, pets, and rodents. Packaging materials are extensively used in many types of ready-to-eat foods. Any failure to produce microbiologically acceptable products can reduce the quality of food. Flies, vermin, birds, and rodents in food processing and food preparation facilities should be viewed with concern as they can carry pathogenic microorganisms.

REFERENCES

1. Krieg, N. R., Ed., *Bergey's Manual of Systematic Bacteriology,* Vol. 1, Williams & Wilkins, Baltimore, 1984.
2. Sneath, P. H. A., Ed., *Bergey's Manual of Systematic Microbiology,* Vol. 2, Williams & Wilkins, Baltimore, 1988.
3. Holt, J. G., Ed., *Bergey's Manual of Determinative Bacteriology,* 9th ed., Williams & Wilkins, Baltimore, 1993.
4. Vanderzant, C. and Splittstoesser, D. F., Eds., *Compendium of Methods for the Microbial Examination of Foods,* 3rd ed., American Public Health Association, Washington, DC, 1992.
5. Doyle, M., Ed., *Foodborne Bacterial Pathogens,* Marcel Dekker, New York, 1989.

CHAPTER 3 QUESTIONS

1. Briefly discuss how an understanding of the microbial sources in food can be helpful to a food microbiologist.

2. A *Salmonella enteritidis* outbreak from the consumption of a national brand of ice cream, involving more than 50,000 consumers, was found to be related to contamination of heat-treated ice cream mix with liquid egg containing the pathogen. As a food microbiologist, how can this information help determine the original source of *Salmonella enteritidis*?

3. A batch of turkey rolls (10 lb each) was cooked to 165°F internal temperature in bags, opened, sliced, vacuum-packaged, and stored at 40°F. The product is expected to have a refrigerated shelf-life of 50 d. However, after 40 d, the packages contained gas and about 10^7 cells per gram of meat. The bacterial species involved in the spoilage were found to be *Leuconostoc carnosum,* which is killed at 165°F. What could be the source(s) of these bacterial species in this cooked product?

4. About 20,000 lb of beef patties were found to be contaminated with *Escherichia coli* 0157:H7. The presence of this pathogen is suspected in the gastrointestinal tract of cows, but less than 1% of cows are carriers of this organism. Discuss how such a large volume of finished product can be contaminated from a source of very low-level contamination.

5. Vacuum-packaged beef (5- to 10-lb size) was roasted to an internal temperature of 168°C, cooked, and stored at about 10°C without opening the original bag. In about 2 weeks, some of the bags accumulated large amounts of gas and purge (liquid). Examination of the purge showed the presence of cells of a psychrotrophic *Clostridium* species in large numbers (10^{6-7}/ml). Explain the possible sources of this species in the vacuum-packaged roasted beef.

CHAPTER 4

Significance of Microorganisms in Food

CONTENTS

I. INTRODUCTION

It is evident from previous discussion that although there are many types (genera, species, strains) of microorganisms present in nature, under normal conditions generally a food harbors only a few types. They constitute those that are naturally present in raw foods (which provide the ecological niche) and those that get in from outside sources to which the foods are exposed from the time of production until consumption. The relative numbers (population level) of a specific type of microorganism normally present in a food will depend upon the intrinsic and extrinsic conditions to which the food is exposed. The predominant type(s) will be the ones for which the optimum growth condition is present. This aspect is discussed in Chapter 6. The objective of this discussion is to develop an understanding of the microbial types (and their levels where possible) that can be expected in different foods under normal conditions. Instead of each food, the information for several food groups is presented.

II. RAW AND READY-TO-EAT MEAT PRODUCTS

Following slaughtering and dressing, the carcasses of animals and birds contain many types of microorganisms, predominantly bacteria, coming from themselves (skin, hair, feathers, gastrointestinal tract, etc.), the environment of the feedlot and pasture (feed, water, soil, and manure), and the environment at the slaughtering facilities (equipment, air, water, and humans). Normally, the carcasses contain an average of 10^{1-3} cells per square inch. Different enteric pathogens, *Salmonella* spp., *Yersinia enterocolitica, Campylobacter jejuni, Escherichia coli, Clostridium perfringens,* and *Staphylococcus aureus*, both from animals/birds and humans can be present, but normally in low levels. Carcasses of birds, as compared to animals, generally have a higher incidence of *Salmonella* contamination coming from fecal matter.[1-4]

Following boning, the chilled raw meat and ground meat contain microorganisms coming from the carcasses as well as from different equipment used during processing, personnel, air, and water. Some of the equipment used can be important sources of microorganisms, such as conveyors, grinders, slicers, and similar types that can be difficult to clean. The chilled meat will have mesophiles, such as *Micrococcus, Enterococcus, Staphylococcus, Bacillus, Clostridium, Lactobacillus*, coliforms, and other *Enterobacteriaceae,* including enteric pathogens. However, because the meats are kept at low temperature (–1 to 5°C), the psychrotrophs constitute major problems. The predominant psychrotrophs in raw meats are some lactobacilli and leuconostocs, *Brochothrix thermosphacta, Clostridium laramie*, some coliforms, *Serratia, Pseudomonas, Alteromonas, Achromobacter, Alcaligenes, Acinetobacter, Moraxella, Aeromonas,* and *Proteus*. The psychrotrophic pathogens include *Listeria monocytogenes* and *Yersinia enterocolitica*. The microbial load of fresh meat varies greatly, with bacteria being predominant. Ground meat can contain $10^{4,5}$ microorganisms per gram; *Salmonella* can be present at about 1 cell per 25 g. As indicated before, the frequency of the presence of *Salmonella* is higher in chicken than in red meats. If the products are kept under aerobic conditions, the psychrotrophic aerobes will grow rapidly, especially Gram-negative rods such as *Pseudomonas, Alteromonas, Proteus,* and *Alcaligenes* as well as yeasts. Under anaerobic packaging, growth of psychrotrophic, facultative anaerobes and anaerobes (e.g., *Lactobacillus, Leuconostoc, Brochothrix, Serratia*, some coliforms, and *Clostridium*) predominate. The pH of the meat (which is low in beef, about 5.6, but high in birds, about 6.0), high protein content, and low carbohydrate level, along with the environment, will determine which type or types will predominate.

Low-heat processed red meat and poultry products include perishable cured or uncured products that have been subject to heat treatment to about 160°F, packaged aerobically or anaerobically, and stored at refrigerated temperature. They include products like franks, bologna, lunch meats, and hams. The products, especially those packaged anaerobically and cured, are expected to have long storage lives (50 d or more). The microbial sources prior to heat treatment include the raw meat, ingredients used in formulation, equipment, air, and personnel. Heat treatment, especially at an internal temperature of 160°F or higher, will kill most microorganisms, except some thermodurics and spores of *Bacillus* and *Clostridium*, which include *Micrococcus*, some *Enterococcus*, and maybe some *Lactobacillus*. The microbial level can be 10^{1-2} per gram. Following heating, the products, some of which are further processed (such as removing casing, slicing) come in contact with equipment, personnel, air, and water prior to final packaging. Depending upon the conditions of the processing plants, different types of bacteria, yeasts, and molds, including pathogens, get into these products. Although the initial bacterial level normally does not exceed 10^2 per gram, they can be psychrotrophic facultative anaerobic and anaerobic bacteria (*Lactobacillus, Leuconostoc*, some coliforms, *Serratia, Listeria, Clostridium* spp.). During extended storage in vacuum or controlled-air packages, even from a low initial level, they can reach a high population and adversely affect the safety and shelf-life of the products. This is aggravated by fluctuation in storage temperature and in products having low fat, high pH, and high A_w.[1-4]

III. RAW AND PASTEURIZED MILK

Raw milk may come from cows, buffalo, sheep, and goats, although the largest volume comes from cows. Pasteurized or market milk includes whole, skim, low-fat, and flavored milks, as well as cream, which are heat treated (pasteurized) according to regulatory specifications.[1,2,5] Milk is high in proteins and carbohydrates that microorganisms can utilize for growth. Since both raw and pasteurized milk contain many types of bacteria as principal microorganisms, they are refrigerated; yet they have a limited shelf-life.

In raw milk, microorganisms come from inside the udder, animal body surfaces, feed, air, water, and utensils and equipment used for milking and storage. The predominant types from inside a healthy udder are *Micrococcus, Streptococcus,* and *Corynebacterium*. Normally, raw milk contains $<10^3$ microorganisms per milliliter. In the case of mastitis, *Streptococcus agalactiae, Staphylococcus aureus,* coliforms, and *Pseudomonas* can be excreted in relatively high numbers. Contaminants from animals, feed, soil, and water predominantly constitute lactic acid bacteria, coliforms, *Micrococcus, Staphylococcus, Enterococcus, Bacillus, Clostridium* spores, and Gram-negative rods. Pathogens such as *Salmonella, Listeria monocytogenes, Yersinia enterocolitica,* and *Campylobacter jejuni* can also come from some of these sources. Equipment can be a major source of Gram-negative rods, namely *Pseudomonas, Alcaligenes,* and *Flavobacterium,* as well as Gram-positive *Micrococcus* and *Enterococcus*.

During refrigerated storage prior to pasteurization, only the psychrotrophs can grow in raw milk. These include *Pseudomonas, Flavobacterium, Alcaligenes,* and some coliforms and *Bacillus* spp. They can affect the acceptance quality of raw milk (e.g., flavor and texture). Some of them are also capable of producing heat-stable enzymes (proteinases and lipases) that can affect the product quality even after pasteurization of raw milk. Psychrotrophic pathogens (*Lis. monocytogenes* or *Yer. enterocolitica*) can multiply in refrigerated raw milk.

The microbiological quality of raw and pasteurized milk is monitored in many countries by regulatory agencies. In the United States the standard plate counts of raw milk for use as market milk can be 1 to 3 × 10^5 per milliliter, and for use in product manufacturing can be 0.5 to 1 × 10^6 per milliliter. Grade A pasteurized milk can have standard plate counts of 20,000 per milliliter and ≤ 10 coliforms per milliliter.

The microorganisms of pasteurized milk constitute those that survive pasteurization (i.e., the thermodurics) of raw milk and those that enter after heating and before packaging (i.e., post-pasteurization contaminants). The thermodurics surviving pasteurization include *Micrococcus*, some *Enterococcus* (e.g., *Ent. faecalis*), *Streptococcus*, some *Lactobacillus* (e.g., *Lab. viridescens*), and spores of *Bacillus* and *Clostridium*. Post-heat contaminants can be coliforms as well as *Pseudomonas, Alcaligenes,* and *Flavobacterium*. Some heat-sensitive pathogens can also enter pasteurized milk following heat treatment; among them, psychrotrophs can grow during refrigerated storage.[1,2,5]

IV. SHELL EGG AND LIQUID EGG

Shell eggs are contaminated with microorganisms on the outer surface from fecal matter, nesting materials, feeds, air, and equipment. Each shell, depending upon the contamination level, can have 10^7 bacteria. Washing helps reduce their level considerably. Egg shells can harbor different types of bacteria, namely *Pseudomonas, Alcaligenes, Proteus, Citrobacter, Escherichia coli, Enterobacter, Enterococcus, Micrococcus,* and *Bacillus*. They may also have *Salmonella* from fecal contamination. Infected ovaries have been suspected to be the source of *Salmonella enteritidis* inside eggs. Liquid egg can be contaminated with bacteria from the shell of washed eggs as well as from the breaking equipment, water, and air. Pasteurization can reduce the numbers to 10^3 per milliliter. Bacteria, especially motile Gram-negative bacteria, can enter through pores of the egg shells, particularly if the shells are wet. Several antimicrobial factors present in egg albumin, such as lysozyme, conalbumin (binds iron), avidin (binds biotin), and alkaline pH (8.0 to 9.0), can control bacterial growth. However, they can grow in yolk that is rich in nutrients and has pH 7.0. Pasteurization of liquid egg has been designed to destroy the pathogens (especially *Salmonella*) and other Gram-negative rods. The thermoduric bacteria, namely *Micrococcus, Enterococcus,* and *Bacillus*, present in the raw liquid egg will survive.[1,2,6]

V. FISH AND SHELLFISH

This group includes fin fish, crustaceans (shrimp, lobster, crabs), and mollusks (oysters, clams, scallops) harvested from aquatic environments (marine and freshwater).[1,2,7] They are harvested from both natural sources and from aquacultures. In general, they are rich in protein as well as nonprotein nitrogenous compounds, their fat content varies with type and season, and except for mollusks, they are very low in carbohydrates; mollusks contain about 3% glycogen.

The microbial population in these products varies greatly with the pollution level and temperature of the water. Bacteria from the major groups, as well as viruses and protozoa, can be present in the raw materials. Muscles of fish and shellfish are sterile, but they harbor microorganisms in scales, gills, and intestines. Fin fish and crustaceans can have 10^{3-8} bacterial cells per gram. Mollusks, during feeding, filter large volumes of water and thus can concentrate bacteria and viruses. The products harvested from marine environments have halophilic vibrios as well as *Pseudomonas, Alteromonas, Flavobacterium, Enterococcus, Micrococcus,* coliforms, and pathogens such as *Vibrio parahaemolyticus, Vibrio vulnificus,* and *Clostridium botulinum* type E. Freshwater fish generally have *Pseudomonas, Flavobacterium, Enterococcus, Micrococcus, Bacillus*, and coliforms. Fish and shellfish harvested from water polluted with human and animal waste may contain *Salmonella, Shigella, Clostridium perfringens, Vibrio cholerae*, and hepatitis A and Norwalk viruses. They may also contain opportunistic pathogens, such as *Aeromonas hydrophila* and *Plesiomonas shigelloides*. The harvesting of seafoods, especially shellfish, is controlled by regulatory agencies in the United States. Water with high coliform populations is closed to harvest.

Following harvest, microorganisms are capable of growing rapidly in fish and crustaceans due to the availability of large amounts of nonprotein nitrogenous compounds, high A_w, and high pH. As many of the bacterial species are psychrotrophs, they are capable of growth at refrigerated temperatures. The pathogens can remain viable for a long time during storage. Microbial loads are greatly reduced during their subsequent heat processing to produce different products.

VI. VEGETABLES, FRUITS, AND NUTS

Vegetables include edible plant components that constitute leaves, stalks, roots, tubers, bulbs, and flowers. In general, they are relatively high in carbohydrates, with pH values of 5.0 to 7.0. Thus different types of bacteria, yeasts, and molds can grow if other conditions are favorable.[1,2,8] Fruits are high in carbohydrates, have pH 4.5 or below due to organic acids, and some also have antimicrobial essential oil.[1,2,8] Nuts can be from the ground (peanuts) or from trees (pecans) and have protective shells and low A_w (0.7).[1,2,8] They are converted to nutmeats for further use or to products such as peanut butter.

Microorganisms in vegetables can come from several sources (soil, water, air, animals, insects, birds, or equipment) and vary with the types of vegetables. While a leafy vegetable will have more microorganisms from the air, a tuber will have more from the soil. The microbial levels and types in these products also vary greatly depending upon the environmental conditions and the conditions of harvesting. Generally, vegetables can have 10^{3-5} microorganism per square centimeter or 10^{4-7} per gram. Some of the predominant bacterial types are: lactic acid bacteria, Coryneforms, *Enterobacter, Proteus, Pseudomonas, Micrococcus, Enterococcus,* and sporeformers. They will also have different types of molds, such as *Alternaria, Fusarium,* and *Aspergillus*. Vegetables can have enteric pathogens, especially if animal and human wastes and polluted water are used for fertilization and irrigation. These include *Listeria monocytogenes, Salmonella, Shigella, Campylobacter, Clostridium botulinum*, and *Clostridium perfringens*. If the vegetables are damaged, then plant pathogens, (e.g., *Erwinia*) can also be predominant. Many of the microorganisms can cause different types of spoilage (different types of rot) of the raw products. Pathogens can cause foodborne diseases (e.g., listeriosis or botulism). Lactic acid bacteria have important roles in the natural fermentation of vegetables (e.g., sauerkraut). Different methods used for the processing of vegetables and vegetable products greatly reduce the microbial population.

Fruits, due to high carbohydrate content and low pH, favor the growth of different types of molds, yeasts, and lactic acid bacteria. The bacteria generally come from air, soil, insects, and harvesting equipment. In general, microbial populations can vary between 10^{3-6} per gram. Improperly harvested and processed fruits can have pathogens that survive and cause foodborne disease. Molds, yeasts, and bacteria can cause different types of spoilage. Natural flora, especially yeasts in fruits, can be important in alcohol fermentation.

Microorganisms get into nuts from soil (peanuts) and air (tree nuts). During processing, air, equipment, and water can also be the sources. They are protected by shells, but damage on the shell can facilitate microbial contamination. Raw nuts and nutmeats can have 10^{3-4} microorganisms per gram, with *Bacillus* and *Clostridium* spores, *Leuconostoc, Pseudomonas,* and *Micrococcus* predominating. Due to a low A_w, bacteria will not grow in the products. However, when used as ingredients, they can cause microbiological problems in the products. Molds can grow in nuts and nutmeats and produce mycotoxins (from toxin-producing strains such as aflatoxins by *Aspergillus flavus*).

VII. CEREAL, STARCHES, AND GUMS

Cereal includes grains, flour, meals, breakfast cereals, pasta, baked products, dry mixes, and frozen and refrigerated products of cereal grains (also beans and lentils). Starches include flours of cereals (corn and rice), tapioca (from plant), potato, and other tubers. Gums are used as stabilizers, gelling agents, and film, and are obtained from plants, seaweeds, and microorganisms (tragacanth, pectin, xanthan, agar, and carrageenan) and as modified compounds (carboxymethyl cellulose). They are rich in amylose and amylopectin but can also have simple sugars (e.g., in grains) and protein (e.g., in lentils). Microbial sources are mainly the soil, air, insects, birds, and equipment.[1,2]

The unprocessed products (grains) may contain high bacterial levels (aerobic plate count $\simeq 10^4$ per gram, coliform $\simeq 10^2$ per gram, yeasts and molds $\simeq 10^3$ per gram). They may also harbor mycotoxins produced by toxicogenic molds. Processed products may also contain a wide variety of yeasts, molds, and bacteria. Flours and starches may possess higher microbial counts similar to grains while processed products (such as breakfast cereals and pasta) may contain, per gram: aerobic plate count 10^{2-3}, coliform $<10^{1-2}$, yeasts and molds $<10^{1-2}$. They can possess bacterial spores and psychrotrophs. Some pathogens, such as *Salmonella, Staphylococcus aureus,* and *Clostridium perfringens,* have also been isolated. Depending upon the product, they can either grow (such as in dough) or can be the source (when used as ingredients) for both spoilage and pathogenic microorganisms as well as mycotoxins.

Gums also may be the source of yeasts, molds (also mycotoxins), bacterial spores, and lactic acid bacteria.

VIII. CANNED FOODS

This group of foods includes those packed in hermetically sealed containers and given high heat treatment. The products with pH 4.5 and above are given heat treatments to obtain commercial sterility, but those with pH below 4.5 are given heat treatments around 100°C.[1,2]

Canned foods prepared and processed to obtain commercial sterility can have spores of thermophilic spoilage bacteria, namely *Bacillus stearothermophilus, Clostridium thermosaccharolyticum,* and *Desulfotomaculum nigrificans*. Their major sources in the products are soil and blanching water, as well as sugar and starches used as ingredients. In canned products stored at 30°C or below, the spores do not germinate to cause spoilage; but if the cans are temperature abused to 40°C or higher, the spores germinate, multiply, and spoil the products.

If the canned products are given lower heat treatment (close to 100°C), spores of mesophilic bacteria that include both spoilage (*Bacillus coagulans, Bac. licheniformis, Clostridium sporogenes, Clo. butyricum*) and pathogenic types (*Bac. cereus, Clo. perfringens, Clo. botulinum*), along with the spores of thermophiles, will survive. In low pH products, particularly in tomato products, *Bac. coagulans* spores can germinate, grow, and cause spoilage. Other sporeformers can germinate and grow in high pH products. *Staphylococcus aureus* toxins, if present in raw products, will not be destroyed by heat treatment of the canned products and thus can cause food poisoning following consumption of the products.[1,2]

IX. SUGARS AND CONFECTIONERIES

Refined sugar is obtained from sugar cane and beets. Sugar can have thermophilic spores of *Bacillus stearothermophilus, Bac. coagulans, Clostridium thermosaccharolyticum,* and *Desulfotomaculum nigrificans,* and mesophilic bacteria (such as *Lactobacillus* and *Leuconostoc*), yeasts, and molds.[1,2] When sugars are used as ingredients in food products, the spores can survive and cause product spoilage. Pathogens are not present in refined sugar unless contaminated. In liquid sugar, the mesophiles can grow. Refined sugar, used in canned products or to make liquid sugar, has strict microbiological (for spores) standards.

Confectioneries include a large variety of products with a sweet taste. In general, these products have low A_w (0.84 or less) and some have low pH. They may contain many types of bacteria, yeasts, and molds, but their microbiological standards are well regulated. Although they may harbor *Lactobacillus, Leuconostoc*, spores of *Bacillus* and *Clostridium*, and yeasts and molds, except for a few osmotolerant yeasts and molds, others cannot grow. However, when used as additives in other foods, confectioneries can be a source of these microbes. If the ready-to-consume products are contaminated with pathogens, either from raw materials, environment, or personnel, they can cause foodborne diseases.[1,2]

X. SOFT DRINKS, FRUIT AND VEGETABLE DRINKS, JUICES, AND BOTTLED WATER

Soft drinks are nonalcoholic beverages containing water, sweeteners, acids, flavoring, coloring and emulsifying agents, and preservatives. Some may contain fruit juices and be carbonated or noncarbonated, with pH 2.5 to 4.0. Fruit juices (100%) have pH 4.0 and below. Vegetable juices (e.g., tomato) can have pH 4.5 or above. Bottled water is obtained from either natural springs or drilled wells under conditions that prevent contamination.

Soft drinks can have different types of microorganisms, but only aciduric microorganisms (molds, yeasts, lactic acid bacteria, and acetic acid bacteria) can multiply. In carbonated beverages, yeasts being facultative anaerobic can grow; in beverages with fruit juices, *Lactobacillus* and *Leuconostoc* species can grow. [1,2] In noncarbonated beverages, molds (*Geotrichum*) and *Acetobacter* and *Gluconobacter* spp. can also grow. Most of these come from the processing environment and equipment. In fruit juices, molds, yeasts, *Lactobacillus* spp. (*Lab. fermentum, Lab. plantarum*), *Leuconostoc* spp. (*Leu. mesenteroides*), and acetic acid bacteria can grow. Pathogens (*Salmonella* spp., *Escherichia coli* 0157:H7 in apple cider) can remain viable for a long time (30 d) in the acid products.[1,2] Vegetable juices can have molds, yeasts, and lactic acid bacteria along with *Bacillus coagulans, Clostridium butyricum,* and *Clostridium pasteurianum*.[1,2]

Bottled water should not contain more than 10 to 100 bacteria and >10 coliforms per 100 ml. The indigenous flora are mainly *Flavobacterium, Alcaligenes,* and *Micrococcus*. They may also have some *Pseudomonas* as contaminants from outside. They should not have pathogens unless produced under poor sanitation.[1,2]

XI. MAYONNAISE AND SALAD DRESSINGS

These products constitute water-in-oil emulsion products formulated with oil, water, vinegar (about 0.25% acetic acid) or lemon juice, sugar, salt, starch, gum, egg, spices, vegetable pieces, and have a pH between 3.5 and 4.0. Some low-calorie and less sour products containing less acid, less oil, and more water and may have pH 4.5 or above.

Microorganisms are introduced into the products through the ingredients, equipment, and air. However, except for aciduric microorganisms, most others will die, especially when stored for a long time at room temperature. Among aciduric molds (*Geotrichum* and *Aspergillus* spp.), yeasts (*Saccharomyces* spp.), and several species of *Lactobacillus* (*Lab. fructivorans, Lab. brevis*) and some *Bacillus* spp. (*Bac. subtilis, Bac. mesentericus*) have been isolated.[1,2,9] Normally, their numbers should not exceed 10 per gram. If pathogens are introduced (e.g., *Salmonella* through eggs), they are expected to be killed rapidly; however, they may survive longer in low-calorie, high-pH products kept at refrigerated temperature.

XII. SPICES AND CONDIMENTS

Spices are plant products (seed, flower, leaf, bark, roots, or bulb) used whole or ground, singly or mixed. Condiments are spices blended with components and have a saucelike consistency (catsup, mustard). They are used in relatively small amounts for aroma and color.[1,2,10]

Some spices, unless given antimicrobial treatments (irradiation, ethylene oxide: now not permitted), may contain microorganisms as high as 10^{6-7} per gram. The most important are spores of molds, *Bacillus,* and *Clostridium* spp. Also, micrococci, enterococci, yeasts, and several pathogens such as *Salmonella* spp., *Staphylococcus aureus,* and *Bacillus cereus* have been found. They may also harbor mold toxins. Although used in small amounts, they can be the source of spoilage and pathogenic microorganisms in food. Some spices, such as cloves, allspice, and garlic, have antimicrobial properties.[1,2,10]

REFERENCES

1. Silliker, J. H., Ed., *Microbial Ecology of Foods,* Vol. II, Academic Press, New York, 1980.
2. Vanderzant, C. and Splittstoesser, D. F., Eds., *Compendium of Methods for the Microbiological Examination of Foods,* American Public Health Association, Washington, DC, 1992.
3. Sofos, J., Microbial growth and its control in meat, poultry and fish products, *Advances in Meat Research,* Vol. 9, Pierson, A. M. and Dutson, T. R., Eds., Chapman and Hall, New York, 1994, 359.
4. Kalchayanand, N., Ray, B., and Field, R. A., Characteristics of psychrotrophic *Clostridium laramie* causing spoilage of vacuum-packaged refrigerated fresh and roasted beef, *J. Food Prot.,* 56, 13, 1993.
5. Richardson, G. H., Ed., *Standard Methods for the Examination of Marine Food Products,* American Public Health Association, Washington, DC, 1985.
6. Mayes, F. M. and Takeballi, M. A., Microbial contamination of hen's eggs: a review, *J. Food Prot.,* 46, 1092, 1983.
7. Ward, D. R. and Hackney, C. R., *Microbiology of Marine Food Products,* Van Nostrand Reinhold, New York, 1990.
8. Dennis, C., *Postharvest Pathology of Fruits and Vegetables,* Academic Press, New York, 1983.
9. Smittle, R. B. and Flowers, R. M., Acid tolerant microorganisms involved in the spoilage of salad dressings, *J. Food Prot.,* 45, 977, 1982.
10. Kneifel, W. and Berger, E., Microbiological criteria of random samples of spices and herbs retailed on the Austrian market, *J. Food Prot.,* 57, 893, 1994.

CHAPTER 4 QUESTIONS

1. List the mesophilic and psychrophilic microorganisms in raw meat. Discuss the significance of aerobic and facultative anaerobic microorganisms in raw chilled meats.

2. Discuss the post-heat contamination of low heat-processed meat products by bacteria and their significance in these products.

3. List and discuss the significance of: (a) psychrotrophic bacteria in raw milk, (b) high thermoduric bacteria in raw milk to be used for pasteurized grade A milk, and (c) >10 coliforms per milliliter in pasteurized grade A milk.

4. Discuss the significance of *Salmonella* contamination, with emphasis on *Salmonella enteritidis* in shell eggs.

5. "A large volume of seafoods consumed in the United States and other developed countries is obtained from countries where the level of sanitation is not very stringent." Explain how this situation can affect microbiological quality, with emphasis on microbiological safety of these seafoods.

6. Many vegetables are eaten raw. Discuss what microbiological concerns the consumer should have for these vegetables.

7. Molds can grow on cereal grains, peanuts, and spices at different stages of their production and processing. What concerns should the consumers, regulatory agencies, and food processors have for the use of these products in foods?

8. Discuss the microbial significance of soft drinks, fruit juices, vegetable juices, and bottled water.

9. "Low pH products, such as mayonnaise, can cause foodborne disease." Describe under what condition(s) this is possible.

10. List the microorganisms that are able to survive in properly processed canned foods and discuss their significance.

SECTION II

MICROBIAL GROWTH IN FOOD

Microorganisms present in raw and processed (nonsterile and commercially sterile) foods are important for their involvement in foodborne diseases, food spoilage, and food bioprocessing. This is accomplished, generally, through the growth of microorganisms, except for viruses, in foods. Growth or cell multiplication of bacteria, yeasts, and molds is greatly influenced by the intrinsic and extrinsic environments of the food. Microbial growth is also facilitated through the metabolism of some food components that provide needed energy and cellular materials and substrates for many by-products. Microbial growth in laboratory media is also important for the quantitative and qualitative detection of microbiological quality of a food. In this section, characteristics of and influencing factors associated with microbial growth are discussed. An understanding of these factors is helpful in designing methods to either control (as against spoilage and health hazard) or stimulate (as in bioprocessing and detection) their growth. The following aspects are discussed:

CHAPTER 5

Microbial Growth Characteristics

CONTENTS

I. MICROBIAL REPRODUCTION

A. Binary Fission

An increase in the number or mass of vegetative cells of bacteria, yeasts, and molds is customarily used to reflect growth for microorganisms. True bacteria reproduce by a process called "transverse binary fission" or simply "binary fission." This is a process in which one (mother) cell asexually divides into two (daughter) cells, each being an essentially true replica of the original (mother) cell. In bacteria, a form of sexual recombination can occur that involves transfer of genetic materials from a donor to a recipient cell (e.g., conjugation). However, this is quite different from sexual reproduction, which is facilitated through the union of two cells (gametes of opposite mating type).[1,2]

A bacterial cell has a specific surface area-to-volume (s/v) ratio. A newly divided cell has a higher s/v ratio, which helps in the rapid transport of nutrients from the environment. A young cell predominantly uses nutrients to synthesize energy and cellular components, causing the cell to increase in size. As the cell size increases, the s/v ratio decreases, which adversely affects the transport of nutrients into and by-products out of the cell. In an effort to increase the s/v ratio, the cell initiates division by forming constriction on the cell surface, followed by formation of transverse wall formation, separating the cellular materials equally between two cells (Figure 5.1). The division can occur in one or more planes depending upon the species and the arrangement of the cells. After division, the two cells may separate from each other. In some species, the two cells remain attached to each other and continue to divide in one plane (e.g., in *Streptococcus* and *Streptobacillus* species). If the cells divide in two or more planes and the daughter cells remain close to each other, they form a cluster (e.g., in *Staphylococcus* species). In some species, like *Pediococcus*, cell division occurs in two or more planes and the daughter cells have a tendency to stay together for some time, giving rise to a four- (tetrad) or eight-cell arrangement.

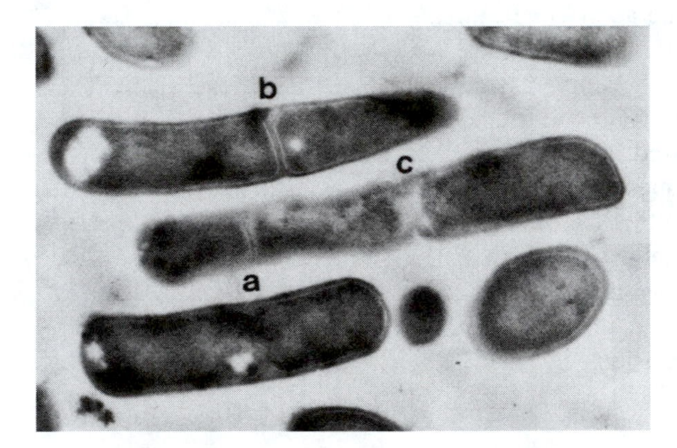

Figure 5.1 Photograph of thin sections of *Lactobacillus* cells by transmission electron microscopy showing cell wall formation during cell division at earlier stage (a), later stage (b), and final stage with partial separation of cells (c).

Yeasts and molds can also reproduce asexually. A yeast (mother) cell produces a bud that initially is much smaller in size and remains attached to the surface of the original cell. As it grows in size, it can also produce a bud, giving an appearance of a chain of buds growing on the surface of the original yeast cell. A yeast cell can have several buds forming on its surface. As the buds mature, they separate out from the original cell. Molds can grow in size by cell division or elongation at the tip of a hypha. They can also form large numbers of asexual spores on a specialized structure. Both yeasts and molds can also reproduce sexually. Viruses do not reproduce by themselves. Instead, bacteriophages attach to the surface and inject their nucleic acid inside the specific host cells, which then replicate the viral nucleic acid and produce viral particles. The viral particles are released into the environment following lysis of the host cells. Viruses associated with foodborne diseases do not increase in number by replication in food.[1,2]

B. Generation Time

The time that a single cell takes to divide is called the generation time.[1,2] However, in practice, generation time is referred to as the doubling time for the entire population. In a population of a microbial species, not all cells divide at the same time or at the same rate. The generation time of a microbial species under different conditions provides valuable information for developing methods to preserve foods. In general, under optimum conditions of growth, bacteria have the shortest generation time, followed by yeasts and molds. Also, among bacterial species and strains, generation times under optimum conditions vary greatly; some species, like *Vibrio parahaemolyticus,* under optimum conditions can have as low as 10 to 12 min generation time. Generally, in food systems, microorganisms have longer generation times than in a rich bacteriological broth.

The generation time of a microbial population can be calculated mathematically from the differences in population during a given time period. Because of large numbers, the calculation is done in logarithmatics ($\log_{10}$) using the formula:

$$G = \frac{0.3t}{\log_{10} z - \log_{10} x}$$

where G is generation time (min), 0.3 is a constant (value of $\log_{10}2$ and indicates doubling), t is the duration of study in min, and $\log_{10}x$ is the initial and $\log_{10}z$ is the final cell number per milliliter (or colony forming units, cfu per milliliter). If, under a given growth condition, the initial population of 10^4 cells per milliliter of a bacterial species increased to 10^6 cells per minute in 120 min, its generation time will be:

$$G = \frac{0.3 \times 120}{6 - 4} = 18 \text{ min}$$

This value for the same bacterial species will change by changing the conditions of growth.

C. Optimum Growth

Many environmental parameters, such as temperature, acidity (pH), water activity (A_w), oxidation-reduction potential (O–R), and nutrients, influence the microbial growth rate. This aspect is discussed in the next chapter. If one of the factors, such as temperature, is varied while keeping all other parameters constant during growth of a microbial species and its growth rate is measured, then it becomes evident that the growth rate is highest (or generation time is shortest) at a certain temperature; this temperature is referred to as the "optimum growth temperature" for the species under a given condition. The growth rate slows down at either side of the optimum growth temperature until the growth stops. The two points on both sides of an optimum growth condition where minimum growth occurs is the growth range. When the cells of a microbial species are exposed to a factor (such as temperature) beyond the growth range, the cells not only stop growing, but depending upon the situation, they may lose viability. The growth range and optimum growth of a microorganism for a specific parameter provide valuable information for its inhibition, reduction, or stimulation of growth in a food.

D. Growth Curve

The growth rate and growth characteristics of a microbial population under a given condition can be graphically represented by counting cell numbers, enumerating cfu, or measuring optical density in a spectrophotometer at a given wavelength (above 300 nm, usually at 600 nm) of a cell suspension. Each method has several advantages and disadvantages. If the cfu values are enumerated at different times of growth and a growth curve is plotted using $\log_{10}$ cfu vs. time ($\log_{10}$ cfu is used because of high cell numbers), a plot similar to the hypothetical plot presented in Figure 5.2 is obtained. It has several features that represent the conditions of the cells at different times. Initially, the population does not change (lag phase). During this time, the cells are assimilating nutrients and increasing in size. Although the population remains unchanged due to changes in size, both cell mass and optical density will show some increase. Following this, the cell number starts increasing, first slowly and then very rapidly. The cells in the population differ in metabolic rate and initially only some multiply and then almost all cells multiply. This is the exponential phase (also called logarithmic phase). Then the growth rate slows down and finally the population enters the stationary phase. At this stage, due to nutrient shortage and accumulation of waste products, a few cells die and a few cells multiply, keeping the living population stable. However, if one counts the cells under a microscope or measures cell mass, both may show an increase, as dead cells may remain intact. After the stationary phase, the population enters into the death phase where the rate of cell death is higher than the rate of cell multiplication. Depending upon a species and the condition of the environment, after a long period of time (a few years), some cells may still remain viable. This information is important for the determination of some microbiological criteria in food, especially in controlling spoilage and pathogenic microorganisms in food. It is important to recognize that by changing environmental parameters (e.g., refrigeration), the growth rate of some microbial species can be slowed down; but after a long time, the population can reach high numbers that cause problems in a food.

II. MICROBIAL GROWTH IN FOOD

A. Mixed Population

Normally, a food harbors a mixed population of microorganisms that can include different species of bacteria, yeasts, and molds.[3,4] Some species can be present in

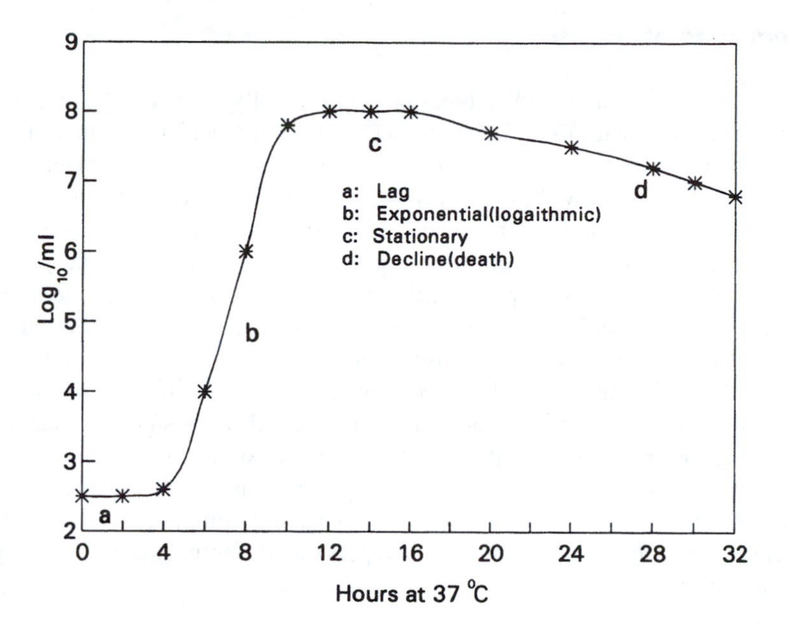

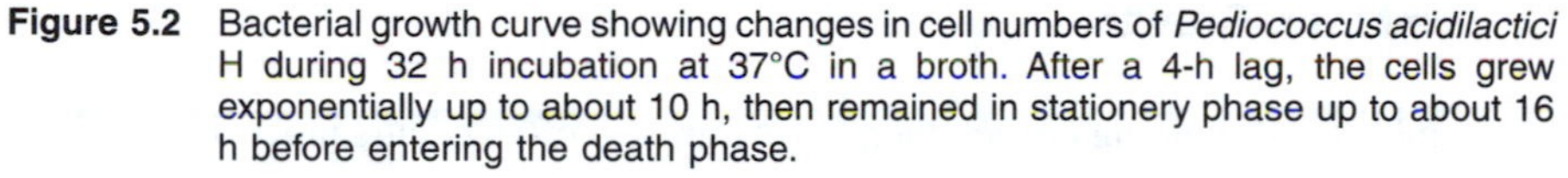

Figure 5.2 Bacterial growth curve showing changes in cell numbers of *Pediococcus acidilactici* H during 32 h incubation at 37°C in a broth. After a 4-h lag, the cells grew exponentially up to about 10 h, then remained in stationery phase up to about 16 h before entering the death phase.

relatively higher numbers than the others. The growth characteristics of the mixed population differ in several respects from that of a pure culture (a single strain of a species). Depending upon the environment, which includes both the food environment (intrinsic) and the environment in which the food is stored (extrinsic), some of the species can be in an optimum or near-optimum growth condition and, due to rapid rate of growth during storage, they will become predominant. This can occur even if they are initially present in low numbers. This is often the case in foods kept for a long time under a specific condition, such as at refrigerated temperature. Initially, if the product is enumerated, it may show that the majority are able to grow at 35°C and only a few grow at 4°C (refrigerated temperature). If the product is enumerated after a few weeks of refrigerated storage, one usually finds that the populations that grow at 4°C have outnumbered those that grow at 35°C but not at 4°C. Another situation can arise if a food contains, among the mixed population, two species both present in equal numbers initially and both grow optimally under the specific intrinsic and extrinsic environments of the food, but one has a shorter generation time than the other. After a storage period, the one with the shorter generation time will become predominant. Many foods most often get spoiled by bacteria rather than by yeasts and molds because, in general, bacteria have shorter generation times. Thus, in a mixed population, the intrinsic and extrinsic environments dictate which one, two, or a few in the mixed population will become predominant and produce specific changes in a food. These aspects are very important in the microbial spoilage of foods and in the production of bioprocessed (or fermented) foods.[3,4]

B. Sequence of Growth

Among the different microbial types present normally in a food, the predominant type(s) can change with time during storage.[3,4] Initially, depending upon the environment, one or two types may grow optimally and create an environment in which they can no longer grow rapidly. However, another type can find this changed environment to be favorable for growth and grow rather rapidly. This shift in predominance can occur several times during the storage of a food. The sequential microbial growth can be seen particularly in foods stored for a long time. If a food is packaged in a bag with a little bit of air (e.g., ground meat), the aerobes will grow first and utilize the oxygen. The environment will become anaerobic, in which the anaerobes (or facultative anaerobes) grow favorably. In the natural fermentation of some foods, such as sauerkraut, four different bacterial species grow in succession, one creating the favorable growth condition for the next one. The desirable characteristics of the final product are dependent upon the growth of all four species in specific sequence. To identify the existence of such a situation, it is necessary to enumerate and determine the microbial type(s) at different stages of storage or fermentation.[3,4]

C. Symbiotic Growth

Symbiosis, or helping one another, during growth often occurs in food containing two or more types of microorganisms.[3,4] One type may produce metabolic product(s) that the second type needs for proper growth, but cannot produce by itself. In turn, the second species produces a nutrient that stimulates the first one to grow better. This is found in the production of some fermented foods such as yogurt. *Streptococcus thermophilus*, hydrolyzes milk proteins by its extracellular proteinases and generates amino acids, which are necessary for good growth of *Lactobacillus delbrueckii* subsp. *bulgaricus*. *Lactobacillus,* in turn, produces formate, which stimulates the *Streptococcus* species to grow quickly.[3,4]

D. Synergistic Growth

This is observed during symbiotic growth of two or more microbial types in a food.[3,4] In such a condition, each type is capable of growing independently and producing some metabolites at lower rates. However, when they are allowed to grow in a mixed population, both their growth rate and the level of by-product formation greatly increase. The increase is more than the additive of the amounts produced by growing the two separately. Both *Streptococcus thermophilus* and *Lactobacillus delbrueckii* subsp. *bulgaricus,* when growing in milk independently, produce about 8 to 10 ppm acetaldehyde, the desirable flavor component of yogurt. However, when growing together in milk, as high as 30 ppm or more acetaldehyde is produced, which is much higher than the additive amounts produced independently by the two species. In the production of a desirable fermented food, two separate strains and species can be used to induce synergistic growth.[3,4]

E. Antagonistic Growth

Two or more types of microorganisms present in a food can adversely affect the growth of each other, or one can interfere with the growth of one or more types; sometimes one can kill the other. This has been found among many bacterial strains, between bacteria and yeasts, between yeasts and molds, and between bacteria and molds. This occasionally occurs due to the production of one or more antimicrobial compounds by one or more strains in the mixed population. Some Gram-positive bacteria produce antibacterial proteins or bacteriocins that can kill many other types of Gram-positive bacteria. Similarly, some yeasts can produce wall-degrading enzymes and can reduce the growth of molds. Some strains have probably developed these specific traits for growth advantage in a mixed population situation. There is interest now to use this phenomenon to control growth and enhance viability loss of undesirable spoilage and pathogenic microorganisms in food.

REFERENCES

1. Buffaloe, N. D. and Ferguson, D. V., *Microbiology: Microbial Growth,* 2nd ed., Part 10, Houghton Mifflin, Boston, 1981.
2. Tortora, G. J., Funke, B. R., and Case, C. L., *Microbiology: An Introduction,* 4th ed., Benjamin/Cummings, Menlo Park, CA, 1992, 155.
3. Sinell, H. J., Interacting factors affecting mixed populations, in *Microbial Ecology of Foods,* Vol. 1, Silliker, J. H., Ed., Academic Press, New York, 1989, 215.

CHAPTER 5 QUESTIONS

1. Describe the process of microbial growth. Use diagram(s).

2. Define generation time as related to bacterial growth. If a pure culture of a bacterial population during incubation in a nutritionally rich broth increased to 3.5×10^6/ml from an initial population of 2.5×10^2/ml in 210 min, what is the generation time of the strain?

3. "In foods, microorganisms are present as mixed population." What disadvantage does this situation impose in applying pure culture study results in food systems?

4. Explain synergistic growth, antagonistic growth, and symbiotic growth of microorganisms in food. What are their advantages and disadvantages?

CHAPTER 6

Factors Influencing Microbial Growth in Food

CONTENTS

I. INTRINSIC FACTORS OR FOOD ENVIRONMENT

The ability of microorganisms (except viruses) to grow in a food is determined by the food environment as well as the environment in which the food is stored. These two are also designated, as the intrinsic and extrinsic environment of a food, respectively. In this chapter, the contributions of the intrinsic factors are discussed. These factors include nutrients, growth factors and antimicrobials, water activity, pH, and oxidation-reduction potential. Although these factors are discussed separately here, one needs to recognize that in a food system, they are present together and exert effects on microbial growth in combination, either favorably or adversely.

A. Nutrients and Growth

Microbial growth is accomplished through the synthesis of cellular components and energy.[1] The necessary nutrients for this purpose are derived from the immediate environment of a microbial cell and, if the cell is growing in a food, it supplies the nutrients. These nutrients include carbohydrates, proteins, lipids, minerals, and vitamins. Water is not considered a nutrient, but it is essential as a medium for the biochemical reactions necessary for the synthesis of cell mass and energy. All foods contain these five major nutrient groups, either naturally or added, and the amount of each nutrient varies greatly with the type of food. In general, meat is rich in protein, lipids, minerals, and vitamins but poor in carbohydrates, while foods from plant sources are rich in carbohydrates, but can be poor sources of proteins, minerals, and some vitamins. In contrast, milk is rich in all five nutrient groups necessary for microbial growth.

Microorganisms normally found in food vary greatly in nutrient requirements, with bacteria requiring the most, followed by yeasts and molds. Microorganisms also differ greatly in their ability to utilize large and complex carbohydrates (e.g., starch and cellulose), large proteins (e.g., casein in milk), and lipids. Microorganisms capable of using these molecules do so by producing specific extracellular enzymes and hydrolyzing the complex molecules to simpler forms before transporting them inside the cell. Molds have the highest capability to do this. However, this provides an opportunity for a species to grow in a mixed population even when it is incapable of metabolizing the complex molecules.[1]

1. Carbohydrates in Foods

Major carbohydrates present in different foods, either naturally or due to addition as ingredients, can be grouped on the basis of chemical nature as follows:

Monosaccharides
  Hexoses:glucose, fructose, mannose, galactose
  Pentoses:xylose, arabinose, ribose, ribulose, xylulose
Disaccharides
  Lactose (galactose + glucose)
  Sucrose (fructose + glucose)

Maltose (glucose + glucose)
Oligosaccharides
Raffinose (glucose + fructose + galactose)
Stachyose (glucose + fructose + galactose + galactose)
Polysaccharides
Starch (glucose units)
Glycogen (glucose units)
Cellulose (glucose units)
Insulin (fructose units)
Hemicellulose (xylose, galactose, mannose units)
Dextrins (glucose units)
Pectins
Gums and mucilages

Lactose is found only in milk and thus can be present in foods made from or with milk and milk products. Glycogen is present in animal tissues, especially in the liver. Pentoses, most oligosaccharides, and polysaccharides are naturally present in foods of plant origin.

All microorganisms normally found in food metabolize glucose, but their ability to utilize other carbohydrates differs considerably. This is because of several major factors, such as the inability of some to transport the monosaccharides and disaccharides and the inability to hydrolyze polysaccharides. Molds have the most capability of using polysaccharides.

Food carbohydrates are metabolized by the microorganisms principally to supply energy through several metabolic pathways. Some of the metabolic products can be used to synthesize cellular components of microorganisms (e.g., to produce amino acids by amination of some keto acids). They also produce metabolic by-products that are associated with food spoilage (CO_2 to cause gas defect) or food bioprocessing (lactic acid in fermented foods). Some of these metabolic pathways are discussed in Chapters 8 and 10. Microorganisms can also polymerize some carbohydrates to produce complex carbohydrates, such as dextrins, capsular materials, and cell wall (or outer membrane and middle membrane in Gram-negative bacteria). Some of these carbohydrates from pathogens may cause health hazards (as complex proteins), some may cause food spoilage (such as slime defect), while some may be used in food production (such as dextrins as stabilizers). Carbohydrate metabolism profiles are extensively used in the laboratory for the biochemical identification of microorganisms isolated from foods.

2. *Proteins in Foods*

The major proteinaceous components in foods are simple proteins, conjugated proteins, peptides, and nonprotein nitrogenous (NPN) compounds (amino acid, urea, ammonia, creatinine, trimethylamine). The proteins and peptides are polymers of different amino acids without or with other organic (e.g., a carbohydrate) or inorganic (e.g., iron) components and contain about 15 to 18% nitrogen. Simple food proteins are polymers of amino acids, such as albumins, globulins, glutelins (glutenin in cereal), prolamins (zein in grains), and albuminoids (collagen in muscle). They differ greatly in their solubility, which determines the ability of microorganisms to utilize a specific protein. Many microorganisms can hydrolyze albumin, which is soluble in water, while collagens that are insoluble in water, or weak salt and acid solutions, are hydrolyzed only by limited microorganisms. As compared to simple proteins, conjugated proteins of food on hydrolysis produce metals (metaloproteins such as hemoglobin and myoglobin), carbohydrates (glycoproteins such as mucin), phosphate (phosphoproteins such as casein), and lipids (lipoproteins such as some in liver). Proteins are present in higher quantities in foods of animal origin than in foods of plant origin. Some plant foods, such as nuts and legumes, are rich in proteins. Proteins as ingredients can also be added to foods.

Microorganisms differ greatly in their ability to metabolize food proteins. Most transport amino acids and small peptides in the cells; small peptides are then hydrolyzed to amino acids inside the cells, such as in some *Lactococcus* spp. Microorganisms also produce extracellular proteinases and peptidases to hydrolyze large proteins and peptides to small peptides and amino acid before they can be transported inside the cells. Soluble proteins are more susceptible to this hydrolytic action than the insoluble proteins. Hydrolysis of food proteins can be either undesirable (texture loss in meat) or desirable (flavor in cheese). Microorganisms are also capable of metabolizing different NPN compounds found in foods.

Amino acids inside the microbial cells are metabolized via different pathways to synthesize cellular components, energy, and various by-products. Many of these by-products can be undesirable (NH_3 and H_2S production to cause spoilage of food and health hazards from toxins and biological amines) or desirable (some sulfur compounds in cheddar cheese flavor). Production of specific metabolic products is used for the laboratory identification of microbial isolates from food. An example of this is the ability to produce indole by *Escherichia coli,* which is used to differentiate this species from non-indole-producing microorganisms (*Enterobacter* spp.).

3. Lipids in Foods

The lipids in foods include those compounds that can be extracted by organic solvents, some of which are free fatty acids, glycerides, phospholipids, waxes, and sterols. Foods of animal origin are relatively higher in lipids than foods of plant origin, although nuts, oil seeds, coconuts, and olives have high amounts of lipids. Fabricated foods can also vary greatly in lipid content. Cholesterols are present in foods of animal origin or food containing ingredients from animal sources. Lipids are, in general, less preferred nutrients for the microbial synthesis of energy and cellular materials. Many microorganisms are capable of producing extracellular lipases that can hydrolyze glycerides to fatty acids and glycerol. Fatty acids can be transported in cells and used for energy synthesis, while glycerol can be metabolized separately. Some microorganisms also produce extracellular lipid oxidases that can oxidize unsaturated fatty acids to produce different aldehydes and ketones. In general, molds are more capable of producing these enzymes. However, certain bacterial groups, such as *Pseudomonas, Achromobacter,* and *Alcaligenes* are capable of producing these enzymes. Lysis of dead microbial cells causes the release of lipases and oxidases, which then can carry out these reactions. In many foods, the action of these enzymes is associated with spoilage (such as rancidity); while in other foods, the enzymes are credited with desirable flavors (such as in mold ripening cheeses). Some beneficial intestinal microorganisms, such as *Lactobacillus acidophilus* strains, are capable of metabolizing cholesterol and are believed to reduce serum cholesterol levels in humans.

4. Minerals and Vitamins in Foods

Microorganisms need several elements such as phosphorous, calcium, magnesium, iron, sulfur, manganese, and potassium, in small amounts, and most foods have these in sufficient amounts. Many microorganisms are capable of synthesizing B vitamins, and also foods contain most of them.

In general, most foods contain different carbohydrates, proteins, lipids, minerals, and vitamins in sufficient amounts to supply necessary nutrients for the growth of molds, yeasts, and bacteria, especially Gram-negative bacteria found in foods. Some foods may have limited amounts of one or a few nutrients for rapid growth of some Gram-positive bacteria, especially some fastidious *Lactobacillus* species. For their growth, some essential amino acids and B vitamins may be added to a food where their growth is desired (e.g., adding glucose to make fermented sausages). It is not possible or practical to control microbial growth in a food by restricting nutrients.

B. Growth Factors and Inhibitors in Food

Foods can also have some factors that either stimulate growth or adversely affect the growth of microorganisms.[2] The exact nature of the growth factors is not known, but they are naturally present in some foods. An example would be the growth factor(s) in tomatoes that stimulate growth of some *Lactobacillus* species. These factors can be added to raw materials during food bioprocessing or to media for the isolation of some fastidious bacteria from foods.

Foods also contain many chemicals, either naturally or added, that adversely affect microbial growth. Some of the natural inhibitors are lysozyme in egg, agglutinin in milk, and eugenol in cloves. The inhibitors, depending upon their mode of action, are capable of either preventing or reducing growth of and killing microorganisms. This aspect is discussed in Chapter 37.

C. Water Activity and Growth

1. Principle

Water activity (A_w) is a measure of the availability of water for biological functions and relates to water present in a food in "free" form. In a food system, total water or moisture is present in "free" and "bound" forms. Bound water is necessary to hydrate the hydrophilic molecules and dissolve the solutes, and is not available for biological functions; as well, it does not contribute to A_w. The A_w of a food can be expressed by the ratio of water vapor pressure (P) of the food to that of pure water (P_o which is 1), i.e., P/P_o. It ranges between 0 to 1 or more accurately

>0 to <1, as no food can have either 0 or 1 water activity. The A_w of a food can be determined from its equilibrium relative humidity (ERH) by dividing ERH by 100 (as ERH is expressed as a percentage).[3-6]

2. A_w of Food

Foods differ in A_w between about 0.1 and 0.99. The A_w of some food groups are as follows: cereals, crackers, sugar, salt, dry milk, 0.10–0.20; noodles, honey, chocolate, dried egg, <0.60; jam, jelly, dried fruits, parmesan cheese, nuts, 0.60–0.85; fermented sausage, dry cured meat, sweetened condensed milk, maple syrup, 0.85–0.93; evaporated milk, tomato paste, bread, fruit juices, salted fish, sausage, processed cheese, 0.93–0.98; and fresh meat, fish, fruits, vegetables, milk, eggs, 0.98–0.99. The A_w of foods can be reduced by removing water (desorption) and increased by the adsorption of water, and these two parameters can be used to draw a sorption isotherm graph for a food (Figure 6.1). The desorption process gives relatively lower A_w values than the adsorption process at the same moisture content of a food. This has important implications in the control of a microorganism by reducing A_w in a food. The A_w of a food can be reduced by several means: namely, by adding solutes, ions, hydrophilic colloids, and by freezing and drying.[3-6]

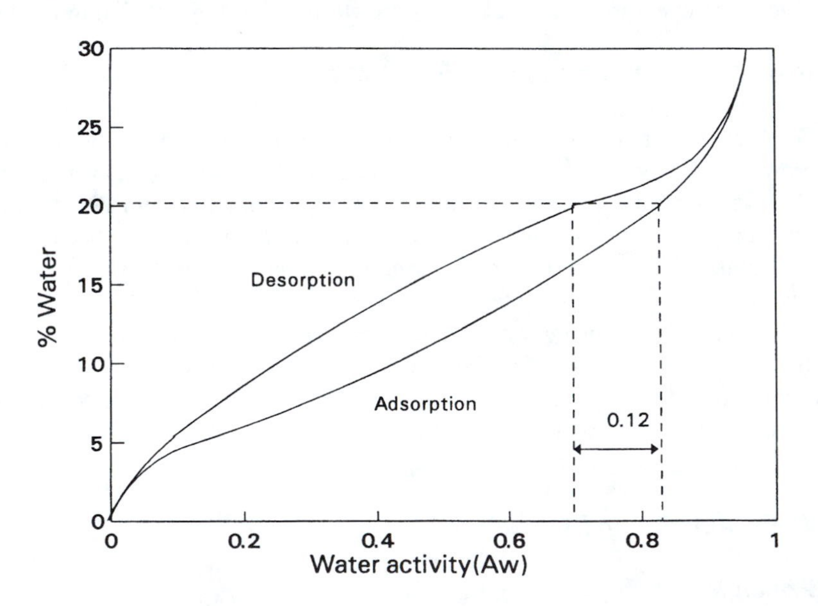

Figure 6.1 Water sorption isotherm showing hysterisis. At the same percentage of water, A_w is lower by desorption than by adsorption.

3. A_w and Microbial Growth

The free water in a food is necessary for microbial growth. It is necessary for the transport of nutrients and removal of waste materials, to carry out enzymatic reactions, to synthesize cellular materials, and to take part in other biochemical reactions, such as hydrolysis of a polymer to monomers (proteins to amino acids). Each microbial species (or group) has an optimum, maximum, and minimum A_w level for growth. In general, the minimum A_w values for growth of microbial groups are: most molds, 0.8, with xerophilic molds as low as 0.6; most yeasts, 0.85, with osmophilic yeasts between 0.6 and 0.7; most Gram-positive bacteria, 0.90; and Gram-negative bacteria, 0.93, with *Sta. aureus* at 0.85 and halophilic bacteria at 0.75. The A_w needed for sporeforming bacteria to sporulate and germinate and the toxin-producing microorganisms to produce toxins is generally higher than the minimum A_w for their growth. Also, the minimum A_w for growth in an ideal condition is lower than in a nonideal condition. As an example, if the minimum A_w for growth of a bacterial strain at pH 6.8 is 0.91, then at pH 5.5, it can be 0.95 or more. When the A_w is reduced below the minimum level for growth of a microorganism, the cells remain viable for awhile. But if the A_w is reduced drastically, microbial cells will lose viability, generally rapidly at first and then more slowly. This information is used to both control spoilage and pathogenic microorganisms in food, as well as to enhance the growth of desirable types in food bioprocessing (such as adding salt in processing of cured ham; see Chapter 33) and in laboratory detection of microorganisms (adding salt to media to enumerate *Sta. aureus*).[3-6]

D. pH and Growth

1. Principle

pH indicates the hydrogen ion concentrations in a system and is expressed as $-\log [H^+]$, the negative logarithm of the hydrogen ion or proton concentration. It ranges from 0 to 14, with 7.0 being neutral pH. $[H^+]$ concentrations can differ in a system depending upon what acid is present. Some strong acids used in foods, such as HCl and phosphoric acid, dissociate completely. Weak acids, such as acetic or lactic acids, remain in equilibrium with the dissociated and undissociated forms.

$$[\mathrm{HCl}] \rightarrow [\mathrm{H}^+] + [\mathrm{Cl}^-], \text{ pH of 0.1 } N \text{ HCl is 1.1}$$

$$\mathrm{CH_3COOH} \rightleftarrows [\mathrm{H}^+] + [\mathrm{CH_3COO}^-], \text{ pH of 0.1 } N\ \mathrm{CH_3COOH} \text{ is 2.9}$$

Acidity is inversely related to pH, so that a system with high acidity has a low pH, and vice versa.[7,8]

2. pH of Food

Depending upon the type, the pH of a food can vary greatly. On the basis of pH, foods can be grouped as high acid foods (pH below 4.6) and low acid foods (pH 4.6 and above). Most fruits, fruit juices, fermented fruits, vegetables, meat and dairy products, and salad dressings are high acid (low pH) foods, while most vegetables, meat, fish, milk, and soups are low acid (high pH) foods. Tomatoes, however, are a high acid vegetable (pH 4.1 to 4.4). The higher pH limit of most low acid foods remains below 7.0; only in a few foods, such as clams (pH 7.1) and egg albumen (pH 8.5), does the pH exceed 7.0. Similarly, the low pH limit of most high acid foods remains at pH 3.0, except in some citrus fruits and fruit juices (lemon, lime, and grapefruit) and cranberry juice, where the pH can be as low as 2.2. The acid in the foods can either be present naturally (as in fruits), produced during fermentation (as in fermented foods), or added during processing (as in salad dressings). Foods can also have compounds that have a buffering capacity. A food such as milk or meat, due to good buffering capacity, will not show a pH reduction when compared to a vegetable product in the presence of the same amount of acid.[7,8]

3. pH and Microbial Growth

The pH of a food has a profound effect on the growth and viability of microbial cells. Each species has an optimum and a range of pH for growth. In general, molds and yeasts are able to grow at lower pH than bacteria. The pH range of growth for molds is between 1.5 and 9.0; for yeasts, between 2.0 and 8.5; for Gram-positive bacteria, between 4.0 and 8.5; and for Gram-negative bacteria, between 4.5 and 9.0. Individual species differ greatly in the lower pH limit for growth; for example, *Pediococcus acidilactici* can grow at pH 3.8 and *Staphylococcus aureus* can grow at pH 4.5, but *Salmonella* spp. cannot. The lower pH limit of growth of a species can be a little higher if the pH is adjusted with strong acid instead of weak acid (due to undissociated molecules).

When the pH is reduced below the lower limit for growth of a microbial species, not only will the cells stop growing, but they will lose viability, the rate of which depends upon the extent of pH reduction. This is more apparent with weak acids, especially with those that have higher pK values, such as acetic acid vs. lactic acid with pK values of 4.8 vs. 3.8, respectively. This is because, at the same pH, acetic acid will have more undissociated molecules than lactic acid. The undissociated molecules, being lipophilic, enter into the cell and dissociate to generate H^+ in the cytoplasm. This causes a reduction in internal pH, which ultimately destroys the proton gradient between the inside and the outside of the cells and dissipates proton motive force as well as the ability of the cells to generate energy. The information on the influence of pH on growth and viability of microbial cells is important for the development of the methods to prevent the growth of undesirable microorganisms in food (e.g., in acidified foods; see Chapter 34) and used to produce some fermented foods (e.g., sequential growth of lactic acid bacteria in sauerkraut fermentation) or to selectively isolate aciduric microorganisms from food (e.g., yeasts and molds in a medium with pH 3.5).[7,8]

E. Redox Potential, Oxygen, and Growth

1. Principle

The redox or oxidation-reduction (O–R) potential measures the potential difference in a system generated by a coupled reaction in which one substance is oxidized and a second substance is reduced simultaneously. The process involves the loss of electrons from a reduced substance (thus it is oxidized) and the gain of electrons by an oxidized substance (thus it is reduced). The electron donor, since it reduces an oxidized substance, is also called a reducing agent. Similarly, the electron recipient is called an oxidizing agent. The redox potential, designated as Eh, is measured in electrical units of millivolts (mV). In the oxidized range, it is expressed in +mV; and in reduced range, in –mV. In biological systems, the oxidation and reduction of substances is the primary means of generating energy. If free oxygen is present in the system, then it can act as an electron acceptor. In the absence of free oxygen, oxygen bound to some other compound, such as NO_3 and SO_4, can accept the electron. In a system where no oxygen is present, other compounds can accept the electrons. Thus, the presence of oxygen is not a requirement of oxidation-reduction reactions.[9]

2. *Redox Potential in Food*

The redox potential of a food is influenced by its chemical composition, specific processing treatment given, and storage condition (in relation to air). Fresh foods of plant and animal origin are in a reduced state due to the presence of reducing substances, such as ascorbic acid, reducing sugars, and the –SH group of proteins. Following stoppage of respiration of the cells, oxygen will diffuse inside and change the redox potential. Processing, such as heating, can increase or decrease reducing compounds and alter the Eh. A food stored in air will have a larger Eh (+MV) than when it is stored under vacuum or in modified gas (such as CO_2 or N_2). Oxygen can be present in a food in the gaseous state (on the surface, trapped inside) or in dissolved form.[9]

3. *Redox Potential and Microbial Growth*

On the basis of the growth of microorganisms in the presence and absence of free oxygen, they may be grouped as aerobes, anaerobes, facultative anaerobes, and microaerophiles. Aerobes need free oxygen for energy generation, as the free oxygen acts as the final electron acceptor through aerobic respiration. Facultative anaerobes can do that if free oxygen is available, or they can use bound oxygen in compounds like NO_3 or SO_4 as final electron acceptor through anaerobic respiration. If oxygen is not available, then other compounds are used to accept the electron (or hydrogen) through (anaerobic) fermentation. An example of this is the acceptance of hydrogen from $NADH_2$ by pyruvate to produce lactate. Anaerobic and facultative anaerobic microorganisms can only transfer electrons through fermentation. Many anaerobes (obligate or strict anaerobes) cannot grow in the presence of even small amounts of free oxygen as they lack the superoxide dismutase necessary to scavenge the toxic oxygen free radicals. Addition of scavengers, such as thiols (e.g., thiolglycolate), helps overcome the sensitivity to these free radicals. Microaerophiles grow better in the presence of less oxygen.

Growth of microorganisms and the specific metabolic reactions to generate energy are extremely dependent on the redox potential of the foods. The range of Eh at which different groups of microorganisms can grow are: aerobes, between

+500 and +300 mV; facultative anaerobes, between +300 and –100 mV; and anaerobes, between +100 and –250 mV or less. However, this varies greatly with food and the presence of oxygen. Molds, yeasts, and *Bacillus, Pseudomonas, Moraxella,* and *Micrococcus* genera are some examples that have aerobic species. Some examples of facultative anaerobes are the lactic acid bacteria and those in the family *Enterobacteriaceae*. The most important anaerobes in food are the *Clostridium*. The Eh range indicates that in each group there are species that are more strict in their Eh need than the others. While most molds are strict aerobes, there are a few that can tolerate less aerobic conditions. Similarly, yeasts are basically aerobic, but some can grow under low Eh (below +300 mV). Many clostridial species can grow at Eh +100 MV, but some need –150 mV or less.

The presence or absence of oxygen and the Eh of food determine the growth of a particular microbial group in a food and the specific metabolic pathways used during growth to generate energy and produce by-products. This is important in microbial spoilage of a food (such as putrification of meat by *Clostridium* species) and in desirable characteristics of fermented foods (such as growth of *Penicillium* species in blue cheese). This information is also important in the isolation of microorganisms of interest from foods (such as *Clostridium laramie*, a strict anaerobe from spoiled meat). This aspect is discussed further in Chapter 35.

II. EXTRINSIC FACTORS

Extrinsic factors important in microbial growth in a food include the environmental conditions in which it is stored. These are temperature, relative humidity, and gaseous environment. The relative humidity and gaseous condition of storage, respectively, influence the A_w and Eh of the food. The influence of these two factors on microbial growth has been discussed before. In this section, the influence of storage temperature of food on microbial growth is discussed.

A. Temperature and Growth

1. Principle

Microbial growth is accomplished through enzymatic reactions. It is well known that with every 10°C rise in temperature within the reaction range, the catalytic rate of an enzyme doubles. Similarly, the enzymatic reaction rate is reduced to half by decreasing the temperature by 10°C. Temperature, due to its influence on enzyme reactions, has an important role in microbial growth in food.

2. Food and Temperature

Foods are exposed to different temperatures from the time of production until consumption. Depending upon processing conditions, a food can be exposed to high heat, ranging between 60 and 65°C (roasting of meat) and over 100°C (in ultra-high temperature processing). For long-term storage, a food can be kept at 5°C (refrigeration) to –20°C or below (freezing). Some stable foods are also kept between 10 and 35°C (cold to ambient temperature). Different temperatures are also used to stimulate desirable microbial growth in food processing.[10]

3. Microbial Growth and Viability

Microorganisms important in foods are divided into three groups on the basis of their temperature of growth. Each group has an optimum temperature and a temperature range of growth. They are: thermophiles (grow at relatively high temperature), with optimum around 55°C and range between 45 and 70°C; mesophiles (grow at ambient temperature), with optimum at 35°C and range between 10 and 45°C; and psychrophiles (grow at cold temperature), with optimum at 15°C and range between –5 and 20°C. However, these divisions are not clear-cut and do overlap each other.

Two other terms used in food microbiology are very important with respect to microbial growth at refrigerated temperature and their survival to pasteurization, since both methods are widely used in the storage and processing of foods. The designation "psychrotrophs" is applied to microorganisms that grow at refrigerated temperature (0 to 5°C) irrespective of their optimum temperature and range of growth temperature. They usually grow rapidly between 10 and 30°C. Molds, yeasts, many Gram-negative bacteria from genera *Pseudomonas, Achromobacter, Yersinia, Serratia,* and *Aeromonas,* and Gram-positive bacteria from genera *Leuconostoc, Lactobacillus, Bacillus, Clostridium,* and *Listeria,* are included in this group. Those microorganisms that survive pasteurization temperature are designated as thermoduric. They include species from genera *Micrococcus, Bacillus, Clostridium, Lactobacillus, Pediococcus,* and *Enterococcus*. They have different growth temperatures and many can grow at refrigerated temperatures as well as thermophilic temperatures.[10]

When the foods are exposed to temperatures beyond the maximum and minimum temperatures of growth, the microbial cells die rapidly at higher temperatures and relatively slowly at lower temperatures. Microbial growth and viability are important considerations in reducing food spoilage and enhancing safety against pathogens, as well as in food bioprocessing. Temperature of growth is also effectively used in the laboratory to enumerate and isolate microorganisms from foods.

REFERENCES

1. Potter, N. N., *Food Science,* 2nd ed., AVI Publishing, Westport, CN, 1973, 36.
2. Conner, D. E., Naturally occurring compounds, *Antimicrobials in Foods,* Marcel Dekker, New York, 1993, 441.
3. Sperber, W. H., Influence of water activity of foodborne bacteria — a review, *J. Food Prot.,* 46, 142, 1983.
4. Troller, J. A., Water relations to foodborne bacterial pathogens — an update, *J. Food Prot.,* 49, 656, 1986.

5. Beuchat, L. R., Influence of water activity on growth, metabolic activities and survival of yeasts, *J. Food Prot.,* 46, 135, 1983.
6. Christian, J. H. B., Reduced water activity, in *Microbial Ecology of Foods,* Vol. 1, Silliker, J. H., Ed., Academic Press, New York, 1980, 70.
7. Corlett, D. A., Jr. and Brown, M. H., pH and acidity, in *Microbial Ecology of Foods,* Vol. 1, Silliker, J. H., Ed., Academic Press, New York, 1980, 92.
8. Baird-Parker, A. C., Organic acids, in *Microbial Ecology of Foods,* Vol. 1, Silliker, J. H., Ed., Academic Press, New York, 1980, 126.
9. Brown, M. H. and Emberger, O., Oxidation reduction potential, in *Microbial Ecology of Foods,* Vol. 1, Silliker, J. H., Ed., Academic Press, New York, 1980, 112.
10. Olson, J. C., Jr. and Nottingham, P. M., Temperature, in *Microbial Ecology of Foods,* Vol. 1, Silliker, J. H., Ed., Academic Press, New York, 1980, 1.

CHAPTER 6 QUESTIONS

1. List the intrinsic and extrinsic factors necessary for the growth of microorganisms in a food.

2. What are the major nutrients in food metabolized by the microorganisms? List the major groups of carbohydrates present in foods.

3. Discuss how bacteria are able to metabolize large molecules of carbohydrates, proteins, and lipids. How do molds differ from bacteria in the metabolism of these molecules?

4. Discuss the importance of antimicrobials in foods that can adversely affect microbial growth.

5. Define A_w and the desorption and adsorption processes of moisture in a food. Discuss the importance of A_w in microbial growth. How do halophilic, osmophilic, and xerophilic microorganisms differ in minimum A_w need for growth?

6. Define pH and discuss the factors that influence the pH of a food. Discuss the role of pH on microbial growth. How does a bacterial cell maintain a high intracellular pH (6.0) while growing in a low pH (5.0) environment? Give an example of aciduric bacteria.

7. Define redox potential and discuss how it influences microbial growth in a food. How can microorganisms be grouped on the basis of their growth capabilities at different redox potentials and oxygen availabilities?

8. How are microorganisms grouped on the basis of their temperature of growth and survival? Discuss the significance of psychrotrophic and thermoduric microorganisms in the processing and refrigeration storage of foods.

CHAPTER 7

Microbial Sporulation and Sublethal Injury

CONTENTS

I. MICROBIAL SPORULATION AND GERMINATION

Microorganisms important in food normally divide by binary fission (or elongation, as in nonseptate molds). In addition, molds and some yeasts and bacteria are capable of forming spores. While in molds and yeasts sporulation is associated with reproduction (and multiplication), in bacteria it is a process of survival in an unfavorable environment. In both molds and yeasts, sporulation can occur sexually and asexually, and sexual reproduction provides a basis for strain improvement for those that are used industrially. Among the spores, bacterial spores have special significance in foods due to their resistance to many treatments used in food processing and preservation. Compared to bacterial spores, mold and yeast spores are less resistant to such treatments. Spore formation in molds, yeasts, and bacteria is briefly discussed here.

A. Mold Spores

Molds form spores by both asexual and sexual reproduction and are classified as "perfect" or "imperfect" molds, respectively. Molds form large numbers of asexual spores and, depending upon types, can form conidia, sporangia spores, and arthrospores. Conidia are produced on special fertile hyphae called "conidiophores." Among the important molds in food, *Aspergillus* and *Penicillium* species form conidia. Sporangiospores are formed in a sack (sporangium) at the tip of a fertile hypha (sporangiophores). *Mucor* and *Rhizopus* species are examples of molds that form sporangiospores. Arthrospores, formed by the segmentation of a hypha, are produced by *Geotrichum*. An asexual spore in a suitable environment germinates to form a hypha and resumes growth to produce the thallus. Sexual spores form from the union of the tips of two hyphae, two gametes, or two cells. However, among the molds important in food, sexual reproduction is very rarely observed. Some examples include *Mucor* and *Neurospora*.

B. Yeast Spores

On the basis of sporeforming ability, the yeasts that are important in food are divided into two groups. Those capable of producing sexual ascospores are designated as *Ascomycetes* (true yeasts), and those that do not form spores are called "false yeasts." Examples of some yeasts important in food that form ascospores are *Saccharomyces, Kluyveromyces, Pichia,* and *Hansenula*. Species in the genera *Candida, Torulopsis,* and *Rhodotorula* do not form spores. Ascospores form by the conjugation of two yeast cells; in some cases, this can result from the union of the mother cell and a bud (daughter cell). The number of spores developed in an ascus varies with species. In a suitable environment, each spore develops into a yeast cell.

C. Bacterial Spores

The ability to form spores is confined to only a few bacteria genera that include Gram-positive *Bacillus, Clostridium, Sporolaclobacillus,* and *Sporosarcina* and

Gram-negative *Desulfotomaculum* species. Among these, *Bacillus, Clostridium,* and *Desulfotomaculum* are of considerable interest in food as they include species implicated in food spoilage and foodborne diseases.[1-3] Several *Bacillus* and *Clostridium* species are used to produce enzymes important in food bioprocessing.

In contrast to mold and yeast spores, bacterial cells produce endospore and one spore per cell. During sporulation and until a spore emerges following cell lysis, a spore can be located terminally, centrally, or off-center, causing bulging of the cell and under a microscope appears as a refractile spheroid or oval structure. Its surface is negatively charged and hydrophobic. The spores, as compared to the vegetative cells, are much more resistant to physical and chemical antimicrobial treatments, many of which are employed in the processing and preservation of food. This is because the specific structure of the bacterial spores is quite different from that of the vegetative cells from which they are formed. From inside to outside a spore has the following structures (Figure 7.1): a protoplasmic core containing important cellular components such as DNA, RNA, enzymes, dipicolinic acid (DPN), divalent cations, and very little water; an inner membrane that is the forerunner of the cell cytoplasmic membrane; surrounding this membrane is the germ cell wall, the forerunner of the cell wall in the emerging vegetative cell; around the cell wall is the cortex, composed of peptides, glycan, and an outer fore-spore membrane; outside the cortex and membrane are the spore coats, composed of layers of proteins that provide resistance to the spores. Spores of some species can have a structure called "exosporium" outside the coat. During germination and outgrowth, the cortex is hydrolyzed, and the outer fore-spore membrane and spore coats are removed by the emerging cell.

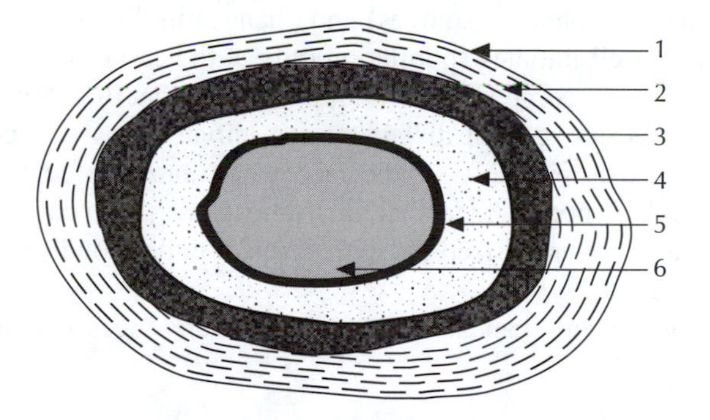

Figure 7.1 Schematic section of a bacterial spore showing different structures: (1) exosporium, (2) coat, (3) outer membrane, (4) cortex, (5) inner membrane, and (6) core.

The spores are metabolically inactive or dormant, can remain in dormant form for years, but are capable of emerging out as vegetative cells (one cell per spore) in a suitable environment. As opposed to nonsporeforming bacteria, the life cycle of a sporeforming bacteria has a vegetative cycle (by binary fission) and a spore cycle (Figure 7.2). The spore cycle also goes through several stages in sequence during which a cell sporulates and a vegetative cell emerges out of a spore. These stages are controlled by different environmental parameters and biochemical events.

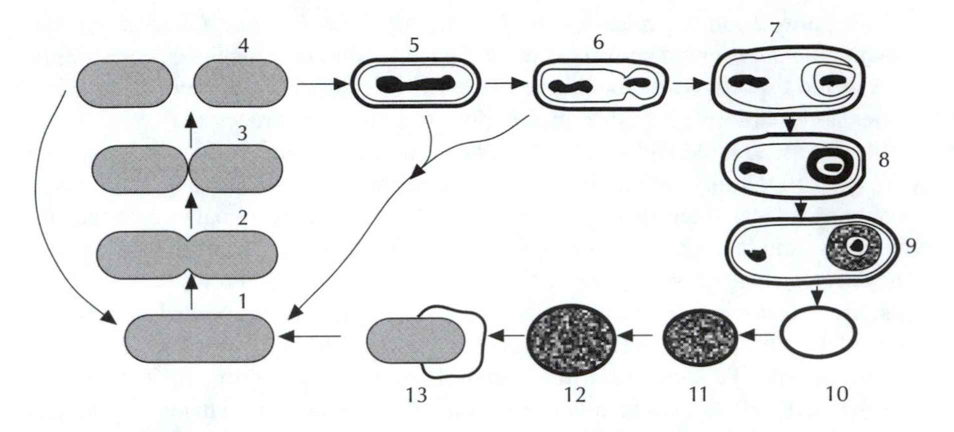

Figure 7.2 Schematic presentation of the cycles of (A) cell multiplication (1–4) and (B) endospore formation, germination and outgrowth of spore forming bacteria. Different steps are: (5) formation of axil filament, (6) septation, (7) prespore formation, (8) cortex formation, (9) coat formation, (10) free spore, (11) germination following activation, (12) swelling of spore, (13) outgrowing cell. Cells from step 4 can either divide (1–4) or sporulate. Cells from steps 5 and 6 can reverse back to cell division; from step 7, the process is irreversible.

D. Sporulation

The transition from normal vegetative cell cycle to sporulation in sporeforming bacteria is triggered by the changes in the environmental parameters in which the cells are growing. These include reduction in nutrient availability (particularly carbon, nitrogen, and phosphorous sources) and changes in the optimum growth temperature and pH. A cell initiates sporulation only at the end of completion of DNA replication. A triggering compound may be involved at that time for a cell to decide to either go through normal cell division or to initiate steps for sporulation. The triggering compound is probably synthesized when nutrition depletion and other unfavorable conditions occur. Adenosine bistriphosphate (Abt) could be the triggering compound, as it is synthesized by sporeformers under conditions of carbon or phosphorous depletion.

The sporulation events can be divided into seven stages (Figure 7.2). These include:[1,2]

1. Termination of DNA replication, alignment of chromosome in axial filament, and formation of mesoseme
2. Invagination of cell membrane near one end and completion of septum
3. Engulfment of prespore or forespore
4. Formation of germ cell wall and cortex, accumulation of Ca^{2+}, and synthesis of DPN
5. Deposition of spore coats
6. Maturation of spore: dehydration of protoplast, resistance to heat, and refractile appearance
7. Enzymatic lysis of wall and liberation of spore

Prior to stage 3, the process is reversible. However, once the process has entered stage 3, the cell is committed to sporulation.[1,2]

E. Dormancy

The spores are formed in such a manner as to remain viable in unfavorable conditions. This is achieved by increasing the resistance to extreme environments and reducing the metabolic activity to dormancy. Dehydration of the core and reduced molecular movement have been attributed to dormancy.

In a suitable environment, the dormancy of a spore can be ended by a series of biochemical reactions involved in spore activation, germination, outgrowth, and growth. Some spores may need a long time before they go through the sequences of germination and are designated "superdormant" spores. This is quite common among *Bacillus* and *Clostridium* spores. Superdormancy is thought to be the consequence of the inherent nature of a spore, spore injury, and environmental factors. Some spores can have stringent germination needs and will not germinate with other spores. Injured spores need to repair their injury before they can germinate and outgrow. Some component in the media could prevent germination of some spores. In food, superdormant spores could cause problems. Following processing, they may not be detected in a food; but during storage, they can germinate and outgrow, causing spoilage of a food or making it unsafe for consumption.[1,2]

F. Activation

Spore activation prior to germination is accompanied by reorganization of macromolecules in the spores. Spores can be activated in different ways, such as sublethal heat treatment, radiation, treatment with oxidizing or reducing agents, exposure to extreme pH, and sonication. These treatments probably accelerate the germination process by increasing the permeability of spore structures to germinating agents for the macromolecular reorganization. This process can be reversible, i.e., a spore does not have to germinate if the environment is not suitable.[1,2]

G. Germination

Several structural and functional events occur during this stage. The dormant stage is irreversibly terminated. The structural changes involve hydration of core, excretion of Ca^{2+} and DPN, and loss of resistance and refractile property. The functional changes include initiation of metabolic activity, activation of specific proteases and cortex-lytic enzymes, and release of cortex-lytic products. Generally, it is a metabolically degradative process.

Germination can be initiated (triggered) by many factors: pH, temperature, high pressure, lysozyme, nutrients (amino acids, carbohydrates), calcium-DPN, and others. The process can be inhibited by D-alanine, ethanol, EDTA, NaCl (high concentrations), NO_2, and sorbate.[1,2]

H. Outgrowth

Outgrowth constitutes the biosynthetic and repair processes during the period following germination of a spore and growth of a vegetative cell. The events during this phase include swelling of the spore due to hydration and nutrient uptake, repair and synthesis of RNA, proteins, and materials for membrane and cell wall, dissolution of coats, cell elongation, and DNA replication. The factors that can enhance the process include nutrients, pH, and temperature.

With the termination of the outgrowth stage, the vegetative cells emerge from the spores and enter the vegetative cell cycle of growth by binary fission.[1,2]

I. Importance of Spores in Food

As indicated before, species of *Bacillus, Clostridium,* and *Desulfotomaculum* are associated with food spoilage and foodborne diseases. Due to high heat resistance, the spores are of special interest and importance in food processing. Special attention must be given to the processing and preservation of the foods so that the spores are either destroyed or prevented from undergoing germination and outgrowth. The superdormant spores of the spoilage and pathogenic species pose another problem. As they are not detected with the other spores, a processing condition could be adopted with the idea that it will eliminate all spores. Subsequently, these surviving dormant spores can germinate, outgrow, and grow, which will either cause the food to spoil or make it unsafe. As destruction of all spores in a food is almost impossible to achieve, a combination of processing and preservation methods can be developed to overcome problems of spores in food.[1,2]

II. MICROBIAL INJURY AND REPAIR

A. Sublethal Injury

Microbial cells and spores exposed to different physical and chemical sublethal treatments (or stresses) suffer injury that is reversible in nature. Many of these treatments include those used directly during food processing and storage, as well as microbial detection from foods. These treatments include low heat (such as pasteurization), low temperature (freezing, refrigeration, and chilling), low A_w (different types of drying, adding high solutes like sugar or salt), radiation (UV or X-ray), high hydrostatic pressure, electric pulse, low pH (both organic and inorganic acids), preservatives (sorbates or benzoates), sanitizer (chlorine or quaternary ammonium compounds), and heated microbiological media (especially selective agar media above 48°C). This phenomenon has been observed in many bacterial cells and spores, yeasts, and molds and includes those that are important in foodborne diseases, food spoilage, food bioprocessing, and as indicators of sanitation (Table 7.1). From this list it becomes evident that other microorganisms that have not been studied most likely will also be injured by the sublethal stresses. In general, Gram-

Table 7.1 Cells and Spores of Some Microorganisms Important in Food in Which Sublethal Injury Has Been Detected

Gram-positive pathogenic bacteria:
Staphylococcus aureus, Clostridium botulinum, Listeria monocytogenes, Clostridium perfringens, Bacillus cereus
Gram-negative pathogenic bacteria:
Salmonella spp., *Shigella* spp., Enteropathogenic *Esc. coli, Vibrio parahaemolyticus, Campylobacter jejuni, Yersinia enterocolitica, Aeromonas hydrophila*
Gram-positive spoilage bacteria:
Clostridium sporogenes, Clostridium bifermentum, Bacillus subtilis, Bacillus stearothermophilus, Bacillus megaterium, Bacillus coagulans
Gram-negative spoilage bacteria:
Pseudomonas spp.
Gram-positive bacteria used in food bioprocessing:
Lactococcus lactis spp., *Lactobacillus delbrueckii* subsp. *bulgaricus, Lactobacillus acidophilus*
Gram-positive indicator bacteria:
Enterococcus faecalis
Gram-negative indicator bacteria:
Esc. coli, Enterobacter aerogenes, Klebsiella spp.
Bacterial spores:
Clostridium botulinum, Clostridium perfringens, Clostridium bifermentum, Clostridium sporogenes, Bacillus cereus, Bacillus subtalis, Bacillus stearothermophilus, Desulfotomaculum nigrificans
Yeasts and molds:
Saccharomyces cerevisiae, Candida spp., *Aspergillus flavus*

negative bacteria are more susceptible to injury than Gram-positive bacteria, and bacterial spores are much more resistant than vegetative cells to a particular stress.[4-8]

Microbial injury and growth of injured cells have been studied with both bacterial cells and spores. The material discussed here mainly covers bacterial cell and spore injury.

B. Manifestation of Bacterial Sublethal Injury

A bacterial population, following exposure to a sublethal stress, contains three physiologically different subpopulations: the uninjured (normal cells), reversibly injured (or injured cells), and irreversibly injured (or dead cells). Their relative percentages vary greatly and are dependent upon the species and strains, nature of suspending media, nature and duration of a stress, and the methods of detection. The injured cells show differences in several characteristics as compared to their normal counterparts. One of these is increased sensitivity to many compounds for which normal cells are resistant, such as surface-active compounds (bile salts, deoxycholate, or SDS), NaCl, some chemicals (LiCl, bismuthsulfite, or tetrathionate), hydrolytic enzymes (lysozyme or RNase), antibiotics, dyes (Crystal Violet or Brilliant Green), and low pH and undissociated acids. They also lose cellular materials, such as K^+, peptides, amino acids, and RNA. When exposed to or maintained in an unfavorable environment, the injured cells die. Also, the injured cells cannot multiply unless the injury has been repaired. In a nonselective and preferably nutritionally balanced medium, the cells are able to repair their injury, which can extend

1 to 6 h, depending upon the nature of stress and degree of injury. After an extended repair phase, the cells regain their normal characteristics and initiate multiplication. The injured cells in a population can be detected by a suitable enumeration technique. The results in Table 7.2 show that normal *Escherichia coli* cells grew almost equally in both the nonselective and selective media, indicating that the cells are not sensitive to the surface-active agent, deoxycholate. Following freezing and thawing, 93.7% cells died. The survivors formed colonies in the nonselective TS agar (they repaired and multiplied), but 80.1% of the survivors failed to form colonies in the selective TSD agar (due to their injury) and developed sensitivity to deoxycholate. Among the survivors, however, 19.9% were normal or uninjured cells, inasmuch as they grew equally well both in the TS and TSD agar media and were not sensitive to deoxycholate. Many food systems can reduce both the cell death and injury incurred from a specific stress.

Table 7.2 Effect of Freezing and Thawing on the Viability Loss and Injury of *Escherichia coli* NCSM

Enumeration media[a]	Media type	cfu/ml Before freezing	cfu/ml After freezing	Subpopulations
TS agar	Nonselective	276×10^6	17.5×10^6	*Original population* Dead: $276 \times 10^6 – 17.5 \times 10^6$ (93.7%)
TSD agar	Selective	267×10^6	3.5×10^6	*Among survivors* Uninjured: 3.5×10^6 (19.9%) Injured: $17.6 \times 10^6 – 3.5 \times 10^6$ (80.1%)

[a] *Esc. coli* cells in water suspension were enumerated before and after freezing (–20°C for 16 h) and thawing simultaneously in TS (tryptic soy) agar and TSD (TS + 0.075% deoxycholate) agar media by pour plating followed by incubation at 37°C for 24 h.

Bacterial spore injury has been observed following heating, UV and ionizing radiation, and treatment by hydrostatic pressure and some chemicals (e.g., hypochlorite, H_2O_2, ethylene oxide, and probably nitrite) that are important in foods. The injured spores developed a sensitivity to NaCl, low pH, NO_2, antibiotics, redox potential, gaseous atmosphere, and temperature of incubation. They also have delayed germination and longer lag for outgrowth, and develop a need for some specific nutrients. These manifestations also vary with the nature of stress to which the spores were exposed.

C. Sites and Nature of Injury

Altered physiological characteristics of the injured bacterial cells have been used to determine the site of damage in the cellular structural and functional components. From the evidence, it is now recognized that some cell components are damaged by almost all stresses studied. In addition, specific components could be damaged by specific stresses. The structural and functional components known to be damaged by sublethal stresses are the cell wall (or outer membrane, OM), cytoplasmic membrane (or inner membrane, IM), ribosomal RNA (rRNA) and DNA, as well as some enzymes. While the damages in the cell wall (or OM) and cytoplasmic membrane (or IM) could be more evident in injury caused by freezing and drying, damage to rRNA is more extensive in sublethal heating and DNA damage following radiation of cells.[4-8]

In Gram-positive and -negative bacteria, freezing and drying cause changes in cell surface hydrophobicity, and the inability to form compact pellets and to adsorb some phages. In Gram-positive bacteria, surface layer proteins are also lost. In sublethally stressed Gram-negative bacteria, the lipopolysaccharide (LPS) layer undergoes conformational alteration and loses its barrier property against many chemicals (such as SDS, bile salts, antibiotics, lysozyme, and RNase), which could easily enter the injured cells. Very little LPS is found outside the cells. The alteration in conformation of LPS is due to the loss of divalent cations, which are necessary for the stability of LPS. In both Gram-positive and -negative cells, the cytoplasmic membrane (or IM) remains intact in injured cells, but they lose their permeability barrier function. The cells become sensitive to NaCl and also lose different cellular materials. There is a suggestion that protein molecules in this structure probably undergo conformational changes in the injured cells. rRNA is extensively degraded by the activated RNase. DNA can undergo single- and double-strand breaks. In some strains, autolytic enzymes can be activated due to stress, causing lysis of the cells.

In bacterial spores, depending upon the type of sublethal stress, different structural and functional components can be injured. High heat was shown to cause damage to the lytic enzymes necessary in causing lysis of cortex prior to spore germination and to the spore membrane structures, causing the loss of permeability barrier functions. Damages by irradiation (UV and γ) are mainly confined to DNA in the form of single-strand breaks, and by some chemicals (H_2O_2, antibiotics, or chlorine) to the lytic enzymes of the germination system. Hydrostatic pressure, in combination with heat, damages cortex, while γ-irradiation damages both cortex and DNA. Milder to strong acid treatments can cause the spores to become dormant by removing Ca^{2+} from the spores and making them sensitive to heat.[4-8]

D. Repair of Reversible Injury

One of the most important characteristics of the injured bacterial cells is to repair the injury in a suitable environment and become similar to the normal cells. The repair process and the rate of repair can be measured using specific methods. One of them is by measuring regain in resistance of injured cells to the surface-active agents due to repair in the cell wall or the outer membrane. Suspending a sublethally stressed population in a repair medium and simultaneously enumerating the colony-forming units during incubation in nonselective and selective plating media help to determine the rate of repair (Figure 7.3). Initially, the injured survivors fail to form colonies in the selective, but not in the nonselective, media. However, as they repair and regain resistance to the selective agent, they form colonies on both media, as indicated by the increase in counts only in the selective media.[4-8]

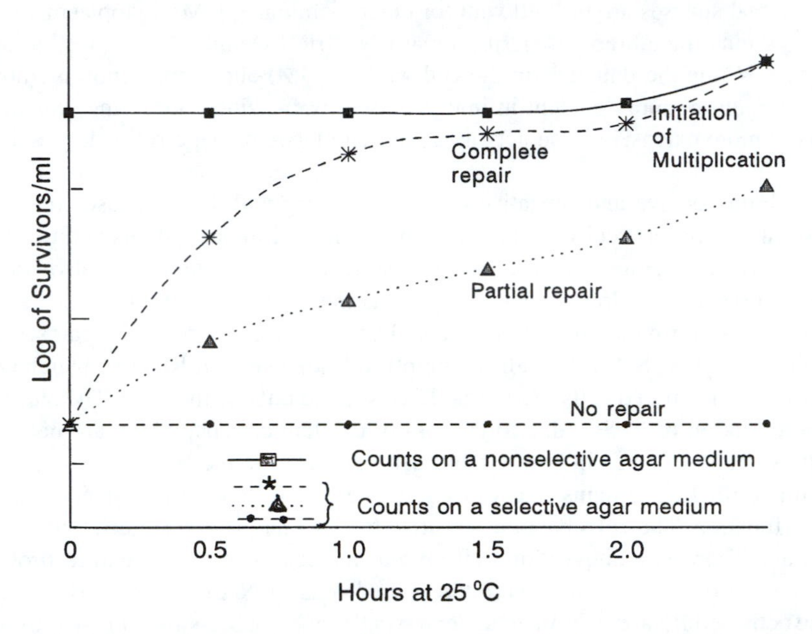

Figure 7.3 A hypothetical repair curve of injured bacteria. Repair is indicated by an increase in counts only in selective agar media during incubation in nonselective repair broth. Cell multiplication is indicated by a simultaneous increase in counts on both selective and nonselective agar media.

The injured cells can repair in a medium devoid of selective compound but containing the necessary nutrients during incubation at optimum pH and temperature. In general, the cells repair well in a medium rich in metabolizable carbon and nitrogen sources and several vitamins. Supplementation with catalase and pyruvate also enhances repair. However, a simple medium with a suitable energy source can also enable some cells to repair damage in the outer membrane (or cell wall). Depending on the sublethal stress, complete repair can be achieved in 1 to 6 h at 25 to 37°C. For freezing and drying injuries, the rate is very rapid; while for heat injuries, the rate can be slow. Specific studies have shown that the metabolic processes during repair vary with the nature of a stress and involve synthesis of ATP, RNA, DNA, and mucopeptide. Reorganization of the existing macromolecules can also be an important event during the repair process. The cell wall (or OM) regains the ability to prevent entrance of many chemicals to the cells, the cytoplasmic membrane (or IM) regains its permeability barrier function as well as the enzymes, and RNA and DNA regain their original characteristics. Finally, the cells regain their ability to multiply.

Repair conditions for injured spores vary with the type (aerobic or anaerobic species) of spores. The composition of media is very important and, in addition to good carbon and nitrogen sources, the addition of special compounds, such as starch, reducing compound (such as cysteine), lysozyme, and divalent cations, may be necessary. Time, temperature, and gaseous environment also have to be optimal for the species. Generally, a longer repair time is required by spores damaged by heat, radiation, and chemicals, than by low temperature.[4-8]

E. Injury in Yeasts and Molds

Very limited studies on injury and repair in yeasts and molds have been conducted. Freezing, low heat treatment, and irradiation have been reported to cause injury in vegetative cells of *Saccharomyces, Kluyveromyces, Candida, Aspergillus, Penicillium,* and *Rhizopus* spp. Spores of the molds have been found to be damaged by irradiation. The main characteristic of the injured cells and spores is their increased sensitivity to many selective environments. The cell membrane seems to be the major structure implicated in injury. Repair in nutritionally rich, nonselective media occurs prior to multiplication.[9]

F. Importance of Injured Microorganisms in Food

Many of the physical and chemical treatments able to induce sublethal injury in microbial cells and spores are used in the processing, storage, and preservation of foods and the sanitation of the facilities. Thus, it is quite likely that the foods and the facilities will harbor injured microorganisms. Microbial injury is important in food microbiology for several reasons.[10]

1. Detection of Undesirable Microorganisms

The injured microorganisms are potentially capable of multiplying. Thus, injured pathogens can cause foodborne disease, and injured spoilage microorganisms can reduce the shelf-life of a product. It is important that if they are present in a food, they should be detected. For detection of many microorganisms in food, several types of selective media are used. Injured microorganisms may not be detected in these media. As a result, foods containing viable but injured pathogens and indicators above the regulatory or acceptable limits, and high numbers of spoilage microorganisms, can be sold. These products can be hazardous and have a short shelf-life, yet they will meet the regulatory standards and specifications. To overcome these problems, a short repair phase has been incorporated prior to the selective detection procedures of important microorganisms in foods. This information is also important in designing processing parameters to obtain proper reduction of undesirable microorganisms in finished products (such as heating temperature and time).

2. Enhancing the Shelf Life of Foods

The injured cells are susceptible to many physical and chemical environments. These conditions, where possible, could be incorporated into the preservation of foods (such as low temperature or preservatives) to kill the injured cells and spores. In this manner, they will be unable to repair, and their potential ability to grow and cause product spoilage can be reduced.

3. Enhancing the Viability of Starter Cultures

In the bioprocessing of foods, starter cultures are used as frozen concentrates or freeze-dried products. However, both of these conditions are known to cause death and injury to the cells. By studying the mechanisms responsible for cell death or cell injury, it may be possible to stop those events and reduce death and injury. This will help produce starter concentrates that can be stored for a long time without a reduction in their desirable characteristics.

REFERENCES

1. Gould, G. W., Ed., Germination, in *The Bacterial Spores,* Academic Press, New York, 1969, 397.
2. Gombas, D. A., Bacterial sporulation and germination, in *Food Microbiology,* Vol. 1, Montville, T. J., Ed., CRC Press, Boca Raton, FL, 1985, 131.
3. Sneath, P. H. A., Ed., Endospore-forming Gram-positive rods and cocci, in *Bergey's Manual of Systematic Bacteriology,* Vol. 2, Williams & Wilkins, Baltimore, 1983, 1104.
4. Ray, B., Ed., *Injured Index and Pathogenic Bacteria,* CRC Press, Boca Raton, FL, 1989.
5. Andrew, M. H. E. and Russell, A. D., Eds., *The Revival of Injured Microbes,* Academic Press, New York, 1984.
6. Busta, F. F. and Foegeding, P. M., Bacterial spore injury — an update, *J. Food Prot.,* 44, 776, 1981.
7. Ray, B., Sublethal injury, bacteriocins and food microbiology, *ASM News,* 59, 285, 1992.
8. Ray, B. and Foegeding, P. M., Repair and detection of injured microorganisms, in *Compendium of Methods for the Examination of Foods,* 3rd ed., Vanderzant, C. and Splittstoesser, D. F., Eds., American Public Health Association, Washington, DC, 1992, 121.
9. Beuchat, L. R., Injury and repair of yeasts and molds, in *The Revival of Injured Microbes,* Andrew, M. H. E. and Russell, A. D., Eds., Academic Press, New York, 1984, 293.
10. Busta, F. F., Importance of injured foodborne microorganisms in minimal processing, in *Minimal Processing of Foods and Process Optimization and Interface,* Singh, R. P. and Oliveira, F. A. R., Eds., CRC Press, Boca Raton, FL, 1994, 227.

CHAPTER 7 QUESTIONS

1. List the differences between mold, yeast, and bacterial spores.
2. Draw and label the structure of a bacterial spore and discuss the functions or characteristics of each structural component.
3. List the stages included between the formation of a bacterial spore and its emergence as a vegetative cell. Also, list the major events that occur in each stage.
4. Discuss the triggering mechanisms in sporulation and spore germination in bacteria.

5. Discuss the importance of bacterial spores in food.
6. Define sublethal injury in microorganisms and discuss the importance of sublethal injury in bacterial cells, bacterial spores, and yeasts and molds in food.
7. List the manifestations of injury in microorganisms.
8. Discuss the site and nature of injury in bacterial cells.
9. Briefly describe the mechanisms by which injured cells and spores of bacteria repair injury.
10. Calculate (%): Dead, uninjured, injured (among survivors), and repaired (among injured) cells in a bacterial population subjected to a sublethal stress (data will be provided by the instructor).

CHAPTER 8

Microbial Metabolism of Food Components

CONTENTS

I. RESPIRATION AND FERMENTATION DURING GROWTH

During growth in a food, microorganisms synthesize energy and cellular materials. A large portion of the energy is used to synthesize the cellular components.

This is achieved through linking the energy-producing reactions with the reactions involved in the synthesis of cell materials. The energy-generating reactions are of the nature of oxidation, and in microorganisms they are organized in sequences (metabolic pathways) for gradual liberation of energy from an organic substrate. The energy is then either used directly in an endergonic reaction or stored for release during a later reaction. The energy can be stored by forming energy-rich intermediates capable of conserving the free energy as biochemical energy. Some of these are derivatives of phosphoric acid (nucleotide triphosphates, acylphosphates, or inorganic polyphosphates) and derivatives of carboxylic acids (acetyl-coenzyme A). The most important of these is ATP, which is formed from ADP in coupled reactions, either through oxidative phosphorylation or through substrate level phosphorylation.

The energy-liberating oxidation reactions of the substrates generate electrons ($H_2 \rightarrow 2H^+ + 2e^-$), which are then accepted by the oxidizing agents. This aspect has briefly been discussed under redox potential (Chapter 6). In a food system, the substrates are mainly the metabolizable carbohydrates, proteins, and lipids. The microorganisms important in foods are heterotrophs (i.e., they require organic carbon sources, substances more reduced than CO_2) and chemoorganotrophs (i.e., they use organic compounds as electron donors to generate energy). On the basis of the nature of the terminal electron acceptors, the energy-generating reactions are differentiated as: aerobic respiration that requires molecular oxygen as the electron acceptor; anaerobic respiration in which inorganic compounds act as the electron acceptors; and fermentation that uses organic compounds as electron acceptors.[1,2] The methods used for electron transfer by different groups of microorganisms important in food can be illustrated by the following scheme:[1]

Organic compounds Inorganic compounds (NO_3, SO_4)

$$\begin{array}{ccccccccc} & & \uparrow 5 & & \uparrow 6 & & & & \\ \text{Substrates} & \xrightarrow{1} & \text{(2H:NAD/FD)} & \xrightarrow{2} & \text{cytochrome(s)} & \xrightarrow{3} & \text{cyt.a} & \xrightarrow{4} & O_2 \\ & & \downarrow 7 & & & & & & \\ & & H_2O_2 \text{ or } O_2 & & & & & & \end{array}$$

Aerobes (some *Bacillus* spp., *Pseudomonas*, molds, and yeasts): 1,2,3,4
Anaerobes (cyt. independent; some *Clostridium* spp.): 1,5
(cyt. dependent, *Desulfotomaculum* spp.): 1,2,6
Facultative anaerobes
(cyt. independent; lactic acid bacteria, some yeasts): aerobic — 1,7
anaerobic — 1,5
(cyt. dependent; such as *Enterobacteriaceae* group): aerobic — 1,2,3,4, or 1,7
anaerobic — 1,5 or 1,2,6

The energy-generating metabolic pathways also produce (from the substrates) many metabolic products that the microbial cells either use for the synthesis of cellular components or release into the environment. The nature of these metabolites differs greatly and is dependent upon the nature of the substrates, the type of microorganisms with respect to their aerobiosis nature, and the oxygen availability (more correctly, redox-potential) of the environment. The metabolism and growth of microorganisms in food are important for several reasons. Microbial spoilage of foods with the loss of acceptance qualities (e.g., flavor, texture, color, and appearance) is directly related to microbial growth and metabolism. Toxin production in food by food-poisoning microorganisms also results from their growth in a food. Many microbial metabolites are also important for their ability to produce desirable characteristics in foods, such as texture, flavor, and long shelf-life. Microbial metabolic products are also used in foods for processing (enzymes), preservation (bacteriocins and acids), and improving texture (dextran) and flavor (diacetyl).

Among the food components, microbial metabolisms of carbohydrates, proteins, and lipids are of major importance. Some of these metabolic pathways are briefly presented in this chapter. Foods, depending upon the source, can contain many types of carbohydrates, proteins, and lipids. This has been discussed briefly before. Depending upon the type and source, foods differ greatly in the amounts of these three groups of nutrients. Plant foods are, in general, rich in carbohydrates, although some (e.g., nuts, lentils, and beans) are also rich in protein, while some others (e.g., oilseeds) are rich in lipids. Foods of animal origin are rich in proteins and lipids, while some (e.g., meat and fish) are low in carbohydrates, others, such as milk, organ meats (liver), and mollusks (oysters), are rich in proteins as well as carbohydrates. Fabricated foods can have all the nutrients in enough quantities for microbial growth. In general, microorganisms preferentially metabolize carbohydrates as an energy source over proteins and lipids. Thus microorganisms growing in a food rich in metabolizable carbohydrates will utilize carbohydrates; but in a food low in metabolizable carbohydrates yet rich in metabolizable proteins will metabolize proteins. In a food rich in both carbohydrates and proteins, microorganisms will usually utilize the carbohydrates first, produce acids, and reduce the pH. Subsequent microbial degradation of proteins can be prevented at low pH, causing a protein sparing effect. In the formulation of processed meat products, added carbohydrates can provide this benefit.

II. METABOLISM OF FOOD CARBOHYDRATES

Food carbohydrates comprise a large group of chemical compounds that include monosaccharides (tetroses, pentoses, and hexoses), disaccharides, oligosaccharides, and polysaccharides.[1-5] Although carbohydrates are the most preferred source of energy production, microorganisms differ greatly in their ability to degrade individual carbohydrates. Carbohydrates are degraded at the cellular level as monosaccharides; disaccharides and trisaccharides can be transported inside the cell and hydrolyzed to monosaccharide units before further degradation. Polysaccharides are broken down to mono- and disaccharides by extracellular microbial enzymes secreted in the environment before they can be transported and metabolized.

A. Degradation of Polysaccharides

Molds, some *Bacillus* spp. and *Clostridium* spp., and several other bacterial species are capable of degrading starch, glycogen, cellulose, pectin, and other polysaccharides. The mono- and disaccharides are then transported and metabolized in the cell. Breakdown of these polysaccharides, especially pectins and cellulose, in fruits and vegetables by microorganisms can affect the texture and reduce the acceptance quality of the products.[1-5]

B. Degradation of Disaccharides

Disaccharides of foods, either present in food (lactose, sucrose) or produced during microbial growth (maltose), are hydrolyzed to monosaccharides inside the cell by these specific enzymes: lactose by lactase to galactose and glucose, sucrose by sucrase to glucose and fructose, maltose by maltase to glucose. Many microbial species are not capable of metabolizing one or more disaccharides.[1-5]

C. Degradation of Monosaccharides

Monosaccharides are degraded (catabolized) by aerobic, anaerobic, and facultative anaerobic microorganisms via several pathways that generate many types of by-products. The metabolic pathways are dependent upon the type and amount of monosaccharides, type of microorganisms, and redox potential of the system. Although all microorganisms important in foods are capable of metabolizing glucose, they differ greatly in their ability to utilize fructose, galactose, tetroses, and pentoses. The fermentable monosaccharides are metabolized by five major pathways, and many microbial species have more than one pathway. These include the Embden-Meyerhoff-Parnas (EMP) pathway, the hexose monophosphate shunt (HMS) or pathway, the Entner-Doudroff (ED) pathway, and two phosphoketolase (PK) pathways (pentose phosphoketolase and hexose phosphoketolase). Pyruvic acid produced via these pathways is subsequently metabolized by microorganisms in several different pathways through fermentation, anaerobic respiration, and aerobic respiration.[1-5]

D. Fermentation

The monosaccharides are fermented by the anaerobic and facultative microorganisms by the five major pathways mentioned above.[1-3] In addition, several other pathways are used, especially for the metabolism of pyruvate, by some specific microbial species and groups. In general, the terminal electron acceptors are organic compounds, and energy is produced at the substrate level. Some of these aspects have been presented in Section IV, dealing with food bioprocessing.[1-3]

The overall reactions of the five main pathways and the end-products are briefly listed here and in Table 8.1.[1-3] These metabolic pathways are discussed in more detail in Chapter 10.

Table 8.1 End-products of Carbohydrate Metabolism by Some Microorganisms

Some microbial types	Fermentation pattern[a]	Major end-products
Yeasts	Alcohol[a]	Ethanol, CO_2
Lactic acid bacteria	Homofermentative[a]	Lactate
	Heterofermentative[a]	Lactate, acetate, ethanol, CO_2, diacetyl, acetoin
Bifidobacteria	Bifidus (hexose ketolase)[a]	Lactate, acetate
Propionibacteria	Propionic acid[a]	Propionate, acetate, CO_2
Enterobacteriaceae, Pseudomonas	Mixed acid	Lactate, acetate, formate, CO_2, H_2, succinate
Bacillus, Pseudomonas	Butanediol	Lactate, acetate, formate, 2,3-butanediol, CO_2, H_2
Clostridium	Butyric acid, butanol, acetone	Butyrate, acetate, H_2, CO_2, butanol, ethanol, acetone, isopropanol

[a] See Chapter 10.

1. EMP Pathway

[Homofermentative lactic acid bacteria, *Enterococcus faecalis, Bacillus* spp., yeasts]

Glucose phosphate → Fructose diphosphate → 2 Pyruvate

Lactate fermentation: Pyruvate → Lactate

Alcohol fermentation: Pyruvate → Acetaldehyde → Ethanol
↓
CO_2

Other hexose monophosphates enter the EMP pathway at different steps, mostly before fructose diphosphate.

2. HMP Pathway (Other names: HMP shunt, pentose cycle, Warburg-Dickens-Horecker pathway)

[Heterofermentative lactic acid bacteria, *Bacillus* spp., *Pseudomonas* spp.]

Glucose-phosphate → Phosphogluconate → Ribulose phosphate
↓
CO_2

→ Ribose phosphate → Acetyl phosphate + Pyruvate

Acetyl-P → Acetate or Ethanol

Pyruvate → Lactate

Ribose phosphate can be used for the synthesis of ribose and deoxyribose moieties in the nucleic acids.

3. ED Pathway

[*Pseudomonas* spp.]

Glucose phosphate → Phosphogluconate → Pyruvate + Glyceraldehyde phosphate

Pyruvate → Acetaldehyde → Ethanol
↓
CO_2

Glyceraldehyde phosphate → Pyruvate → Acetaldehyde → Ethanol
↓
CO_2

4. Pentose Phosphoketolase Pathway

[*Esc. coli, Enterobacter aerogenes, Bacillus* spp., some lactic acid bacteria]

Ribose phosphate → Xylulose phosphate → Acetyl-phosphate + Pyruvate → Acetate or Ethanol + Lactate

5. Hexose Phosphoketolase Pathway

[Bifidus pathway; *Bifidobacterium* spp.]

2 Glucose phosphate → Acetyl phosphate + 2 Xylulose phosphate
→ 3 Acetate + 2 Pyruvate
→ 3 Acetate + 2 Lactate

6. Some Specific Pathways

a. Mixed Acid Fermentation

[*Enterobacteriaceae*]

Pyruvate produced from monosaccharide fermentation can be used to produce different end-products.

Pyruvate → Lactate
Pyruvate → Formate + Acetyl ~ CoA
Formate → H_2 + CO_2
Acetyl ~ CoA → Acetate and Ethanol
Pyruvate → Succinate

b. Propionic Acid Fermentation

[Propionic bacteria, *Clostridium* spp.]

Lactate → Pyruvate → Propionyl ~ CoA → Propionate (propionibacteria)

Lactate → Acrylyl ~ CoA → Propionyl ~ CoA → Propionate (*Clostridium* spp).

c. Butyrate, Butanol, and Acetone Fermentation

[*Clostridium* spp.]

$$\text{Pyruvate} \rightarrow H_2 + CO_2 + \text{Acetyl} \sim \text{CoA}$$
$$\text{Acetyl} \sim \text{CoA} \rightarrow \text{Acetate, Butyrate, Butanol, and Acetone}$$
Acetyl ~ CoA is used to produce different compounds

d. Diacetyl, Acetoin, and Butanediol Fermentation

[some lactic acid bacteria, some *Enterobacteriaceae*]

$$2\ \text{Pyruvate} \underset{\substack{\downarrow \\ 2\ CO_2}}{\rightarrow} [\text{Acetaldehyde} + \text{Acetaldehyde} \sim \text{TPP}]$$

$$\rightarrow \text{Diacetyl} \overset{\substack{2\ H^+ \\ \downarrow}}{\rightarrow} \text{Acetoin} \overset{\substack{2\ H^+ \\ \downarrow}}{\rightarrow} 2,3\text{-Butanediol}$$

E. Anaerobic Respiration

Sulfate-reducing *Desulfatomaculum nigrificans* metabolizes glucose as the energy source, primarily through the EMP pathway to produce pyruvate, which is then decarboxylated to generate acetate (or ethanol) and CO_2. Sulfate, acting as an electron acceptor, is reduced to generate H_2S. NO_3-reducing bacteria, containing nitrate reductase (such as the species in the family *Enterobacteriaceae*, some *Bacillus* spp., and *Staphylococcus* spp.) degrade metabolizable carbohydrates through EMP, HMS, and ED (also mixed acid fermentation) pathways. Pyruvate produced through these pathways can act as an effective electron donor and, depending upon the species, may be converted to lactate, acetate, ethanol, formate, CO_2, H_2, butanediol, acetoin, and succinate.

F. Aerobic Respiration

Aerobes (*Bacillus* spp., *Pseudomonas* spp., molds, and yeasts) and many facultative anaerobes (*Enterobacteriaceae, Staphylococcus* spp.) under aerobic conditions are capable of using molecular oxygen as the terminal electron acceptor during metabolism of carbohydrates to produce pyruvate by one or more of the major pathways mentioned earlier. Pyruvate, as well as other carboxylic acids, can be oxidized completely through oxidative decarboxylation to generate CO_2, H_2O, and large quantities of ATP. The pathway (designated as the Krebs cycle, the tricarboxylic acid cycle, or the citric acid cycle) also generates large numbers of intermediates that are utilized for the synthesis of cell materials. Initially, pyruvate is decarboxylated to generate acetyl ~ CoA and CO_2. Acetyl ~ CoA then combines with oxaloacetate (4C compound) to produce citrate (6C compound). Through successive reactions, citrate is metabolized to a 5C compound (and CO_2) and 4C compounds (and CO_2). The 4C succinate is then, through several steps, changed to oxalacetate for reuse. During these reactions, reducing compounds are generated that, in turn, enter the electron transport system, thereby generating $2H^+$ and $2e^-$. The terminal cytochrome, cytochrome oxidase (cyt.a), releases the electron for its acceptance by oxygen. If the cyt.a transfers only two pairs of electrons to molecular oxygen, the end-product is H_2O. However, if one pair of electrons is transferred, the product is H_2O_2, which is subsequently hydrolyzed by catalase or peroxidase to H_2O and O_2. Each pyruvate is potentially capable of generating 15 ATP molecules.

G. Synthesis of Polymers

Leuconostoc mesenteroides cells growing on sucrose hydrolyze the molecules and predominantly metabolize fructose for energy production. Glucose molecules are polymerized to form dextrin. Polymers are also formed from carbohydrates by some *Lactococcus lactis* and *Lactobacillus* strains, *Alcaligenes faecalis* strains, and *Xanthomonas* spp. Some of these polymers are useful as food stabilizers and to give viscosity in some fermented foods; they can also cause quality loss in some foods.

Metabolism of food carbohydrates by microorganisms is undesirable when it is associated with spoilage. Fermentation of carbohydrates is desirable in food bioprocessing and production of metabolites for use in foods (such as lactate and diacetyl). Several end-products are also used for the identification of microorganisms; for example, 2,3-butanediol production by *Enterobacter* spp., helps to differentiate them from the nonproducer *Escherichia coli* spp. (Voges Proskauer test). The microbial ability to metabolize different polysaccharides, disaccharides, and monosaccharides is also used to identify unknown isolates.

III. METABOLISM OF FOOD PROTEINS

The proteinaceous compounds present in foods include different types of simple proteins (e.g., albumin, globulin, zein, keratine, and collagen), conjugated proteins (e.g., myoglobin, hemoglobin, and casein), and peptides containing two or more amino acids. Amino acids, urea, creatinine, trimethyl amine, and others form the nonprotein nitrogenous group. In general, microorganisms can transport the amino acids and small peptides (about 6 to 8 amino acid in length) in the cells. The proteins and large peptides in a food are hydrolyzed to amino acids and small peptides by the microbial extracellular proteinases and peptidases. Species from genera *Alcaligenes, Bacillus, Clostridium, Enterococcus, Enterobacter, Flavobacterium, Klebsiella, Lactococcus, Micrococcus, Pseudomonas*, and *Serratia* are among those capable of producing extracellular proteinases and peptidases. The small peptides are transported in the cell and converted to amino acids before being metabolized further.[1,4,5]

A. Aerobic Respiration (Decay)

Many aerobic and facultative anaerobic bacteria are capable of oxidizing amino acids and use them as their sole source of carbon, nitrogen, and energy. The L-amino acids generally undergo either oxidative deamination or trans-amination to produce respective keto acids. The keto acids are then utilized through different pathways. Several amino acids can also be oxidized in different pathways by many bacterial species. Some examples are conversion of L-threonine to acetaldehyde and glycine, L-tryptophan to anthranilic acid, L-lysine to glutaric acid, L-valine to ketoisovalerate, L-leucine to ketoisocaproate, L-arginine to citrulline, and L-histidine to urocanic acid.

B. Fermentation (Putrefaction)

Degradation of L-amino acids by anaerobic and facultative anaerobic bacteria is carried out either with single amino acid or two amino acids in pairs. Metabolism of single amino acids is carried out through different types of deamination (with the production of C-skeletone and NH_3), decarboxylation (with the production of amines and CO_2), and hydrolysis (with production of C-skeletone, CO_2, NH_3, and H_2). The C-skeletones (fatty acids, α-keto acids, and unsaturated acids) are then used to supply energy and other metabolic products. The metabolism of amino acids in pairs involves simultaneous oxidation-reduction reactions between suitable pairs in which one acts as a hydrogen donor (oxidized) and the other acts as a hydrogen acceptor (reduced). Alanine, leucine, and valine can be oxidized, while glycine, proline, and arginine can be reduced by this type of reaction (Stickland reaction). The products in this reaction are fatty acids, NH_4, and CO_2.

The products of microbial degradation of amino acids vary greatly with the types of microorganisms and amino acids and the redox potential of the food. Some of the products are: keto acids, fatty acids, H_2, CO_2, NH_3, H_2S, amines, and others. Metabolic products of several amino acids are of special significance in food because many of them are associated with spoilage (foul smell) and health hazards. These include indole and skatole from tryptophan; putrescine and cadaverine from lysine and arginine; histamine from histidine; tyramine from tyrosine; and sulfur-containing compounds (H_2S, mercaptans, sulfides) from cysteine and methionine. Some of these sulfur compounds, as well as proteolytic products of proteinases and peptidases (both extra- and endocellular) of starter culture microorganisms, are important for desirable and undesirable (bitter) flavor and texture in several cheeses. The breakdown of threonine to acetaldehyde by *Lactobacillus acidophilus* is used to produce the desirable flavor in acidophilus yogurt. Indole production from tryptophan is used to differentiate *Escherichia coli* from other coliforms. Also, the amino acid metabolism profile is used in species identification of unknown bacterial isolates. In addition to degradation (catabolism) of proteinaceous compounds of foods, the syntheses (anabolism) of several proteins by some foodborne pathogens while growing in foods are important due to the ability to produce proteins that are toxins. These include the thermostable toxins of *Staphylococcus aureus*, thermolabile toxins of *Clostridium botulinum*, and toxins produced by some bacteria associated with foodborne infections (such as Shiga toxin). The ability of some microbial species to synthesize essential amino acids (such as L-lysine), antibacterial peptides (such as nisin and pediocin), and enzymes (such as amylases and proteinases) in relatively large amounts has been used for beneficial purposes in foods.

IV. METABOLISM OF FOOD LIPIDS

The main lipids in food are mono-, di-, and triglycerides, free saturated and unsaturated fatty acids, phospholipids, sterols and waxes, with the glycerides being the major lipids. Microorganisms have low preference for metabolizing lipids. Being hydrophobic, lipids are difficult to attack when present in a large mass. In emulsion, they can be attacked by the microorganisms at the oil/water interphase. The glycerides are hydrolyzed by the extracellular lipases to release glycerol and fatty acids. The fatty acids then can be transported inside the cells and metabolized by β-oxidation to initially generate acetyl ~ CoA units before being utilized further. Fatty acids, if produced at a rapid rate, will accumulate in the food. Unsaturated fatty acids can be oxidized by microbial oxidases to initially produce hydroperoxides and then carbonyl compounds (aldehydes and ketones).

Some of the microorganisms that are important in food and capable of releasing lipases are found in the following genera: *Alcaligenes, Enterobacter, Flavobacterium, Micrococcus, Pseudomonas, Serratia, Staphylococcus, Aspergillus, Geotrichum,* and *Penicillium*. Oxidative enzymes are produced mainly by the molds. While both groups of enzymes are associated with food spoilage, oxidative enzymes are also important for desirable flavor in mold-ripened cheeses.[4,5]

REFERENCES

1. Doelle, H. W., *Bacterial Metabolism*, 2nd ed., Academic Press, New York, 1975, 208, 442, and 559.
2. Gottschalk, G., *Bacterial Metabolism*, 2nd ed., Academic Press, New York, 1986, 13, 96, 141, and 210.
3. Rose, A. H., *Chemical Microbiology*, Butterworths, London, 1965, 79, 85, 94.
4. Holt, J. G., Ed., *Bergey's Manual of Systemic Bacteriology*, Vol. 1 and Vol. 2, Williams & Wilkins, Baltimore, 1984.
5. Gunsalus, I. C. and Stanier, R. Y., Eds., *The Bacteria*, Vol. II, Academic Press, New York, 1961, 59, and 151.

CHAPTER 8 QUESTIONS

1. Discuss the major differences between aerobic respiration, anaerobic respiration, and fermentation of food nutrients by microorganisms.

2. Discuss how the aerobes, anaerobes, and facultative anaerobes differ from each other in their ability to transfer electrons by different acceptors.

3. A facultative bacterial species is growing anaerobically in three foods: one rich in carbohydrates but low in proteins, one rich in proteins but low in carbohydrates, and one rich in both. Also, the carbohydrates and proteins in all three foods can be metabolized by the bacterial species. Suggest how the bacterial species metabolize the two nutrients.

4. How are the polysaccharides and disaccharides metabolized by microorganisms?

5. List the five major pathways microorganisms use to metabolize monosaccharides found in foods.

6. List some metabolites produced by *Escherichia coli* and *Enterobacter* spp. by mixed acid fermentation, and *Clostridium* spp. by butyric acid fermentation.

7. Discuss briefly the significance of microbial metabolism of carbohydrates by aerobic and anaerobic respiration.

8. How are food proteins metabolized by bacteria? What is the importance of protein synthesis during growth of microorganisms in food?

9. Discuss the differences and importance of amino acid degradation via microorganisms by fermentation and aerobic respiration in food.

10. What is the significance of lipid metabolism by microorganisms in food?

SECTION III

BENEFICIAL USES OF MICROORGANISMS IN FOOD

The major concern of microbial presence in food is due to their undesirable properties. Most are able to spoil foods, and several are associated with foodborne health hazards. However, there are other microorganisms that have beneficial properties in food production, controlling the undesirable spoilage and pathogenic bacteria in food and maintaining normal health of the gastrointestinal tract of humans. The beneficial attributes of the desirable microorganisms are briefly discussed in this section through the following topics:

CHAPTER 9

Microorganisms Used in Food Fermentation

CONTENTS

I. INTRODUCTION

Beneficial microorganisms are used in foods in several ways. These include actively growing microbial cells, non-growing microbial cells, and by-products and cellular components of microorganisms. An example where growing microbial cells are used is the conversion of milk to yogurt by bacteria. Non-growing cells of some bacteria are used to increase the shelf-life of refrigerated raw milk or raw meat. Many by-products, such as lactic acid, acetic acid, some essential amino acids, and bacteriocins produced by different microorganisms, are used in many foods. Finally, microbial cellular components, such as single cell proteins, dextran, and many enzymes, are all used in food for different purposes. These microorganisms or their by-products or cellular components have to be safe, food grade, and approved by the regulatory agencies. When the microbial cells are used in a way so that they are consumed live with the food (as in yogurt), it is very important that they and their metabolites have no detrimental effect on the health of the consumers. Where a by-product (such as an amino acid) or a cellular component (such as an enzyme) is used in a food, the microorganisms producing them have to be regulatory approved and the by-product and cellular component have to be safe. Thus the microorganisms used for these purposes have to meet some commercial and regulatory criteria. In this chapter, characteristics of some microorganisms that are used in the processing of foods, designated as fermented foods, are discussed. Many of these microorganisms are also used to produce several by-products and cellular components that are used in foods.

II. MICROBIOLOGY OF FERMENTED FOODS

Food fermentation involves a process in which raw materials are converted to fermented foods by the growth and metabolic activities of the desirable microorganisms. The microorganisms will utilize some components present in the raw materials as substrates to generate energy and cellular components, to increase in population, and to produce many usable by-products that are excreted in the environment. The unused components of the raw materials and the microbial by-products, together, constitute fermented foods. The raw materials can be milk, meat, fish, vegetables, fruits, cereal grains, and seeds and beans, fermented individually or in combination. Worldwide, over 2000 types of fermented foods are produced. Many ethnic types are being produced and used in small localities by small groups of people. Many of the fermented foods consumed today have been produced and consumed by humans for thousands of years. The old city civilizations, as far back as 3000 to 5000 B.C. in Mesopotamia, Egypt, and the Indus Valley, developed exceptional skills in the production of fermented foods from milk, fruits, cereal grains, and vegetables.

The basic principles developed by these ancient civilizations are used even today to produce many types of fermented foods by the process known as natural fermentation. In this method, either the desirable microbial population naturally present in the raw materials, or some products containing the desirable microbes from a previous fermentation (called "back slopping"), are added to the raw materials. Then the conditions of fermentation are set in a way that favor growth of the desirable types, but either prevent or retard growth of undesirable types that could be present in the raw materials. In another type of fermentation, designated as "controlled" or "pure culture fermentation," the microorganisms associated with fermentation of a food are first purified from the food, identified, and maintained in the laboratory. When required for the fermentation of a specific food, the microbial species associated with this fermentation are grown in large volume in the laboratory and then added to the raw materials in very high numbers. Then the fermentation conditions are set such that these microorganisms grow to produce a desired product. Characteristics of some of the microorganisms used in fermentations are discussed here. These microbial species, when used in controlled fermentation, are also referred to as "starter cultures." Many of the microbial species are present in foods that are naturally fermented, along with other associated microorganisms, some of which also contribute to the desirable characteristics of the products.

III. LACTIC STARTER CULTURES

At present, bacterial species from 10 genera are included in a group designated as lactic acid bacteria, due to their ability to metabolize relatively large amounts of lactic acids from carbohydrates.[1-3] The genera include *Lactococcus, Leuconostoc, Pediococcus, Streptococcus, Lactobacillus, Enterococcus, Aerococcus, Vagococcus, Tetragenococcus,* and *Carnobacterium*. Many of the genera have been created recently and include one or a few species that were once with other genera. Thus, *Lactococcus* and *Enterococcus* were previously classified as *Streptococcus* group N and group D, respectively. *Vagococcus* is indistinguishable from *Lactococcus*, except that these bacteria are motile. *Tetragenococcus* includes a single species that was previously included with *Pediococcus (Ped. halophilus)*. *Carnobacterium* was created to include a few species that were previously in genus *Lactobacillus* and are obligatory heterofermentative. Species from the last five genera, except for maybe *Tet. halophilus*, are, at present, not used as starter cultures and will not be discussed further.

IV. GENUS *LACTOCOCCUS*

This genus includes several species, but only one species, *Lactococcus lactis*, has been widely used in dairy fermentation. It has three subspecies: *Lac. lactis* subspecies *lactis*, *Lac. lactis* subsp. *cremoris,* and *Lac. lactis* subsp. *hordniae;* but only the first two are used in dairy fermentation. A biovar, *Lac. lactis* subsp. *lactis* biovar diacetilactis is also used in dairy fermentation.[1,2]

The cells are ovoid, about 0.5 to 1.0 μm in diameter, present in pairs or short chains, nonmotile, nonsporulating, facultative anaerobic to microaerophilic (Figure 9.1). In general, they grow well between 20 and 30°C, but do not grow in 6.5% NaCl or at pH 9.6. In a suitable broth, they can produce about 1% L(+)-lactic acid and reduce the pH to about 4.5. Subsp. *cremoris* can be differentiated from subsp. *lactis* due to its inability to grow at 40°C, in 4% NaCl, ferment ribose, and hydrolyze arginine to produce NH_3. Biovar diacetylactis, as compared to others, produces large amounts of CO_2 and diacetyl from citrate. They are generally capable of hydrolyzing lactose and casein. They also ferment galactose, sucrose, and maltose. Natural habitats are green vegetation, silage, dairy environments, and raw milk.

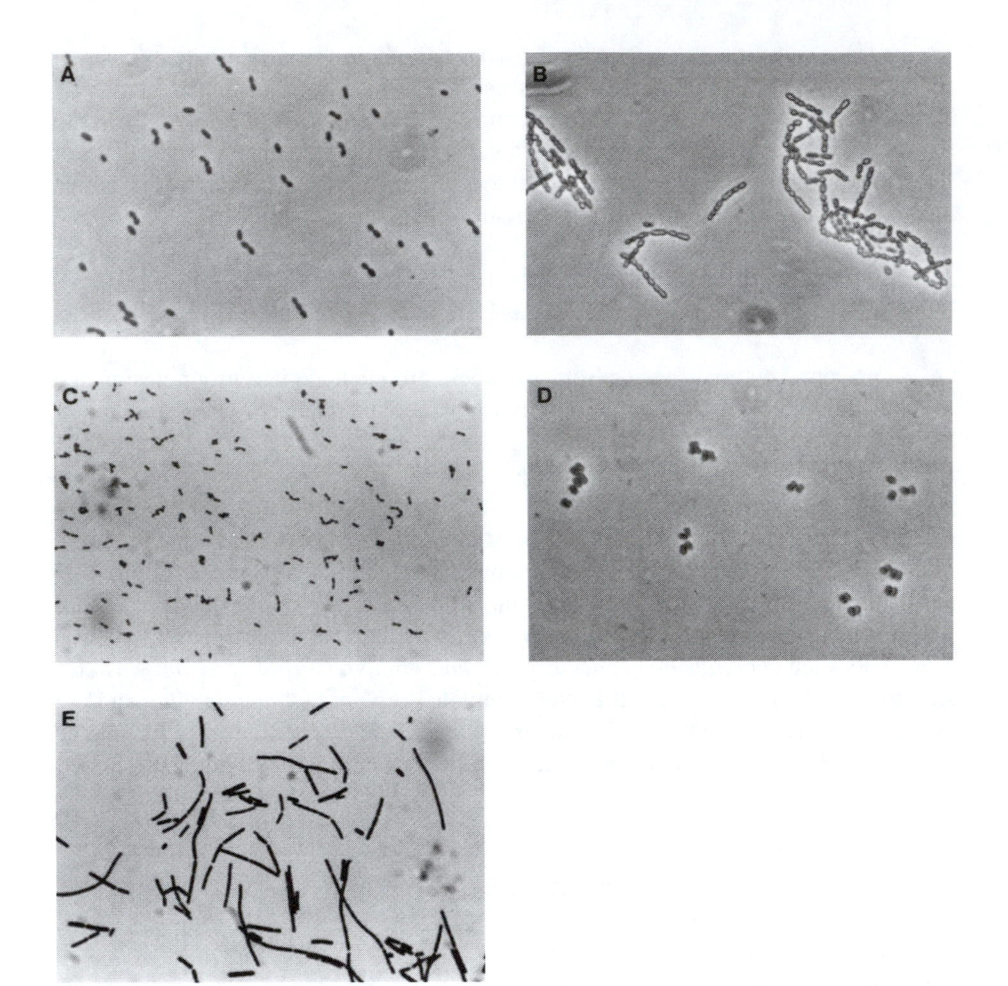

Figure 9.1 Photograph of lactic acid bacteria: (a) *Lactococcus lactis,* (b) *Streptococcus thermophilus,* (c) *Leuconostoc mesenteroides,* (d) *Pediococcus acidilactici,* and (e) *Lactobacillus acidophilus.*

V. GENUS *STREPTOCOCCUS*

Only one species, *Str. thermophilus*, has been used in dairy fermentation. A change in designation to *Str. salivarius* subsp. *thermophilus* was suggested, but has not been used. The Gram-positive cells are spherical to ovoid, 0.7 to 0.9 μm in diameter, and exist in pairs to long chains (Figure 9.1). The cells grow well at 37 to 40°C, but can also grow at 52°C. They are facultative anaerobes and, in glucose broth, can reduce the pH to 4.0. They ferment fructose, mannose, and lactose, but generally not galactose and sucrose. Cells survive 60°C for 30 min. Produces L(+)-lactic acid. Natural habitat is unknown. Found in milk.

VI. GENUS *LEUCONOSTOC*

The Gram-positive cells are spherical to lenticular, arranged in pairs or in chains, nonmotile, nonsporulating, catalase negative, facultative anaerobes (Figure 9.1). The species grow well between 20 and 30°C with a range of 1 to 37°C. Glucose is fermented to D(–)-lactic acid, CO_2, ethanol, or acetic acid with the pH reduced to 4.5 to 5.0. Grows in milk but may not curdle. Also, arginine is not hydrolyzed. Many form dextran while growing on sucrose. Citrate is utilized to produce diacetyl and CO_2. Some species can survive 60°C for 30 min. *Leuconostoc* species are found in plants, vegetables, silage, milk and some milk products, and raw and processed meats. One species, *Leu. oenos*, is found in wine and related habitats.[1,4]

At present, six species are known: *Leu. mesenteroides, Leu. paramesenteroides, Leu. lactis, Leu. oenos, Leu. carnosum,* and *Leu. gelidum*. *Leu. mesenteroides* has three subspecies: subsp. *mesenteroides*, subsp. *dextranicum*, and subsp. *cremoris*. *Leu. mesenteroides* subsp. *cremoris* and *Leu. lactis* are used in some dairy fermentations, while *Leu. oenos* is used in wine for malo-lactic fermentation. Many of these species, particularly *Leu. carnosum* and *Leu. gelidum,* have been associated with spoilage of refrigerated vacuum-packaged meat products.

VII. GENUS *PEDIOCOCCUS*

The cells are spherical and form tetrads, but can be present in pairs. Single cells or chains are absent (Figure 9.1). They are Gram-positive, nonmotile, nonsporulating, facultative anaerobes. They grow well between 25 and 40°C; some species grow at 50°C. They ferment glucose to L(+)- or DL-lactic acid, reducing the pH by some species to 3.6. Depending on the species, they can ferment sucrose, arabinose, ribose, and xylose. Lactose is not generally fermented, especially in milk, and milk is not curdled.[1]

Depending on the species, they are found in plants, vegetables, silage, beer, milk and fermented vegetables, meats, and fish. The genus has 7 to 8 species, of which *Ped. pentosaceus* and *Ped. acidilactici* are used in vegetables, meat, cereal, and other types of fermented foods. They have also been implicated in the ripening and flavor production of some cheeses as secondary cultures. These two species are difficult to differentiate, but as compared to *Ped. acidilactici*, *Ped. pentosaceus* ferments maltose, does not grow at 50°C, and is killed at 70°C in 5 min.[1]

VIII. GENUS *LACTOBACILLUS*

The genus *Lactobacillus* includes a large number of Gram-positive rod-shaped, usually nonmotile, nonsporulating, facultative anaerobic species that vary widely morphologically and in growth and metabolic characteristics (Figure 9.1).[1] Cells vary from very short (almost coccoid) to very long rods, slender or moderately thick, often bent, and can be present as single cells or in short to long chains. While growing on glucose, depending upon a species, they produce either only lactic acid [L(+), D(–), or DL] or a mixture of lactic acid, ethanol, acetic acid, and CO_2. Some also produce diacetyl. Many species utilize lactose, sucrose, fructose, or galactose, and some species can ferment pentoses. Growth temperature can vary from 1 to 50°C, but most that are used as starter cultures in controlled fermentation of foods grow well between 25 and 40°C. Several species involved in natural fermentation of some foods at low temperature can grow well between 10 and 25°C. While growing in a metabolizable carbohydrate, the pH can be reduced to 3.5 to 5.0.

They are distributed widely and can be found in plants, vegetables, grains, seeds, raw and processed milk and milk products, raw, processed, and fermented meat products, and fermented vegetables, and some are found in the digestive tract of humans, animals, and birds. Many have been associated with spoilage of foods.

Among the large number of species, some have been used in controlled fermentation, some are known to be associated with natural fermentation of foods, a few are consumed live for their beneficial effect on intestinal health, while some others can be associated with their undesirable effect on foods. On the basis of their metabolic patterns of hexoses and pentoses (discussed in Chapter 10), the species have been divided into three groups (Table 9.1).[1] Those in group I ferment hexoses (and disaccharides such as lactose and sucrose) to produce mainly lactic acids and do not ferment pentoses (such as ribose, xylose, arabinose, etc.). However, those that belong to group II, depending upon the carbohydrates and the amounts available, either produce mainly lactic acid or a mixture of lactic, acetic, formic acids, ethanol, and CO_2. Group III species, however, ferment carbohydrates to a mixture of lactate, acetate, ethanol, and CO_2.

Table 9.1 Division of *Lactobacillus* Species into Groups

Characteristics	Group I	Group II	Group III
Previous designation	Thermobacterium	Streptobacterium	Betabacterium
Carbohydrate fermentation pattern[a]	Obligately homofermentative	Facultatively heterofermentative	Obligately heterofermentative
End-products of carbohydrate fermentation	Lactate	Lactate or lactate, acetate, ethanol, CO_2, formate	Lactate, acetate, ethanol, CO_2
Ferment pentoses	–	+	+
Representative species	*Lab. delbrueckii* subsp.	*Lab. casei* subsp.	*Lab. fermentum*
	:*delbrueckii*	:*casei*	*Lab. divergens*
	:*bulgaricus*	:*rhamnosus*	*Lab. kefir*
	:*lactis*	:*pseudoplantarum*	*Lab. confusus*
	Lab. leichmannii	*Lab. plantarum*	*Lab. brevis*
	Lab. acidophilus	*Lab. curvatus*	*Lab. sanfrancisco*
	Lab. helveticus	*Lab. sake*	*Lab. reuteri*

[a] Homofermentative (produce mainly lactic acid) and heterofermentative (produce lactic acid as well as large amounts of other products).

The three *Lac. delbrueckii* subspecies are used in the fermentation of dairy products, such as some cheeses and yogurt. They grow well at 45°C and ferment lactose to produce large amounts of D(–) lactic acid. β-galactosidase in these subspecies is constitutive. *Lab. acidophilus* is considered a beneficial intestinal microbe and present in the small intestine. It is used to produce fermented dairy products and also either added to pasteurized milk or made into tablets and capsules for consumption as probiotics. It metabolizes lactose and produces large amounts of D(–) lactic acid. However, in *Lab. acidophilus*, β-galactosidase is generally inducible. *Lab. helveticus* is used to make some cheeses and ferment lactose to lactic acid (DL). *Lab. casei* subsp. *casei* is used in some fermented dairy products. It ferments lactose and produces L(+)-lactic acid. *Lab. plantarum* is used in vegetable and meat fermentation. It produces (DL)-lactic acid. *Lab. curvatus* and *Lab. sake* are capable of growing at low temperatures (2 to 4°C) and are associated with fermentation of vegetable and meat products. *Lab. sake* is used for fermentation of sake. *Lab. kefir* is important in the fermentation of kefir, an ethnic fermented sour milk. *Lab. sanfrancisco* is associated with other microorganisms in the fermentation of San Francisco sourdough bread. *Lab. viridescens, Lab. curvatus*, and *Lab. sake* have been associated with spoilage of refrigerated meat products.[1]

IX. GENUS *BIFIDOBACTERIUM*

They are morphologically similar to some *Lactobacillus* spp. and were previously included in the genus *Lactobacillus*. The cells are Gram-positive, rods of various shapes and sizes, present as single cells or in chains of different sizes. They are nonsporeforming, nonmotile, and anaerobic, although some can tolerate O_2 in the presence of CO_2. The species grow optimally at 37 to 41°C, with a growth temperature range of 25 to 45°C. They usually do not grow at pH above 8.0 or below 4.5. They ferment glucose to produce lactic and acetic acid in 2:3 molar ratio without production of CO_2, and also ferment lactose, galactose and some pentoses.[1]

They have been isolated from feces of humans, animals, and birds and are considered to be beneficial for the normal health of the digestive tract. They are present in large numbers in the feces of infants within 2 to 3 d after birth, and usually present in high numbers in breast-fed babies. They are usually found in the large intestine.

Many species have been isolated from the feces of humans and animals. Some of these include *Bif. bifidum, Bif. longum, Bif. infantis,* and *Bif. adolescentis*. All four of these species have been isolated in humans; however, some species are more prevalent in infants than in adults. Some of these species have been added to dairy products to supply live cells in high numbers to humans to restore and maintain intestinal health.

X. GENUS *PROPIONIBACTERIUM*

The genus includes species in the "classical or dairy propionibacterium group" and the "cutaneous or acne propionibacterium group." Here, only the dairy group is discussed.[1]

The cells are Gram-positive pleomorphic thick rods of 1 to 1.5 μm in length, occur as single cells, pairs, or short chains with different configurations. They are nonmotile, nonsporulating, anaerobic (can also tolerate air), catalase positive, and ferment glucose to produce large amounts of propionic acid and acetic acid. They also, depending on the species, ferment lactose, sucrose, fructose, galactose, and some pentoses. They grow optimally at 30 to 37°C. Some species form pigments. They have been isolated from raw milk, some types of cheeses, dairy products, and silage.

At present, four species of dairy propionibacterium are included in the genus: *Pro. freudenreichii, Pro. jensenii, Pro. thoenii,* and *Pro. acidipropionici*. All of them have been associated with natural fermentation of Swiss-type cheeses, but *Pro. freudenreichii* has been used as a starter culture in controlled fermentation.

XI. GENUS *ACETOBACTER*

A species in this genus, *Ace. aceti,* is used for the production of acetic acid from alcohol.[1] The cells are Gram-negative, aerobic, rods (0.5 to 1.5 μm), occurring as single cells, pairs, or chains; they can be motile or nonmotile. They are obligately aerobes, catalase positive, oxidize ethanol to acetic acid and lactic acid to CO_2 and H_2O. Grow well between 25 to 30°C. Found naturally in fruits, sake, palm wine, cider, beer, sugar cane juice, "tea fungus," and soil.

XII. YEASTS

Among many types of yeasts, only a few have been associated with fermentation of foods and alcohol, production of enzymes for use in food, production of single cell proteins, and as an additive to impart desirable flavor in some foods.[5] The most important genus and species used is *Saccharomyces cerevisiae*. It has been used for leavening bread, producing beer, wine, distilled liquors, industrial alcohol, producing invertase (enzyme), and to flavor some foods (soups). However, many strains have been developed to suit a specific need.

The cells are round, oval, or elongated. They multiply by multipolar budding or by conjugation and formation of ascospores. The strains are generally grouped as bottom yeasts or top yeasts. The top yeasts grow very rapidly at 20°C, producing alcohol and CO_2. They also form clumps which, due to rapid CO_2 production, float at the surface. In contrast, the bottom yeasts grow better at 10 to 15°C, produce CO_2 slowly (also grow slow), do not clump, and thus settle at the bottom. The top yeasts and bottom yeasts are used according to the need of a particular fermentation process.

In addition, *Candida utilis* has been used to produce single-cell proteins. It is a false yeast (*Fungi imperfecti*), reproduces by budding (not by conjugation). The cells are oval to elongated and form hyphae with large numbers of budding cells. They have also been involved in food spoilage.

Kluyveromyces marxianus and *Klu. marxianus* var. *lactis* are capable of hydrolyzing lactose and have been associated with the natural fermentation, along with other yeasts and lactic acid bacteria, of alcoholic dairy products such as kefir. They have also been associated with spoilage of some dairy products.[5]

XIII. MOLDS

Although most molds are associated with food spoilage and some form mycotoxins while growing in foods, other species and strains are used in processing of foods and to produce additives and enzymes for use in foods.[5]

In general, the molds are multicellular, filamentous fungi. The filaments (hyphae) can be septate or nonseptate and have nuclei. They divide by elongation at the tip of a hypha (vegetative reproduction) or by forming sexual or asexual spores on a spore-bearing body.

Among many genera, several species from genera *Aspergillus* and *Penicillium,* and a few from *Rhizopus* and *Mucor,* have been used for beneficial purposes in food. One major importance is that the strains to be used for this purpose should not produce mycotoxins. It is difficult to identify a nonmycotoxin producer strain in the case of natural fermentation, but should be an important consideration in the selection of strains for use in controlled fermentation.

Aspergillus oryzae is used in the fermentation of several oriental foods, such as sake, soy sauce, and miso. It is also used as a source of some food enzymes. *Aspergillus niger* is used to produce citric acid and gluconic acid from sucrose. It is also used as a source of the enzymes pectinase and amylase. *Penicillium roquefortii* is used for the ripening of Roquefort, Gorgonzola, and Blue cheeses. Some strains can produce a neurotoxin, roquefortin. In the selection and development of strains for use in cheese, this aspect needs careful consideration. *Pen. camembertii* is used in Camembert cheese and *Pen. caseicolum* is used in Brie cheese. They are also used to produce the enzyme glucose oxidase.

REFERENCES

1. Sneath, P. H. A., Mair, N. S., Sharpe, M. E., and Holt, J. G., Eds., *Bergey's Manual of Systemic Bacteriology,* Vol. 2, Williams & Wilkins, Baltimore, 1985, 1065, 1071, 1075, 1209, 1346, and 1418.
2. Schleifer, K. H., Kraus, J., Dvorak, C., Kilpper-Blaz, R., Collin M. D., and Fisher, W., Transfer of *Streptococcus lactis* and related streptococci to the genus *Lactococcus* genus nov, *System. Appl. Microbiol.,* 6, 183, 1985.
3. Axelsson, L. T., Lactic acid bacteria: classification and physiology, in *Lactic Acid Bacteria,* Salminen, S. and von Wright, A., Eds., Marcel Dekker, New York, 1993, 1–64.
4. Shaw, B. G. and Harding, C. D., *Leuconostoc gelidum* sp. nov. from chill-stored meats, *Int. J. System. Bacteriol.,* 39, 217, 1989.
5. Beneke, E. S. and Stevenson, K. E., Classification of food and beverage fungi, in *Food and Beverage Mycology,* Beauchat, L. R., Ed., AVI Publishing, Westport, CN, 1978, 1.

CHAPTER 9 QUESTIONS

1. Discuss the criteria used to select a microorganism for beneficial purposes in foods.
2. List the different ways microorganisms are used beneficially in foods.
3. List the genera of lactic acid bacteria used as starter cultures in food fermentation.
4. List one species each from *Lactococcus* and *Streptococcus*, and two each from *Leuconostoc, Pediococcus, Bifidobacterium,* and *Propionibacterium,* used in food fermentation.
5. How are the species in genus *Lactobacillus* (the basis) divided into groups? List two species from each group.
6. When are the following terms used: lactococci and *Lactococcus*; streptococci and *Streptococcus*; leuconostocs and *Leuconostoc*; pediococci and *Pediococcus*; and lactobacilli and *Lactobacillus*?
7. In yeast fermentation of different foods and beverages, only one species is used. Name the species, discuss how one species can be effective in so many fermentation processes. Discuss the characteristics of "bottom" and "top" yeasts.
8. How are molds used in different ways in food? Name two species and list their uses. What precautions does one need while using a mold strain in food fermentation?

CHAPTER 10

Biochemistry of Some Beneficial Traits

CONTENTS

I. INTRODUCTION

The beneficial microorganisms metabolize some of the components present in the starting materials (such as milk or meat) to produce energy and cellular materials and to multiply. In this process, they produce some end-products that are no longer necessary for the cells, so these by-products are excreted into the environment. Some of these by-products impart unique characteristics (mostly texture and flavor) to the remaining components of the starting materials. These are fermented products and are considered desirable by the consumers. Some of the by-products of fermentation can also be purified and used as food additives. The production of several of these by-products by some desirable microorganism is discussed in this chapter.

Before describing the metabolic pathways used by these microbes, it will be helpful to review which components of foods (substrates) are used in fermentation. Also, before these substrates are metabolized inside the microbial cells, they have to be transported from the outside environment. It will also be beneficial to recognize, in brief, the cellular components involved in the transport of these substrates.

The important substrates available in the starting materials of fermentation include several carbohydrates, proteinaceous and nonprotein nitrogenous (NPN) compounds, and lipids. The fermentable carbohydrates in foods are: starch, glycogen (in meat), lactose (in dairy products), sucrose, maltose (from breakdown of starch), glucose, and fructose and pentoses (from plant sources). The proteinaceous and NPN components include mainly large proteins (both structural and functional), peptides of different sizes, and amino acids. The lipids could include triglycerides, phospholipids, fatty acids, and sterols. The microorganisms differ greatly in their ability to transport these components from outside and metabolize inside the cells.

II. MECHANISMS OF TRANSPORT OF NUTRIENTS

The nutrient molecules have to pass through the cell barriers, which constitute both the cell wall and the cell membrane. However, in most Gram-positive lactic acid bacteria the main barrier is the cytoplasmic membrane. The cytoplasmic membrane is made up of two layers of lipids in which protein molecules are embedded, some of which span the lipid bilayer from the cytoplasmic side to the cell wall side (Figure 10.1). Many of them are transport proteins involved in carrying nutrient molecules from the outside into the cell (also removing many by-products from the cell into the environment).[1,2]

In general, small molecules, such as mono- and disaccharides, amino acids, and small peptides (up to 8 to 10 amino acids), are transported almost unchanged inside the cell by specific transport systems, either singly or in groups. The fatty acids can dissolve and diffuse through the lipid bilayers. In contrast, the large carbohydrates (such as polysaccharide), large peptides, and proteins cannot be transported as such inside the cell. If a cell is capable of producing specific extracellular hydrolyzing enzymes that are either present on the surface of the cell wall or released into the environment, then the large nutrient molecules can be broken down to small molecules that are then transported by the appropriate transport systems.

The mono- and disaccharides, amino acids, and small peptides are transported through the membrane by the mechanism known as the active transport system. Some characteristics of this system are: a system can be specific for a type of molecule or for a group of similar molecules; can transport against the concentration gradient of a substrate; the transport process requires energy. In the phosphoenol pyruvate phosphotransferase (PEP-PTS) system for PTS-sugars, the energy is derived from PEP; in the permease system (for permease sugars, amino acids, and probably small peptides), energy is derived from the proton motive force.

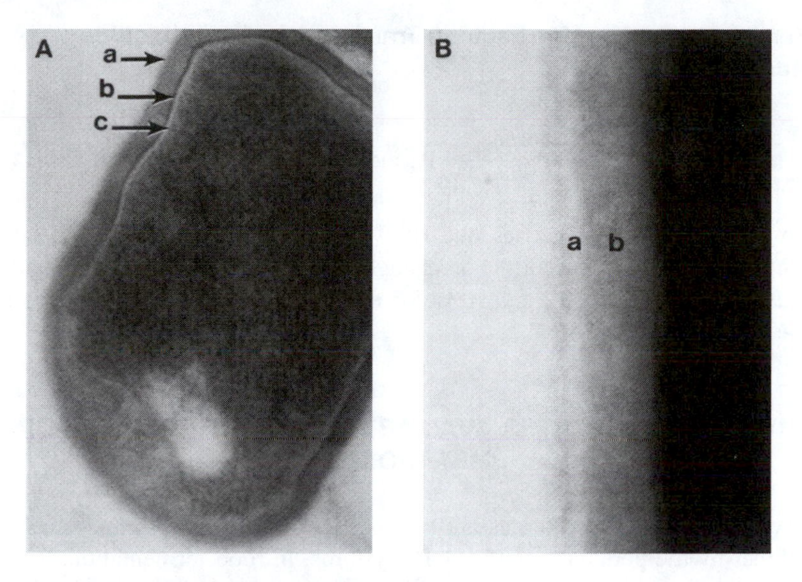

Figure 10.1 A: Photograph of transmission electron microscopy of the thin section of *Lactobacillus acidophilus* cell showing the: (a) anionic (teichoic, teichouronic, lipoteichoic acids) polysaccharide layer, (b) mucopeptide layer, and (c) cytoplasmic membrane. B: Photograph of negatively stained electron microscopy of *Lactobacillus acidophilus*, showing: (a) surface layer protein, and (b) cell wall; cytoplasm is stained dark.

III. PEP-PTS AND PERMEASE TRANSPORT SYSTEMS FOR CARBOHYDRATES

In the lactic acid bacteria and other bacteria used in food fermentation, the disaccharide and monosaccharide (both hexoses and pentoses) molecules can be transported by PEP-PTS as well as by permease systems.[1,2] The same carbohydrate can be transported by the PEP-PTS system in one species, while by the permease system in another species. Similarly, in a species, some carbohydrates are transported by the PEP-PTS system, while others are transported by the permease system.

A. PEP-PTS System for Lactose Transport in *Lactococcus Lactis*

Lactose $\quad$ $EnzII_{Lac}$ / $EnzII_{Lac\sim P}$ $\quad$ $FacIII_{Lac\sim P}$ / $FacIII_{Lac}$ $\quad$ HPr / HPr ~ P $\quad$ EnzI ~ P / EnzI $\quad$ Pyr / PEP

Lactose → Lactose ~ P

The high-energy phosphate from phosphoenol pyruvate (PEP) is transferred sequentially to EnzI, HPr (both are in the cytoplasm and nonspecific for lactose), $FacIII_{Lac}$, and $EnzII_{Lac}$ (both on the membrane and specific for lactose), and finally to lactose. Lactose from the environment is transported in the cytoplasm as lactose-phosphate (galactose-6 phosphate-glucose).

B. Permease System for Lactose Transport in *Lactobacillus Acidophilus*

$$\text{Lactose} + \left[\text{H}^+\right] \xrightarrow{\text{Permease}_{\text{Lac}}} \text{Permease-Lactose-H}^+ \rightarrow \text{Lactose} + \text{H}^+$$

One molecule of lactose carries one H^+ with it in the $\text{permease}_{\text{Lac}}$ (specific for lactose). Once inside, a conformation change occurs in the permease molecule that releases the lactose molecule and H^+ inside the cytoplasm. Lactose is transported as lactose (galactose-glucose).

IV. CARBOHYDRATES AVAILABLE INSIDE THE CELLS FOR METABOLISM

The mono- and disaccharides are transported from the environment inside the cells either by the permease or PEP-PTS systems. In food fermentation, they generally include several pentoses, glucose, fructose, sucrose, maltose, and lactose. The pentoses and hexoses are metabolized by several different pathways, as described later. The three disaccharides are hydrolyzed by enzymes sucrase, maltase, and lactase (β-galactosidase) to hexoses. Lactose-P (galactose-6 phosphate-glucose) is hydrolyzed by phospho-β-galactosidase to yield glucose and galactose-6 phosphate before further metabolism.

V. HOMOLACTIC FERMENTATION OF CARBOHYDRATES

Hexoses in the cytoplasm, either transported as hexoses or derived from the disaccharides, are fermented by homolactic species of lactic acid bacteria to produce mainly lactic acid. Theoretically, one hexose molecule will produce two molecules of lactate. The species include genera from *Lactococcus, Streptococcus, Pediococcus,* and group I and group II *Lactobacillus* (Table 10.1). The hexoses are metabolized through the Embden-Meyerhoff-Parnas (EMP) pathway (Figure 10.2). These species have fructose diphosphate (FDP) aldolase, which is necessary to hydrolyze a 6-carbon hexose to two molecules of 3-carbon compounds. They also lack a key enzyme, phosphoketolase, present in those species that are heterolactic fermentors.[3-5]

In the EMP pathway, with glucose as the substrate, 2 ATP molecules are used to convert glucose to fructose 1,6-diphosphate (Figure 10.2). Hydrolysis of these molecules generates two molecules of 3-carbon compounds. Subsequent dehydrogenation (to produce NADH + H^+ from NAD), phosphorylation, and generation of two molecules ATP lead to production of phosphenol pyruvate (PEP; can be used in PEP-PTS sugar transport). Through the generation of substrate-level ATP, PEP is converted to pyruvate, which, by the action lactate dehydrogenase, is converted to lactic acid. The ability of a lactic acid bacterial species to produce L(+)-, D(–)-, or DL-lactic acid is determined by the type of lactate dehydrogenase (L, D, or a mixture of both, respectively) it contains. The overall reaction is the production of two

Table 10.1 Fermentation of Monosaccharides by Some Starter Culture Bacteria to Produce Different By-products

Genus	Monosaccharide	Fermentation	Pathway[a]	Main product(s)
Lactococcus	Hexoses	Homolactic	EMP	Lactate
Streptococcus	Hexoses	Homolactic	EMP	Lactate
Pediococcus	Hexoses	Homolactic	EMP	Lactate
Leuconostoc	Hexoses	Heterolactic	HMS	Lactate, CO_2, acetate/ethanol (1:1:1)
	Pentoses	Heterolactic	PP	Lactate, acetate/ethanol (1:1)
Lactobacillus				
Group I	Hexoses	Homolactic	EMP	Lactate
Group II	Hexoses	Homolactic	EMP	Lactate
	Pentoses	Heterolactic	PP	Lactate, acetate/ethanol (1:1)
Group III	Hexoses	Heterolactic	HMS	Lactate, CO_2, acetate/ethanol (1:1:1)
	Pentoses	Heterolactic	PP	Lactate, acetate/ethanol (1:1)
Bifidobacterium	Hexoses	Heterolactic	BP	Lactate, acetate (1:1.5)

[a] EMP, Embden-Meyerhoff-Parnas; HMS, hexose monophosphate shunt (also called phosphogluconate-phosphoketolase); PP, pentose-phosphate; BP, bifidus (also called fructose ketolase).

molecules each of lactic acid and ATP from one molecule of hexose. The lactic acid is excreted into the environment.

Other hexoses, such as fructose (transported as fructose or from hydrolysis of sucrose), galactose (from hydrolysis of lactose), and galactose-6-phosphate (from hydrolysis of lactose-phosphate following transport of lactose by PEP-PTS system), undergo different molecular conversion before they can be metabolized in the EMP pathway. Thus, fructose is phosphorylated by ATP to fructose-6-phosphate, and galactose is converted first to galactose-1-phosphate, then to glucose-1-phosphate, and finally to glucose-6-phosphate through the Leloir pathway before entering the EMP pathway. Galactose-6-phosphate is first converted to tagatose-6-phosphate, then to tagatose-1,6-diphosphate, and then hydrolyzed to dihydroacetone phosphate and glyceraldehyde-3-phosphate by the Tagatose pathway before entering the EMP pathway.

VI. HETEROLACTIC FERMENTATION OF CARBOHYDRATES

Hexoses are metabolized to produce a mixture of lactic acid, CO_2, and acetate-ethanol by the heterofermentative lactic acid bacteria (Table 10.1). Species from genera *Leuconostoc* and group III *Lactobacillus* lack fructose diphospho aldolase,

```
              ATP                                                               ATP
(1) Glucose --↓--> (1) Glucose-6-phosphate → (1) Fructose-6-phosphate --↓--> (1) Fructose-1, 6-diphosphate
              ADP                                                               ADP

  *
-----> (1) Dihydroacetone phosphate + (1) Glyceraldehyde-3-phosphate → (2) Glyceraldehyde-3-phosphate

2NAD  2 phosphate                          2ADP
            ↓
-↓---------------> (2) 1,3-diphosphoglycerate ---↓--> (2) 3-phosphoglycerate → (2) 2-phosphoglycerate
2NADH                                      2ATP

                          2ADP                   ** 2NADH+2H+
→ (2) Phosphoenolpyruvate ---↓--> (2) Pyruvate-------↓-----> (2) Lactate
                          2ATP                       2NAD

      [1 Glucose → 2 lactic acid + 2 ATP]

* Fructose diphosphate aldolase, ** L, D or LD-lactate dehydrogenase
```

Figure 10.2 Homolactic fermentation of hexoses through the EMP pathway.

but have glucose phosphate dehydrogenase and xylulose phosphoketolase that enables them to metabolize hexoses through the phospho-gluconate-phosphoketolase pathway (or hexose monophosphate shunt) to generate energy.[3-6]

This pathway has an initial oxidative phase followed by a nonoxidative phase (Figure 10.3). In the oxidative phase, glucose following phosphorylation is oxidized to 6-phosphogluconate by glucose-phosphate dehydrogenase and then decarboxylated to produce one CO_2 molecule and a 5-carbon compound, ribulose-5-phosphate. In the nonoxidative phase, this 5-carbon compound is then converted to xylulose-5-phosphate that, through hydrolysis, produced one glyceraldehyde-3-phosphate and one acetyl-phosphate. Glyceraldehyde-3-phosphate is subsequently converted to lactate. Acetyl-phosphate can be oxidized to yield acetate, or reduced to yield ethanol. Species differ in their ability to produce ethanol, acetate, or a mixture of both. The end-products are excreted into the environment.

VII. METABOLISM OF PENTOSE

The species in genera *Leuconostoc* and group III *Lactobacillus* are capable of fermenting different pentose sugars by the pentose-phosphate pathway to produce ATP, lactate, and acetate, as they have phosphoketolase enzyme. In group II *Lactobacillus,* this enzyme is inducible and is produced only when a pentose is present in the environment. Although *Ped. pentosaceus, Ped. acidilactici,* and *Lac. lactis* can metabolize some pentoses, the pathway(s) are not clearly known.[3-6]

The metabolizable pentoses by the *Leuconostoc* and *Lactobacillus* (group II and III) are first converted to xylulose-5-phosphate by several different ways. Xylulose-5-phosphate is then metabolized to produce lactate and acetate-ethanol by the mechanisms described in the nonoxidizing portion of metabolism of hexoses by the heterofermentative lactic acid bacteria (Figure 10.3). No CO_2 is produced from the metabolism of pentoses through this pathway.

VIII. HEXOSE FERMENTATION BY *BIFIDOBACTERIUM*

Bifidobacterium species metabolize hexoses to produce lactate and acetate by the fructose-phosphate shunt or bifidus pathway.[1,2] For every two molecules of hexoses, two molecules of lactate and three molecules of acetate are produced without generation of any CO_2 (Figure 10.4). From two molecules of fructose-6-phosphate, generated from two molecules of glucose, one molecule is converted to produce one 4-carbon erythrose-4-phosphate and one acetyl-phosphate (which is then converted to acetate). Another molecule of fructose-6-phosphate combines with erythrose-4-phosphate to generate two molecules of 5-carbon xylulose-5-phosphate through several intermediate steps. Xylulose-5-phosphates are then metabolized to produce lactates and acetates by the method described in the nonoxidizing part of heterolactic fermentation (also in pentose fermentation; Figure 10.3).

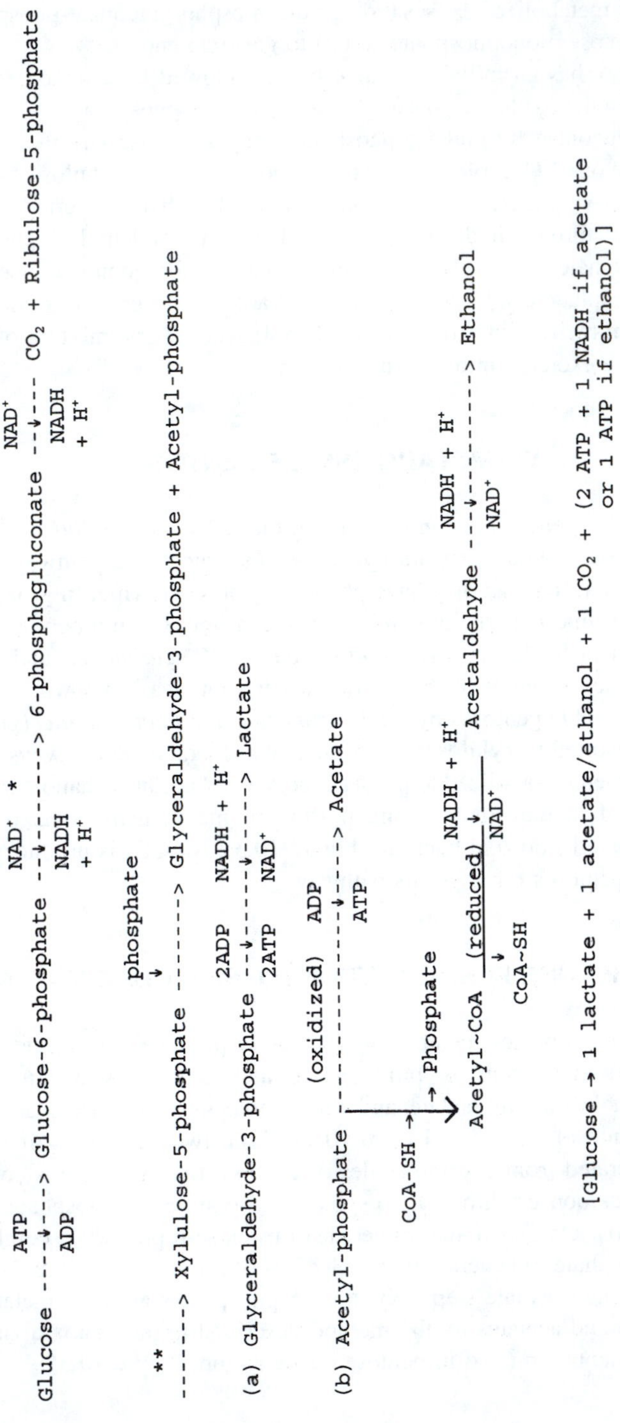

```
                ATP                               NAD+   *                                 NAD+
Glucose ----↓---> Glucose-6-phosphate --↓-------> 6-phosphogluconate --↓--- CO2 + Ribulose-5-phosphate
                ADP                               NADH                                     NADH
                                                  + H+                                     + H+

                                          phosphate
  **                                          ↓
-------> Xylulose-5-phosphate -------> Glyceraldehyde-3-phosphate + Acetyl-phosphate

                                 2ADP    NADH + H+
(a) Glyceraldehyde-3-phosphate --↓------↓-----> Lactate
                                 2ATP    NAD+

                    (oxidized)  ADP
(b) Acetyl-phosphate ----------------↓----> Acetate
                     |          ATP
          CoA-SH →   |
                     | → Phosphate
                     ↓                 NADH + H+                       NADH + H+
               Acetyl~CoA  (reduced)  ↓______ Acetaldehyde ---↓---------> Ethanol
                              ↓        NAD+                            NAD+
                           CoA~SH

[Glucose → 1 lactate + 1 acetate/ethanol + 1 CO2 + (2 ATP + 1 NADH if acetate
                                                    or 1 ATP if ethanol)]
```

Figure 10.3 Heterolactic fermentation of hexoses through HMS.

```
                  2 ATP
    (2) Glucose ---↓----> (2) Glucose-6-phosphate --------> (2) Fructose-6-phosphate
                  2 ADP
                                phosphate
                                   ↓
(a) (1) Fructose-6-phosphate ----------> Erythrose-4-phosphate + Acetyl ~ Phosphate

(b) (1) Fructose-6-phosphate + (1) Erythrose-4-phosphate ------> (2) Xylulose-5-phosphate

    2 phosphate
       ↓
    -----------> (2) Glyceraldehyde-3-phosphate + (2) Acetyl-phosphate

                          3ADP
    (3) Acetyl-phosphate ---↓---> (3) Acetate
                          3ATP

                                           2 ADP
    (2) Glyceraldehyde-3-phosphate -------↓-----> (2) Lactate
                                           2 ATP

     [2 Glucose -----> 2 Lactate + 3 Acetate + 3 ATP]
```

Figure 10.4 Hexose fermentation by *Bifidobacterium*

IX. DIACETYL PRODUCTION FROM CITRATE

Diacetyl, a 4-carbon compound, is important in many fermented dairy products for its pleasing aroma or flavor (butter flavor). Many lactic acid bacteria can produce it in small amounts from pyruvate, generated from carbohydrate metabolism.[6] However, *Lac. lactis* subsp. *lactis* biovar diacetilactis and *Leuconostoc* species are capable of producing large amounts of diacetyl from citrate (Figure 10.5). Citrate, a 6-carbon compound, is transported from outside into the cells by the citrate-permease system. It is then metabolized through pyruvate to acetaldehyde-TPP (thiamin pyrophosphate) with the generation of one acetate (which can be converted to acetyl ~ CoA) and two CO_2. Acetylaldehyde-TPP then combines with acetyl ~ CoA to produce diacetyl. Under a reduced condition, diacetyl can be converted to acetoin with loss of desirable flavor.

X. PROPIONIC ACID PRODUCTION BY *PROPIONIBACTERIUM*

The desirable flavor of some cheeses (such as Swiss) where dairy *Propionibacterium* is used as one of the starter cultures, is from propionic acid. *Propionibacterium* generate pyruvate from hexoses through the EMP pathway, and the pyruvate is used to generate propionic acid (Figure 10.6). Pyruvate and methylmalonyl ~ CoA produce propinyl ~ CoA and oxaloacetate. Propinyl ~ CoA is then converted to propionate. Oxaloacetate recycles to generate methylmalonyl ~ CoA through succinyl ~ CoA. *Propionibacterium* also generates acetate and CO_2 (CO_2 contributes to eye formation in Swiss cheese) from pyruvate.

XI. TRANSPORT AND METABOLISM OF PROTEINACEOUS COMPOUNDS AND AMINO ACIDS

Many lactic acid bacteria used in food fermentation have active transport systems for transporting the amino acids and small peptides (about 8 to 10 amino acids in length).[8] Large peptides and metabolizable proteins in the environment are hydrolyzed by the proteinase and peptidase enzymes, located mainly on the cell wall of these species, to small peptides and amino acids and then transported inside. In the cell, they are all converted to amino acids, which are then metabolized differently to produce many different end-products that are excreted from the cell in the environment. Some of the methods involve decarboxylation (generates amines, some of which are biologically active, such as histamine from histidine), deamination, oxidative reduction, and anaerobic reduction. Amino acid metabolism can produce diverse products, many of which have specific flavor characteristics. Some of these include ammonia, hydrogen sulfide, amines, mercaptans, and disulfides. Many of them, in low concentration, contribute to the desirable flavor of different fermented foods. Proper hydrolysis of food proteins by the proteolytic enzymes of starter microorganisms is important for the desirable texture of many fermented foods (such

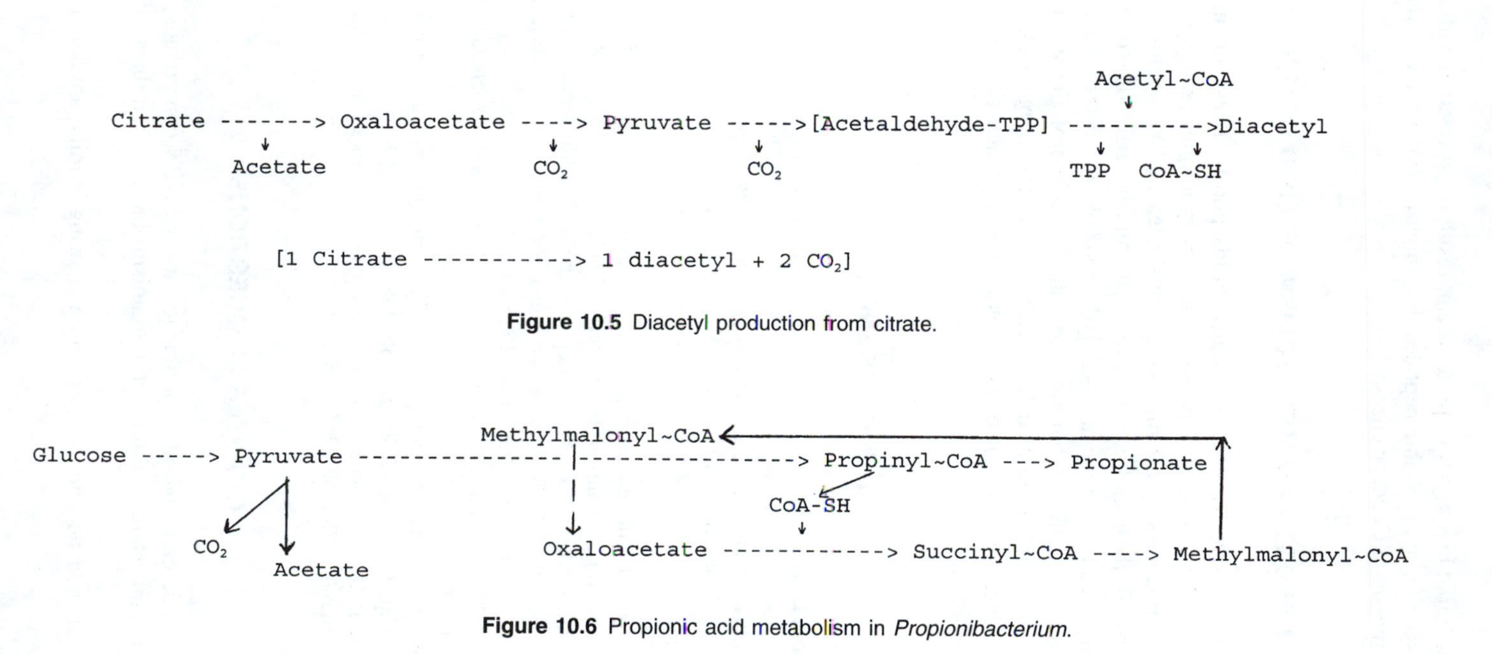

Figure 10.5 Diacetyl production from citrate.

Figure 10.6 Propionic acid metabolism in *Propionibacterium*.

as in cheeses). Rapid hydrolysis of proteins can result in production and accumulation of some specific hydrophobic peptides that impart a bitter taste in the product (such as in some sharp Cheddar cheeses).

XII. TRANSPORT AND METABOLISM OF LIPID COMPOUNDS

Many starter culture bacteria metabolize lipids poorly. However, molds have better lipid metabolism systems. Glycerides and phospholipids are hydrolyzed outside by lipases produced by the microorganisms releasing fatty acids. Fatty acids can diffuse through the membrane into the cells and be metabolized. Some fatty acids are incorporated in the membrane. Hydrolysis of glycerides, especially those with small fatty acids such as butyric acid, can cause hydrolytic rancidity of the products. Oxidation of unsaturated fatty acids by the microorganisms, especially molds, can produce many flavor compounds, some of which are desirable while others are undesirable.

REFERENCES

1. Axelsson, L. T., Lactic acid bacteria: classification and physiology, in *Lactic Acid Bacteria,* Salminen, S. and von Wright, A., Eds., Marcel Dekker, 1993, 1.
2. Sneath, P. H. A., Ed., *Bergey's Manual of Systemic Bacteriology*, Vol. 2, William & Wilkins, Baltimore, 1985, 1209 and 1418.
3. Knadler, O., Carbohydrate metabolism in lactic acid bacteria, *Antonie van Leeuwenhoek,* 49, 209, 1983.
4. Thompson, J., Lactic acid bacteria: novel system for *in vivo* studies of sugar transport and metabolism of Gram-positive bacteria, *Biochimie,* 70, 325, 1988.
5. Ray, B. and Sandine, W. E., Acetic, propionic and lactic acids of starter culture bacteria as biopreservatives, in *Food Biopreservatives of Microbial Origin,* Ray, B. and Daeschel, M. A., Eds., CRC Press, Boca Raton, FL, 1992, 103.
6. Cogan, T. M. and Jordan, K. N., Metabolism of *Leuconostoc* bacteria, *J. Dairy Sci.,* 77, 2704, 1994.
7. Hettinga, D. H. and Reinbold, G. W., The propionic-acid bacteria: a review. II. Metabolism, *J. Milk Food Technol.,* 35, 358, 1972.
8. Thomas, T. D. and Pritchard, G. G., Proteolytic enzymes of dairy starter cultures, *FEMS Microbiol. Rev.,* 46, 245, 1987.

CHAPTER 10 QUESTIONS

1. List the main carbohydrates, proteinaceous and NPN compounds, and lipids in a food system that starter culture microorganisms have in their disposal to metabolize.

2. Discuss how lactic acid bacteria, while growing in milk, are able to transport lactose.

3. What are the different types of food carbohydrates transported inside the starter culture cells? Explain the functions of β-galactosidase and phospho-β-galactosidase.

4. Define homolactic and heterolactic fermentation of carbohydrates by lactic acid bacteria. Give three examples of lactic acid bacteria in each fermentation group.

5. Using glucose as a substrate, list the end-products (including energy generation) of homolactic and heterolactic fermentation. Which enzymes play crucial roles in each pathway?

6. Why is the bifidus pathway of metabolism of hexose also called fructose-phosphate shunt? What are the end-products?

7. How is citrate transported and metabolized to produce diacetyl by some lactic acid bacteria? List two species from two genera that produce diacetyl from citrate. How can other lactic acid bacteria produce diacetyl from carbohydrate metabolism?

8. Pyruvate can be used to produce products other than lactic acid. List four such products and show reactions involved in their production.

9. Briefly discuss how food proteins and food lipids are metabolized by starter culture bacteria.

CHAPTER 11

Genetics of Some Beneficial Traits

CONTENTS

I. INTRODUCTION

Since the 1930s, it has been recognized that many important characteristics (traits or phenotypes) in dairy starter culture bacteria are unstable. Thus, a *Lactococcus lactis* strain once able to ferment lactose while growing in milk-producing lactic acid and coagulating the milk was found to no longer ferment lactose and became commercially not useful. Similar losses of other commercially important traits of starter cultures used in dairy and nondairy fermentations, such as ability to hydrolyze proteins (necessary for some cheese production), ability to utilize citrate (for diacetyl production), loss of resistance to bacteriophages, and loss of hydrolysis of sucrose, were observed. However, the specific mechanism(s) involved in the instability of these important phenotypes were not understood. In the 1960s, the genetic basis of instability of different microbial phenotypes started unfolding. Similar studies, when extended to dairy starter culture bacteria, revealed the genetic basis of the instability of the important traits. In those early days, only a few laboratories were conducting research into the genetics of lactic acid bacteria (notably, Dr. Larry McKay's laboratory at the University of Minnesota). Since the late 1970s, many other laboratories started working in this area and, at present, genetic research of starter culture bacteria has generated a major interest in several laboratories throughout the world. The genetic basis of some of the commercially important phenotypes in some lactic acid bacteria, and methods of transfer of desirable traits from one bacterial strain to another to develop a better strain for use in food fermentation, is discussed briefly in this chapter. To understand this material better, a brief self-review on genetic materials is included at the end of this chapter.

II. PLASMIDS AND PLASMID-LINKED TRAITS IN STARTER CULTURE BACTERIA

Starter culture bacteria, like other bacteria, carry genetic information (genetic code) in the circular chromosomal DNA, circular plasmids, and linear transposons (see "self-review" at the end of this chapter). Chromosomal DNA carries genetic codes for vital functions of a cell (such as a key enzyme in the EMP or HMS pathways in lactic acid bacteria). Although both plasmid DNA and transposons can carry genetic codes, those are only for nonvital functions, i.e., characteristics that are not absolutely necessary for the survival of a cell (such as the ability to hydrolyze a large protein). However, having such a genetic code gives a cell a competitive advantage over the other cells without it, but sharing the same environment. Initial research in the early 1970s revealed that many industrially important phenotypes in different lactic acid bacteria are plasmid linked. Since then, due to the availability of specific techniques, the genetic basis of many plasmid-encoded important phenotypes in several starter culture bacteria, particularly in *Lactococcus lactis*, have been studied. These studies have not only helped to identify the locations of many genes, their structure, and the control systems involved in their expression, but also enabled the researchers to transfer it into a cell lacking a specific phenotype and to create a new desirable strain. Characteristics of plasmids and some plasmid-linked traits in starter culture bacteria are discussed here.[1,2]

A. Important Characteristics of Bacterial Plasmids

- They are double stranded, circular, self-replicating DNA that can differ in size (<1 to >100 kb).
- May not be present in all species or all strains in a species.

- A strain can have more than one type of plasmid that differs in size and the genetic code it carries.
- A plasmid can be present in more than one copy in a cell (copy number; this is in contrast to single copy of chromosome that a cell can carry).
- For some plasmids, copy numbers can be reduced (depressed) or increased (amplified) by manipulation of the control systems.
- Plasmids can differ in their stability in a cell. A plasmid from a cell can be lost spontaneously or by manipulation.
- Two types of plasmids in a cell may be incompatible, resulting in the loss of one.
- Plasmids can be transferred from one cell (donor) to another cell (recipient), spontaneously or through manipulation.
- Plasmid transfer can occur either only between closely related strains (narrow host range), or between widely related strains from different species or genera (broad host range).
- A plasmid can be cryptic (i.e., not known to carry genetic code for a known trait).
- Effective techniques for the isolation, purification, and molecular weight determination of bacterial plasmids have been developed.
- Genetic codes from different sources (different procaryotes and eucaryotes) can be introduced into a plasmid, which then can be transferred in the cell of an unrelated bacterial species in which the phenotype may be expressed.

B. Some Characteristics of Small (about 10 kb) and Large (over 10 to about 150 kb) Plasmids

- Copy number: Small plasmids generally occur in multiple copies (10 to 40 per cell); large plasmids generally present in low copies (for a very large plasmid even one copy per cell)
- Amplification: Many small plasmids can be amplified to a very high copy number; large plasmids, especially very large ones, cannot be amplified
- Conjugal transfer: Small plasmids are nonconjugative (but can be transferred along with a conjugative plasmid); large plasmids are generally conjugative
- Stability: Small plasmids are usually unstable; large plasmids are usually stable; a small plasmid can only encode one or a few phenotypes
- Genetic code: A large plasmid can encode many phenotypes

C. Presence of Plasmids in Some Starter Culture Bacteria

Lactococcus species: Many strains from both subspecies and the biovar have been analyzed. Most strains have 2 to 10 or more types of plasmid, both small and large.

Str. thermophilus: Among the strains examined, a few strains carry plasmids, generally only one to three types; mostly small and not very large.

Leuconostoc species: Many species and strains carry plasmids of different sizes, ranging from 1 to 10 or more types.

Pediococcus species: Limited studies have shown the strains carry from none to two to three plasmids; both small and large types.

Lactobacillus species: Only a limited number of species and strains have been tested. Some species carry plasmids very rarely, such as *Lab. acidophilus*. Some carry usually a few plasmids, such as *Lab. casei*.

Some carry a large number (two to seven) of plasmids of various sizes, such as *Lab. plantarum*.

Bifidobacterium species: Limited studies have revealed the species to harbor two to five different types of plasmids.

Propionibacterium species: Limited studies have shown some species to contain only a few plasmids.

D. Phenotype Assignment to a Plasmid

Following the understanding of plasmid characteristics and the revelation that many starter culture bacteria carry plasmids, studies were conducted to determine that a particular plasmid in a strain carries genetic codes necessary for the expression of a specific phenotype or to determine that a particular phenotype is linked to a specific plasmid in a strain. The loss of lactose fermenting ability (Lac^+ phenotype) among *Lactococcus lactis* strains was suspected to be due to loss of a plasmid that encodes gene(s) necessary for lactose hydrolysis (also for lactose transport). The possible linkage of Lac^+ phenotype to a plasmid in a *Lac. lactis* strain was first studied according to the following protocol:

1. A Lac^+ strain can become Lac^- (inability to hydrolyze lactose) spontaneously, or when grown in the presence of a chemical curing agent (such as acriflavin) or at high temperature (physical curing agent).
2. Analysis of the plasmid profile (types of plasmid present as determined from their molecular weight in kilobase, kb) showed that a 45-kb plasmid, present in the Lac^+ wild strain, was missing in the Lac^- cured variant. Thus, loss of this plasmid was correlated with the loss of Lac^+ phenotype in this strain.
3. To determine further that this 45-kb plasmid was really encoding the Lac^+ phenotype, the wild Lac^+ strain was conjugally mated with a plasmidless Lac^- strain, and several Lac^+ transconjugants were obtained (this aspect is discussed later for mechanisms of DNA transfer). When these transconjugants were analyzed, all were found to contain the 45-kb plasmid (Figure 11.1). Following curing, the transconjugants were converted to Lac^-, and an analysis showed these variants no longer have the 45-kb plasmid.
4. From these series of experiments, it was determined that the 45-kb plasmid in the specific *Lac. lactis* strain used encodes for the Lac^+ phenotype.

Similar studies were conducted to determine plasmid linkage of several other traits in starter culture bacteria.

E. Plasmid-Linked Traits in Starter Culture Bacteria

Many starter culture bacteria, especially *Lac. lactis* strains, have been examined for the plasmid linkage of different phenotypes. These studies have revealed that in these bacteria, many commercially important traits are plasmid-encoded. Some are listed below.

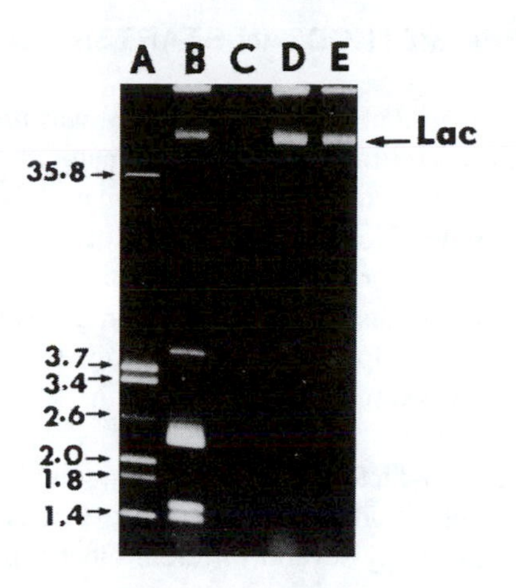

Figure 11.1 Plasmid profiles of *Lactococcus lactis* subspecies *lactis*. Lane A, plasmid standards of different molecular weights in mdal; Lane B, donor strain with *lac*-plasmid (35.5 mda or 53 kb; Lac$^+$ Sm^s); Lane C, plasmidless recipient strain (*Lac*$^-$ Sm^r); Lanes D and E, transconjugants showing *lac*-plasmid (*Lac*$^+$ Sm^r).

- *Lac. lactis:*
  Lac$^+$, lactose hydrolysis (also lactose transport trait);
  Pro$^+$, proteinase activity
  Cit$^+$, citrate hydrolysis (also citrate transport trait)
  Bac$^+$, production of several bacteriocins (also their respective immunity, processing, and translocation traits; bacteriocin, like nisin is encoded in a transposon)
  Phage^r, resistance to specific bacteriophages
  R/M system, restriction/modification
  Resistance to several antibiotics (such as Km^r, resistance to kanamycin)
  Metabolism of several carbohydrates (such as Gal$^+$, galactose utilization)
  Muc$^+$, mucin production
- *Str. thermophilus*: Plasmid linkage of a phenotype is not conclusively known
- *Leuconostoc* species: Bac$^+$, production of different bacteriocins (also immunity against them)
- *Pediococcus* species:
  Suc$^+$, sucrose hydrolysis
  Bac$^+$, production of different bacteriocins (also immunity against them, processing, and translocation)
- *Lactobacillus* species:
  Lac$^+$, lactose hydrolysis
  Mal$^+$, maltose hydrolysis
  Bac$^+$, production of some bacteriocins (also immunity against them)
  Muc$^+$, ability to produce mucin
  Resistance to some antibiotics (such as erythromycin, Em^r)
  R/M system, restriction/modification

The same phenotype in a species can be encoded in different size plasmids.

III. GENE TRANSFER METHODS IN STARTER CULTURE BACTERIA

Once the genetic basis of a phenotype in bacteria was understood, studies were conducted to develop means to transfer the genetic materials from one bacterial cell to another. It is recognized that exchange of genetic materials occurs among bacteria naturally, but at a slower space. However, if a process of introducing genetic materials can be developed under laboratory conditions, the process of genetic exchange not only can be expedited, it can also help in developing desirable strains. In starter culture bacteria, this would help to develop a strain for a specific fermentation process that carries many desirable phenotypes and the least number of undesirable phenotypes.[3,4]

Results of the studies conducted since the 1970s, initially in *Lac. lactis* subspecies and later in other lactic acid bacteria, revealed that genetic materials can be introduced into bacterial cells by several different mechanisms. Some of these are discussed here.

A. Transduction

In this process, a transducing bacteriophage mediates the DNA exchange from one bacterial cell (donor) to another cell (recipient). DNA of some phages (designated as temperate bacteriophages) following infection of a cell can integrate with bacterial DNA and remain dormant (bacteriophages of lactic acid bacteria is presented in Chapter 12). When induced, the phage DNA separates out from the bacterial DNA and, on some occasions, can also carry a portion of the bacterial DNA encoding a gene or genes in it. When the phage-carrying portion of a bacterial DNA infects a bacterial cell and integrates its DNA with bacterial DNA, the phenotype of that gene is expressed by the recipient cell. Initially, the Lac$^+$ phenotype from a lactose-hydrolyzing *Lac. lactis* strain carrying a temperate phage was transduced to a Lac$^-$ *Lac. lactis* strain to obtain a Lac$^+$ transductant. This method has been successfully used to transduce Lac$^+$ phenotype and several other phenotypes (such as Pro$^+$) in different strains of *Lac. lactis* subspecies. Investigation showed that both chromosomal and plasmid-encoded genes from bacteria can be transduced. Transduction has been conducted successfully in some strains of *Str. thermophilus* and in strains of several *Lactobacillus* species.

The transduction process in starter culture bacteria is important in order to determine location of a gene in bacteria for genetic mapping and to study its characteristics. Even though a temperate phage can be induced spontaneously, resulting in lysis of bacterial cells, it is not very useful in commercial fermentation. Also, this method cannot be applied in species that are known not to have bacteriophages, such as some *Pediococcus* species.

B. Conjugation

In this process, a donor bacterial cell transfers a replica of a portion of its DNA to a recipient cell. The two cells have to be in physical contact to effect this transfer.

If the transferred DNA encodes for a phenotype, the transconjugant will have that phenotype. To make the DNA transfer possible, the donor cells should have several other genes, such as a clumping factor (for a physical contact through clumping) and a mobilizing factor (to enable the DNA to move from a donor to a recipient). The process consists of selecting the right donor and recipient strains, mixing the two cell types in a 2:1 to 10:1 (donor:recipient) ratio in several different ways for DNA transfer to occur, and then identifying the transconjugants by appropriate selection techniques.

This technique has been used successfully to transfer several plasmid-linked phenotypes in some lactic acid bacteria. The plasmid-linked Lac^+ phenotype was transferred conjugally between two *Lac. lactis* species. The transconjugant was Lac^+ and had the specific plasmid, loss of which resulted in its phenotype becoming Lac^-. Subsequently, the Lac^+ phenotype located in different plasmids in many *Lac. lactis* subspecies and strains were conjugally transferred to Lac^- strains of the same species. Conjugal transfer of different plasmid-linked traits have also been reported in other lactic acid bacteria, such as the diacetyl production trait in *Lac. lactis* subsp. *lactis* biovar diacetilactis.

The method has several limitations, some of which have been listed previously with the characteristics of plasmids. These include plasmid size, plasmid incompatibility and instability in recipient strains, the inability to express in hosts, the inability to have the proper donors and recipients, and in some cases, the inability to recognize the transconjugant. However, using a broad host range plasmid (e.g., pAMβ1, a plasmid of *Enterococcus* species encoding antibiotic gene), it was possible to show that plasmid transfer by conjugation is possible among lactic acid bacteria between some species, between two different species in the same genus, or between two different species from different genera.

C. Transformation

The method involves extraction and purification of DNA from a donor bacterial strain and mixing the purified DNA with the recipient cells. The expectation is that some DNA can pass through the cell barriers (wall and membrane) and become part of the host DNA expressing the new phenotype. In some Gram-positive bacteria (e.g., *Bacillus* species), this method has been effective for the transfer of certain traits. In lactic acid bacteria, limited studies revealed the technique to be not very effective. However, a modified method was found to be effective. First, *Lac. lactis* cells were treated with lysozyme and/or mutanolysin to remove the cell wall and to form protoplasts in a high osmotic medium. The protoplasts were then exposed to purified DNA (chromosomal, plasmid, or phage DNA) in the presence of polyethylene glycol. The growth conditions were then changed for the protoplasts to regenerate cell wall. The transformants were then detected in a selective medium. By this method, Lac^+ phenotype and Em^r (erythromycin resistance phenotype) and phage DNA (transfection) were transferred to recipient strains of *Lac. lactis* subspecies. Due to limitations in the success rate, this method is not widely studied in lactic acid bacteria.

D. Protoplast Fusion

The technique involves preparation of protoplasts of cells from two different strains and allows them to fuse together in a suitable high osmotic environment. Fusion of cells of the two strains and recombination of the genetic materials may occur. Allowing the protoplasts to regenerate cell wall and using proper selection techniques, recombinants carrying genetic information from both strains can be obtained. It has been used successfully to produce recombinants of *Lac. lactis* subspecies for both Lac^+ and Em^+ phenotypes. However, this technique is not used very much in lactic acid bacteria.

E. Electrotransformation

In this method a suspension of recipient cells in high population levels (10^8 cells per 200 μl) is mixed with purified DNA from a donor strain and then exposed to a high-voltage electric field for a few microseconds. This results in temporary formation of small holes in the cell barrier (membrane) through which purified DNA can pass. Subsequently, the cells were allowed to repair their damage and express the new phenotype to enable their isolation.

This method has been widely used in many lactic acid bacteria to introduce plasmids from the different strains of the same species and from separate species and genera. In addition, vectors carrying cloned gene(s) from other sources have been successfully introduced in several species of lactic acid bacteria. This is currently the most preferred method used to transfer DNA from a source into the recipient cells of lactic acid bacteria.

F. Gene Cloning

In the simplest form in this technique, a DNA segment, carrying a gene or genes, is first obtained by digesting the purified DNA of the donor with the suitable DNA restriction endonuclease(s) and purifying it from the mixture. A suitable plasmid (cloning vector) is selected that has one or several gene markers (such as resistance to an antibiotic) and a site that can be digested with the same restriction enzyme(s). The plasmid is digested with this enzyme, mixed with the donor DNA fragment and incubated for the fragment to align in the opening of the plasmid DNA (Figure 11.2). The open ends are then sealed using suitable enzyme(s). This plasmid carrying the gene(s) from the donor can then be introduced in a bacterial cell in several different ways, most effectively by electroporation in lactic acid bacteria, as described above. This method is now being studied to transfer genes from different sources into lactic acid bacteria.

REFERENCES

1. McKay, L. L., Functional properties of plasmids in lactic *Streptococci, Antoine van Leeuwenhoek,* 49, 259, 1983.

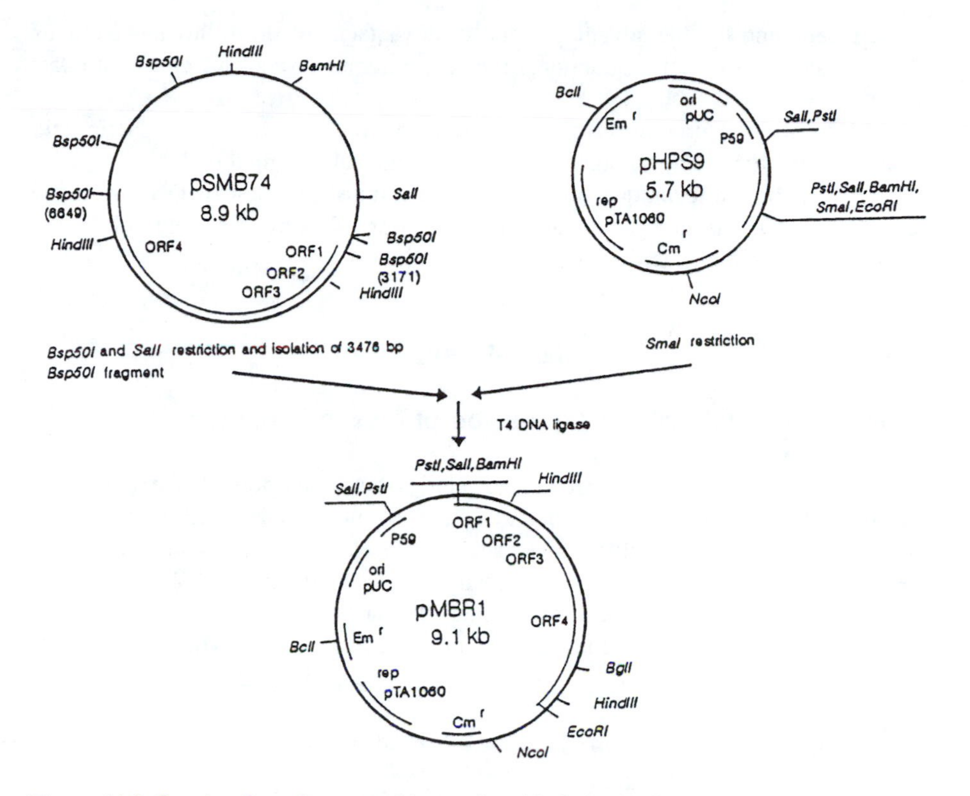

Figure 11.2 Construction of a recombinant plasmid. A 3.5-kb fragment carrying four open reading frames from pSMB74 was derived by digesting with *Bsp* 501 and cloned in the *Sma* I site of pHPS9 to produce 9.1-kb pMBR1.0.

2. von Wright, A. and Sibakov, M., Genetic modification of lactic acid bacteria, in *Lactic Acid Bacteria,* Salminen, S. and von Wright, A., Eds., Marcel Dekker, New York, 1993, 161.
3. Kondo, J. K. and McKay, L. L., Gene transfer systems and molecular cloning in group N streptococci: a review, *J. Dairy Sci.,* 68, 2143, 1985.
4. Gasson, M. J. and Fitzgerald, G. F., Gene transfer systems and transposition, in *Genetics and Biotechnology of Lactic Acid Bacteria,* Gasson, M. J. and de Vos, W. M., Eds., Blackie Academic and Professional, New York, 1994,1–51.

CHAPTER 11 QUESTIONS

1. Discuss the important characteristics of plasmids in lactic acid bacteria.

2. Discuss the characteristic differences between large and small plasmids in lactic acid bacteria.

3. List several commercially important phenotypes that are plasmid-linked in lactic acid bacteria and discuss their importance in relation to plasmid stability.

4. Define and list the advantages and disadvantages of the following methods used in gene transfer: transduction, transfection, transformation, and protoplast fusion in lactic acid bacteria.

5. Describe the process involved in conjugal transfer of DNA in lactic acid bacteria. What are the advantages and disadvantages of this method?

6. Describe the technique involved in electrotransformation of DNA in lactic acid bacteria. Discuss the advantages of transferring DNA by this method over the other methods in lactic acid bacteria.

SELF-REVIEW

From genetic materials to expression of a gene in bacteria.

1. Genetic materials (deoxyribonucleic acid [DNA] and ribonucleic acid [RNA]) are polymers of nucleotides. A nucleotide has a phosphate, a deoxyribose (in DNA) or ribose (in RNA), and a nitrogenous base (2 purines: adenine [A] and guanine [G] and three pyrimidine: thymine [T], cytosine [C], and uracil [U]). While a DNA molecule contains A, G, T, and C nucleotides, the nucleotides in RNA are A, G, U, and C. In a mononucleotide, the base (b) and phosphate (P) are attached to deoxyribose (dR) or ribose (R), respectively, at C–1′ and C–5′ positions with the –OH group free in C–3′ position.

$$\underset{\displaystyle b}{\mathrm{P} - - -\,^{5'}\,\mathrm{dR}^{3'} - - - \mathrm{OH}} \qquad\qquad \underset{\displaystyle b}{\mathrm{P} - - -\,^{5'}\,\mathrm{R}^{3'} - - - \mathrm{OH}}$$

2. Polynucleotide strands

In DNA and RNA, the respective mononucleotide monomers are linked to produce a specific sequence of the bases. The linkage occurs through the phosphodiester bond formation between the 3′–OH group of dR or R in the first nucleotide, with the –OH group of P at 5′ of the next nucleotide. Thus when a strand forms, at its 5′ end there will be a P group; and at the 3′ end, there will be an OH group, giving the strand a 5′ → 3′ polarity. In a strand, the sugar (dR or R) and the P molecules can be imagined to form a backbone with the bases sticking out on one side and the P forming diester bonds between two consecutive nucleotides.

$$\mathrm{P} - - \underset{b_1}{\mathrm{dR}} - - \mathrm{P} - - \underset{b_2}{\mathrm{dR}} - - \mathrm{P} - - \underset{b_3}{\mathrm{dR}} - - \mathrm{P} - - \underset{b_n}{\mathrm{dR}} - \mathrm{OH}\ 3'$$

3. Double-stranded DNA molecule

A DNA molecule has two polynucleotide strands, arranged in opposite polarity, i.e., one in the 5′ → 3′ direction and the other in the 3′ → 5′ direction. Schematically, the bases of two strands can be envisioned to be located inside the two strands. The bases of the opposite strands being close to each other form hydrogen-bonds (H-bonds) in a specific way, i.e., A with T, and G with C. As the bases in one strand form an H-bond with the corresponding bases on the other strand, the base sequences of the two strands are complementary to each other.

An H-bond can form when an "H" is located in between an "N" and an "O" or "N" and "N". Between A and T (or U), two such bonds can form, but between G and C three bonds can form. Large numbers of H-bonds between the complementary bases provide the necessary strength to hold the two strands in a DNA molecule. In a DNA molecule a region rich with A=T has less bonding strength than a region rich in G≡C. A DNA molecule forms a double helix structure, and inside a cell remains in a tightly coiled state (supercoiled).

```
P 5′– dR – – P – – dR – – P – – dR – – – – – P – – dR 3′– OH
       A             G             C                    T
       T             C             G                    A
HO 3′– dR – – P – – dR – – P – – dR – – – – – P – – dR 5′– P
```

4. Bacterial DNA materials

A bacterial cell can have DNA in three forms: the chromosomal, the plasmid, and the transposon DNA. Each cell has only one chromosome, which is relatively large (contains about 10^9 base pairs) and circular (formed by joining 5′-P and 3′-OH of a strand on opposite ends). It carries all the essential or vital genes, together with many other genes (as many as 10^6). A cell can have several types of plasmids of different sizes and a type can have more than one molecule (copy number). The plasmids are also circular DNA, and can carry genes, but only nonvital ones. The transposons are mobile DNA that can be integrated in the chromosome or a plasmid. It is linear and has a DNA segment that can carry gene(s) and is flanked by regions of identical, but opposite, nucleotide sequences. The replication process of a DNA molecule involves synthesis of a complementary strand by each of the two strands, so that at the end of the replication process, each molecule has an old strand and a newly synthesized strand of opposite polarity. The replication process is catalyzed by DNA polymerases, ligase and several other enzymes, as well as RNA polymerase.

5. RNA synthesis (transcription)

Besides replication, the other major function of a DNA molecule is to synthesize messenger RNA (mRNA), transfer RNA (tRNA), and ribosomal RNA (rRNA). About 90% of DNA is used for mRNA synthesis. While the tRNA and rRNA, once synthesized, are conserved for future use, mRNA is destroyed soon after it has been used to synthesize the specific protein (translation). One of two DNA strands at any time is used as a template in the 3′ → 5′ direction. The DNA strands open up in the region to be transcripted, and RNA polymerase catalyzes the synthesis from the start to the stop codon to synthesize an RNA of opposite polarity, i.e., with 5′ → 3′ polarity. If it is mRNA, then the code for the first amino acid of the protein will be at the 5′ end.

6. Protein synthesis (translation)

The three types of RNA have specific functions in translation. An mRNA carries the genetic code to be translated into a protein. A set of three base sequence in the 5′ → 3′ direction in mRNA is a codon and corresponds to a specific amino acid. The codon at the 5′ end corresponds to an NH_2-terminal amino acid of the protein to be translated. At the 3′ end, the mRNA carries a nonsense codon that allows termination of a protein synthesis. The tRNAs carry amino acids to be used in protein synthesis. For each amino acid there is a different type of tRNA. Each tRNA has a three-base sequence of anticodon in the 3′ → 5′ direction to form base pairs with corresponding codon on mRNA. rRNA, in combination with proteins, forms 30 S and 50 S RNA, which then associate together to form 70 S RNA. On 70 S RNA, an mRNA aligns at the 5′ end to begin protein synthesis. A single mRNA molecule can pass through several 70 S RNA so that many molecules of the protein can be synthesized using the same mRNA.

7. Polarity of transcription and translation

DNA template (code)
3′ HO •---------------------- • P 5′

mRNA (codon)
5′ •---------------------- • 3′

Protein (amino acids)
NH_2 •---------------------- • COOH

8. Gene

In bacteria, a gene on the DNA template strand consists of a nucleotide sequence that codes for a protein. It is also called an open reading frame (ORF) or an exon. It can be monocistronic consisting of a noncoding region of promoter followed by a coding region for protein and another noncoding region of terminator. The RNA polymerase binds to the promoter and transcripts the mRNA from the first triplet code to the last triplet code and stops when it reaches the terminator sequence. A gene can also be polycistronic, in which case there will be more than one ORF between a common promoter and a common terminator. The ORFs are separated by noncoding sequences of variable length (also called spacer sequence). During transcription, usually a single mRNA is formed. The mRNA carrying codons for all the proteins can be translated at the same time. Due to the noncoding spacer sequences, specific proteins will form separately. A polycistronic gene can have a regulatory region to control transcription of the genes.

CHAPTER 12

Starter Cultures and Bacteriophages

CONTENTS

I. INTRODUCTION

Starter culture is a generic term and has changed its meaning through the years. Currently, it means selected strain(s) of food-grade microorganisms of known and stable metabolic activities and other characteristics that are used to produce fermented foods of desirable appearance, body, texture, and flavor.[1] Toward the end of the 19th century, the term meant inoculation of a small amount of fermented (sour) cream or milk (starter) to fresh cream or milk to start fermentation in the production of butter and cheese, respectively.[2] This process was found to give better products than the products produced through natural fermentation of the raw materials. These starters were mixtures of unknown bacteria. The processing plant started maintaining a good starter by daily transfer (mother starter) and produced product inoculum from these. However, the bacteriological makeup of these starters (types and proportion of the desirable as well as undesirable bacteria) during successive transfers was continually susceptible to changes as a result of strain dominance among those present initially, as well as from the contaminants during handling. This introduced difficulties in producing products of consistent quality and from product failure due to bacteriophage attack of starter bacteria. Some private companies started supplying the mixed cultures of unknown bacterial composition for cheese manufacture both in the United States and in Europe. Subsequently, the individual strains were purified and examined for their characteristics, and starter cultures with pure strain(s) were produced by these commercial companies. Initially, such starter cultures were developed to produce cheeses. Currently, starter culture for many types of fermented dairy products, fermented meat products, some fermented vegetables, fermented baking products, and for alcohol fermentation are commercially available. In this chapter, a brief discussion of the history, current status, bacteriophage problems, and production of concentrated cultures is presented.

II. HISTORY

Initial development of starter cultures resulted from the need and changes in the cheese industry. Prior to the 1950s, small producers were producing limited amounts of cheese to satisfy local consumers. A plant used to maintain a bottle of "mother culture" by daily transfer, which it received from a culture producer or another neighboring processor. From the mother culture, the plant used to make "bulk culture" through several transfers to meet the inoculation volume (1 to 2% of the volume of milk to be processed for cheese). These starters were a mixture of undefined strains of bacteria, and it was difficult to produce a product of consistent quality. There were also problems with starter failure from bacteriophage attack. To overcome the problem of quality, single-strain starter was introduced (a strain isolated from a mixture). Good sanitation and designing new processing equipment were introduced to overcome the phage problems.[1,2]

Since the 1950s, large cheese operations have replaced the small producers. They needed products of consistent quality and could not afford to have too many starter failures from phage attack. The starter culture producers developed single-strain cultures and supplied these in dried form to the cheese processors, who, in turn, used them to produce mother cultures and bulk cultures. To overcome phage problems, rotation of strains (such as using different strains each day) to prevent the build-up of a particular phage, as well as multiple strain cultures (if one strain is killed by a specific phage, another will work), were practiced. Later, defined media to produce bulk cultures that would reduce phage attack were introduced. Even then, daily production of large amounts of bulk cultures (some cheese processors were handling over a million pounds of milk daily and needed over 10,000 pounds of bulk cultures) and maintaining a large number of defined bacterial strains for use in rotation or multiple strain cultures (some were keeping 30 or more strains) and using them in proper combinations (so that they are compatible) demanded defined and large facilities, expert microbiologists, and a large crew for the operation. This was partially overcome with the introduction in the 1960s of frozen concentrate cultures that could be shipped by air from the culture producers to the cheese processors in dry ice and could be used directly to produce bulk cultures. This, along with the availability of several types of phage inhibitory media (PIM) to produce bulk cultures, helped the cheese processors overcome the cost and labor necessary to maintain a microbiological laboratory and a large number of starter strains. Finally, these dairy-based media contain a high concentration of phosphate to chelate calcium in milk and thus make the divalent cation unavailable for the adsorption of phages to the bacterial cells and cause infection. To obtain high cell density by maintaining a high pH, these media had either internal or external pH control systems. Subsequently, frozen concentrated cultures, containing 10^{11-12} cells per milliliter were introduced; they could be directly inoculated into milk in the cheese vat (direct vat set, DVS), thus eliminating the need to produce bulk starter by a cheese processor.

From the 1970s, the popularity of several types of fermented products (particularly buttermilk and yogurt as well as fermented sausages, some fermented ethnic products, and fermented health products) stimulated their production by large commercial processors. To meet their need, different types of frozen concentrated cultures for direct inoculation into the raw materials were developed.

Efforts have been made to produce freeze-dried concentrated culture. Dried cultures can eliminate the bulk problem in transporting frozen concentrated cultures in dry ice as well as their accidental thawing, which would thus prevent their use. The dried cultures can also be used directly for product manufacture or can be used to produce bulk starters. However, many strains do not survive well in the dried state. Thus, their use as dried cultures in large commercial operations has been limited. They are available in small packages for use directly by the small processors of fermented foods or to produce bulk cultures prior to inoculation in the raw material.

A fairly recent advance has been the availability of custom-designed starter cultures to meet the specific needs of a food processor. An understanding of the genetic basis of some important desirable traits, as well as traits for phage attack inhibition in the starter cultures, has helped to produce the designer cultures.

III. CONCENTRATED CULTURES

In controlled fermentation of food, a starter is added to a raw material at a level of about 10^{6-7} cells per milliliter or cells per gram for the fermentation to proceed at a desired rate. In the conventional process, a bulk culture with about 10^{8-9} cells per milliliter needs to be inoculated at about the 1% level to the raw material. A cheese processor using 100,000 lbs or more milk per day needs 1000 lbs or more bulk starter daily. This large volume is produced from the mother cultures through several intermediate transfers, involving more handling at the processing facilities and possible introduction of phages, as the phages are more abundant in the processing environment (Figure 12.1). In contrast, in the production of concentrate culture, most handling is done by the culture producers under controlled environmental conditions, thereby minimizing phage problems. The strains are grown in suitable media to obtain high cell density (some to over 10^{10} per milliliter). The cells are harvested by centrifugation and resuspended in a liquid at a level of 10^{12-13} cells per milliliter. A 360-ml frozen concentrate culture in DVS can be added to 5000 gal milk vat to get the desired initial concentrations of viable cells (10^{6-7} cells per milliliter). The suspending medium contains cryoprotective agents to reduce cell death and injury during freezing and thawing. The cells in a mixture in metal containers are frozen, either by using dry ice acetone (–78°C) or liquid nitrogen (–196°C), and stored at that temperature. They are transported to the processors, preferably by air in styrofoam boxes with sufficient amounts of dry ice. A processor stores the containers at –20°C or below and uses it within the time specified by the culture producer. Just before use, a container is thawed in warm potable water (45°C) and added to the raw material. The directions for use of these cultures are supplied by the culture producers.[3]

	Conventional cultures		Concentrated cultures
Culture producer	Freeze-dried culture (small volume, single strains)	Culture producer	Stocks (frozen/freeze-dried)
---------	↓		↓
Processors	Stocks (single strains)		Inoculate in liquid media (small volume, single)
	↓		↓
	Mother culture (single strains)		Inoculate in liquid media (large volume, mixed or single)
	↓		↓
	↓		Harvest cells (centrifugation)
	↓		↓
	Intermediate cultures (mixed or single)		Suspend in protective media
	↓		↓
	Bulk cultures (mixed or single)		Freeze (or dry)
	↓		↓
	Processing vat		Transport to processors
			↓
		Processors	Processing vat

Figure 12.1 Production steps and use of conventional and concentrated cultures and changes in culture handling.

To produce freeze-dried concentrate, the liquid cell concentrates are shell frozen (in thin layers or small droplets) in either dry ice-acetone or in liquid nitrogen and then dried under vacuum. The dried materials are packed in plastic bags under vacuum and stored at –20°C or at refrigerated temperature. They are transported to the processors for quick delivery, either at ambient temperature or in boxes containing "ice-packs." A processor stores it at –20°C or in a refrigerator and uses it within the specified time. Just before use, the dried culture is mixed with warm water (preferably boiled and cooled, or sterile) and added to the raw materials.[3]

IV. STARTER CULTURE PROBLEMS

A. Strain Antagonism

In mixed strain cultures, where a starter culture contains two or more strains, dominance of one over the other(s) can change the culture profile quickly. Dominance can result from optimum growth environment or production of inhibitory metabolites. This can affect the product quality and increase starter failure through phage attack. The culture producers test the compatibility between the desirable strains and develop mixed strain cultures only with the compatible strains to avoid strain antagonism.[4]

B. Loss of a Desired Trait

A strain carrying a plasmid-linked desired trait can lose the trait during subculturing and under some growth conditions.[4] Genetic studies are being conducted to understand the mechanisms of the stability of these traits. Several strains with better stability of some traits have been developed and are being used. In the future, more such strains will be available.

C. Cell Death and Injury

This aspect has been discussed in Chapter 7.[4] The effective use of frozen and freeze-dried concentrated cultures, especially for direct use (such as DVS cultures), depends upon two important characteristics. Such cultures need to have large numbers of viable cells, and the cells should have a short lag phase. Freezing and thawing, and freeze-drying and rehydration, are known to cause cell damage leading to cell death and cell injury. This can lead to situations where a concentrate culture will fail to meet the needs of initial viable cell concentrations (10^{6-7} cells per milliliter or per gram) and the relatively short lag in the fermentation process. The culture producers have been successful in reducing cell damage by using cryoprotectants in the suspending menstrua and freezing the cells rapidly at a very low temperature. Some of the causes of cell viability loss are thawing and refreezing, thawing long before using, mixing thawed cultures with or rehydrating dried cultures in concentrated solutions of other constituents (such as curing salt and spice mixtures used in sausage fermentation), and long storage at –20°C or at higher temperatures.[4] For the best performance of concentrated cultures, directions from the culture producers need to be followed.

D. Inhibitors in Raw Materials

The milk may contain either antibiotics, given to the animals to treat some infections (mastitis), or sanitizers (from the equipment).[4] Meat can contain ingredients used to make sausages, such as phosphate or nitrite, and sanitizers from equipment. These factors can prevent or reduce the growth of starters.[4]

E. Bacteriophages

The role of bacteriophages in starter culture failure in food fermentation has been recognized for a long time.[5,6] A short discussion on their life cycle that results in starter failure, as well as some current practices being used to overcome phage problems in food fermentation, is presented here.

1. Morphology

Bacteriophages are filterable viruses of bacteria widely distributed in the environment, especially in food fermentation environments. A phage contains several proteins (that make the head, tail, tail fiber, and contractile sheath) and DNA, which can be linear or circular, double stranded of about 20 to 40 kb in size. The DNA molecule is packed in the head, which could be round or hexagonal.

2. Life Cycle

A phage cannot multiply by itself. Instead, it attaches (needs Ca^{2+} for adsorption) with its tail on the surface of a bacterial cell (specific host) and injects its DNA inside the cell cytoplasm. If it is a lytic phage, the bacterial cell will produce a large number of copies of the phage DNA and the phage proteins. Following assembly of the proteins and phage DNA to produce mature phages, the bacterial cell will lyse, releasing the phages (as many as 200) into the environment (Figure 12.2). They in turn attack other bacterial cells. These are lytic phages. Their growth cycles (lytic cycle) take about 20 to 30 min. There are other phages (called temperate phages) that are nonlytic and their DNA, following injection into the cytoplasm, are integrated with bacterial DNA. The phage DNA (called prophage) is carried by the bacterial cells (lysogeny state), and as the bacterial cells multiply, the prophage also multiplies without showing the presence of the phage in the bacterial strain (lysogenic strain). However, the prophage can be induced by a physical (such as UV) or chemical (such as mitomycin C) agent, causing the phage DNA to separate out of bacterial DNA and resume lytic cycle (Figure 12.2).

3. Host Specificity

Bacteriophages against many species of *Lactococcus, Streptococcus, Leuconostoc, Lactobacillus,* and *Pediococcus oenos* used in food fermentation have been discovered. The phages are host specific and can range from one specific host (strain) for a specific phage to several related strains for a phage. A bacterial strain can also be the host of many different types of phages. A bacterial strain can have restriction enzymes that can hydrolyze and destroy the DNA of a phage. A phage can be either of lytic or temperate type. All require Ca^{2+} for their adsorption on the cell surface of the lactic cultures.

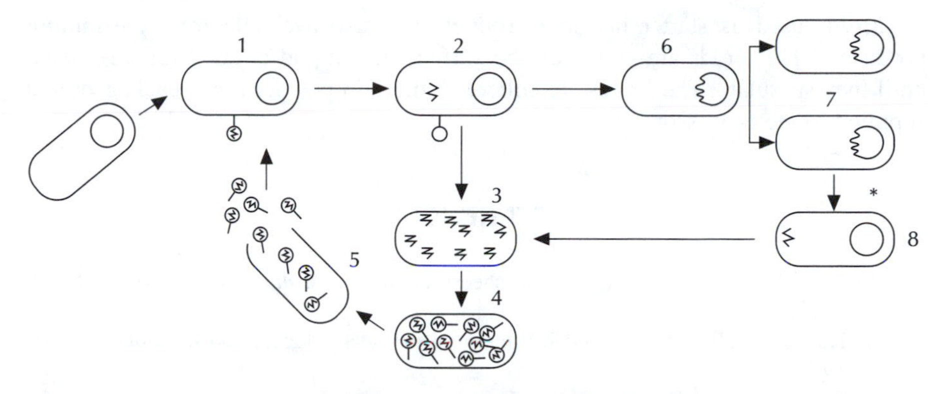

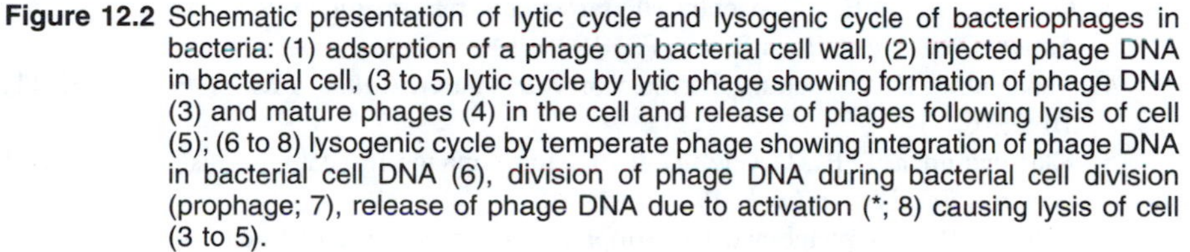

Figure 12.2 Schematic presentation of lytic cycle and lysogenic cycle of bacteriophages in bacteria: (1) adsorption of a phage on bacterial cell wall, (2) injected phage DNA in bacterial cell, (3 to 5) lytic cycle by lytic phage showing formation of phage DNA (3) and mature phages (4) in the cell and release of phages following lysis of cell (5); (6 to 8) lysogenic cycle by temperate phage showing integration of phage DNA in bacterial cell DNA (6), division of phage DNA during bacterial cell division (prophage; 7), release of phage DNA due to activation (*; 8) causing lysis of cell (3 to 5).

4. Control Methods

Following discovery of the bacteriophage-related starter failure in food fermentation, steps were developed aiming at reduction of phage contamination with the starter cultures. These include proper sanitation to reduce phage buildup in the processing facilities, both during bulk starter preparation and product fermentation, use of phage insensitive media, rotation of strains, and use of mixed strains to reduce the buildup of a particular phage to a level to cause a starter failure. Subsequently, phage-resistant starter strains were developed. Initially, by growing a sensitive bacterial strain in the presence of a specific lytic phage, cells that are not killed by this phage are isolated and supplied to the processors. Recent studies have indicated that by using genetic techniques, a bacterial strain can be made insensitive to one or more phages. These include modification of genetic makeup of a starter strain to inhibit phage adsorption, destruction of phage DNA by restriction enzyme systems, or aborting the phages before lysis. By combining these traits through genetic manipulation, a strain can be developed that is resistant to several phages.

V. YEAST AND MOLD CULTURES

Specific strains of yeast cultures (e.g., *Saccharomyces cerevisiae*) for use to leaven dough in bakery products, and to produce alcohol in beer, wine, or distilled liquor, have been developed. These yeasts are produced by culture producers as well as by processors. The culture producers grow the yeast in suitable media, concentrate the cells, and supply in frozen or dried form.

Molds used as starter in some products are also available from the culture producers. The molds are grown on the surface of a liquid or solid (bread) media until they sporulate. The spores are collected, dried in powder form, packaged, and supplied to the processors.[5,6]

REFERENCES

1. Sandine, W. E., Starter systems in cheese making, in *Culture Dairy Product J.,* 10, 6, 1975.
2. Huggins, A. R., Progress in dairy starter culture technology, *Food Technol.,* 6(6), 41, 1984.
3. Gilliland, S. E., Ed., Concentrated starter cultures, in *Bacterial Starter Culture for Foods,* CRC Press, Boca Raton, FL, 1985, 145.
4. Vedamuthu, E. R., Getting the most of your starter, *Cultured Dairy Product J.,* 11, 16, 1976.
5. Klaenhammer, T. R., Development of bacteriophage-resistant strains of lactic acid bacteria, *Food Biotechnol.,* 19, 675, 1991.
6. Hill, C., Bacteriophage and bacteriophage resistance in lactic acid bacteria, *FEMS Microbiol. Rev.,* 12, 87, 1993.

CHAPTER 12 QUESTIONS

1. Define starter culture.
2. Briefly list the major advances in starter culture technology since the 1950s.
3. How are concentrate starter cultures prepared? How does this technology differ from the making of bulk starters?
4. Briefly describe the method used to develop a phage-insensitive medium for the manufacture of bulk starters.
5. List the problems in starter cultures that can affect their performance. What difficulties can occur due to thawing and refreezing a starter culture?
6. How can resistant starter strains against bacteriophages be produced?

CHAPTER 13

Microbiology of Fermented Food Production

CONTENTS

I. INTRODUCTION

In the dawn of civilization humans recognized, probably by accident, that under certain circumstances, when raw foods from plant and animal sources were stored for future consumption, they might change to desirable products with longer storage stability. The possibility of such an event occurring might have been after they learned to produce more foods than they could consume immediately and thus needed storage. It was probably the period during which they learned agriculture and animal husbandry, as well as making baskets and pottery. On this basis, one can assume that fermented food probably originated around 7000 to 8000 B.C. in the tropical areas of Mesopotamia and the Indus Valley. Subsequently, other civilizations also produced fermented foods from different raw materials, particularly to preserve those that were seasonal and thus available in abundance only for a short period. Fermented milk products, alcoholic beverages from fruits and cereal grains, and leavened breads became popular among the early civilizations in the Middle East and in the Indus Valley and later among the Egyptians, Greeks, and Romans.[1]

Currently, more than 2000 different fermented foods are consumed by humans worldwide; many are ethnic and produced in small quantities to meet the need of a group in a particular region. Some are, at present, produced commercially, and only a few are produced by large commercial producers.

At present, there is interest in the consumption of many types of fermented foods besides cheese, bread, pickles, and alcoholic beverages. One reason for this increase is due to consumer interest in natural and healthy foods, which fermented foods have been thought to satisfy. Even countries where many types of fermented foods have been consumed for a long time, but mostly produced in small volumes, have started commercially producing some products in large volumes. It is anticipated that in the future, consumption of many fermented foods will increase worldwide.[1]

II. GENERAL METHOD OF PRODUCTION

The method involves exposing the raw or starting food material(s) to conditions that will favor growth and metabolism of specific and desirable microorganisms. As the desirable microorganisms grow, they utilize some nutrients and produce some end-products. These end-products, along with the unmetabolized components of the starting materials, constitute the fermented foods that have desirable acceptance qualities, many of which are attributed to the metabolic end-products.

A. Raw (or Starting) Materials

A large number of plant and animal products are used to produce fermented foods. These include milk (from cows, buffalo, sheep, goats, and mares), meat (beef, pork, lamb, goat, and fowl), fish (many types), eggs (chicken and duck), different vegetables and vegetable juices, many fruits and fruit juices, cereal grains, tubers, lentils, beans, and seeds. Some are used in combination.

B. Microorganisms Used

Many desirable species and strains of bacteria, yeasts, and molds are associated with fermentation of foods. Depending on the product, fermentation may be achieved by a single predominating species and strain. However, in most fermentations, a mixed population of several bacterial species and strains or even bacteria and yeasts, and bacteria and molds are involved. When a fermentation process involves a mixed population, they should not be antagonistic toward one another; rather, they should preferably be synergistic. Maximum growth of a desirable microorganism and optimum fermentation rate are dependent upon environmental parameters that include nutrients, temperature of incubation, oxidation-reduction potential, and pH. In the fermentation process, if the different species in a mixed population need different environmental conditions (e.g., temperature of growth), a compromise is made to facilitate growth of all the species at a moderate rate. Depending upon a raw or starting material and a specific need, carbohydrates (dextrose in meat fermentation), salts, citrate, and other nutrients are supplemented. In some natural fermentations several species may be involved for the final characteristics of the product. However, instead of growing at the same time, they appear in sequence, with the consequence that a particular species becomes predominant at a certain stage during fermentation. But analyzing the final product to isolate the species involved in the fermentation of such a food will not give the right picture. Instead, samples should be analyzed at intervals to determine predominant types at different times and to know the sequences in their appearance. Finally, some minor flora (secondary flora) may be present in a raw material and the final product, and may not be detected during analysis. However, they may have important contributions for the desirable characteristics, particularly some unique aroma, of the products.

C. Fermentation Process

Foods can be fermented in three different ways, based on the sources of the desirable microorganisms: namely, natural fermentation, back slopping, and controlled fermentation.

1. Natural Fermentation

Many raw materials used in fermentation (usually not heat treated) contain both desirable and associated microorganisms. The conditions of incubation are set to favor rapid growth of the desirable types, and no or slow growth of the associated (many are undesirable) types. A product produced by natural fermentation can have some desirable aroma resulting from the metabolism of the associated flora. However, as the natural microbial flora in the raw materials may not always be the same, it is difficult to produce a product with consistent characteristics over a long period of time. Also, chances of product failure due to growth of undesirable flora and foodborne diseases by the pathogens are high.

2. Back Slopping

In this method, some products from a successful fermentation are added to the starting materials and conditions are set to facilitate the growth of the microorganisms coming from the previous product. This is still practiced in the production of many ethnic products in small volumes. Retention of product characteristics over a long period of time may be difficult due to changes in microbial types. Chances of product failure and foodborne diseases are also high.

3. Controlled Fermentation

The starting materials (maybe heat treated) are inoculated with a high population (10^6 cells per milliliter or more) of a pure culture of single or mixed strains or species of microorganisms (starter culture). The incubation conditions are set for the optimum growth of the starter cultures. Large volumes of products can be produced with consistent and predictable characteristics day after day. Generally, there is less chance of product failure and foodborne diseases. However, growth of desirable secondary flora may not occur. As a result, a product may not have some delicate flavor characteristics.

Microbiological criteria of some fermented dairy, meat, and vegetable products are discussed here to understand the methods involved in controlled and natural fermentation.

III. FERMENTED DAIRY PRODUCTS

Fermented dairy products can be broadly divided into two groups: fermented milk products and cheeses. In fermented milk products, all the constituents of the milk are retained in the final products, with the exception of those partially metabolized by the bacteria. In cheeses, a large portion of milk constituents are removed in whey to obtain the final products.

A. Milk Composition and Quality

The growth of desirable microorganisms and the quality of a fermented dairy product are influenced by the composition and quality of the milk used in a fermentation process. Cow's milk contains approximately 3.2% protein, 4.8% lactose, 3.9% lipids, 0.9% minerals, traces of vitamins, and about 87.2% water. Among the proteins, casein in colloidal suspension as calcium caseinate is present in greater amounts than the other two soluble proteins, albumin and globulin. Lactose is the main carbohydrate and is present in solution, while lipids are dispersed as globules of different sizes in emulsion (fat in water). Minerals are present in solution and with casein. The water-soluble vitamins are present in the aqueous phase, while the fat-soluble vitamins are present in the lipids. The solid components (about 12.8%) are designated as total solids of TS, and TS without lipids is designated as solid-not-fat (SNF; about 8.9%). The whey contains principally the water-soluble components, some fat, and the water.

The growth of desirable microorganisms can be adversely affected by several components either naturally present or entered in the milk as contaminants. The natural antimicrobials are agglutinins and lactoperoxidase-isothiocynate system. The agglutinins can induce clumping of starter culture cells and slow their growth and metabolism. The lactoperoxidase-isothiocynate system can be inhibitory to starter cultures. They can cause problems only when raw milk is used as both are destroyed by heating milk. Milk can also contain antibiotics, either used in the feed or used to treat the animals for some infections, such as mastitis. Their presence can also affect the growth of the starter cultures. Some milk can contain heat stable proteases and lipases produced by some psychrotrophic bacteria, such as *Pseudomonas* species, during storage of raw milk before pasteurization. These enzymes will remain stable after heating and can cause product defects (low yield of cheese, proteolysis, rancidity). Before milk is used for fermentation these aspects need to be considered.

B. Fermented Milk Products

Many types are produced in different parts of the world. A few are produced by controlled fermentation, and the microbial types and their respective contributions are known. In many others, fermented either naturally or by back slopping, the microbial profiles and their contribution are not exactly known. Many types of lactic and bacteria and some yeasts are found to be predominant microbial flora in these products.[2] Some are listed below:

1. Buttermilk: Made with *Lactococcus* species with or without *Leu. cremoris*; some can have biovar diacetilactis in place of *Leu. cremoris* (such as ymer in Denmark); some can have a ropy variant of *Lactococcus* species (langfil in Norway) or mold (*Geotrichum candidum* in villi in Finland).
2. Yogurt: Made with *Str. thermophilus* and *Lab. delbrueckii* subsp. *bulgaricus*; some types can also have *Lab. acidophilus, Bifidobacterium* spp., some may also have *Lactococcus* species and *Lab. plantarum* and lactose fermentating yeasts (dahi in India).
3. Acidophilus milk: *Lab. acidophilus*.
4. Bifidus milk: *Bifidobacterium* spp.
5. Yakult: *Lab. casei* (may contain *Bifidobacterium* spp.).
6. Kefir: *Lab. kefir* (several species of yeasts along with *Leuconostoc, Lactobacillus*, and *Lactococcus* spp.).
7. Kumiss: *Lab. delbrueckii* subsp. *bulgaricus* and yeasts.

Among these, cultured buttermilk and yogurt are discussed here.

C. Microbiology of Cultured Buttermilk Fermentation

It is produced from partially skim milk through controlled fermentation with starter cultures.

1. Product Characteristics

It should have a pleasant acid taste (from lactic acid) and a high degree of aroma (from diacetyl) with slight effervescence (from CO_2). It should have white color with smooth, thick body and pour easily.

2. Process

Skim milk ≥9% SNF + Citrate (0.2%)	Heated 185°F/30 min (kills bacterial cells and phages)	Cooled to 72°F, starter added, agitated 50 min (incorporates air)	Incubated at 72°F/12 h, pH 4.7, acidity 0.9%	Gel broken, cooled to 40°F, salted (packaged)

3. Starter (Controlled Fermentation)

Lac. lactis subsp. *lactis* or *cremoris* for acid; *Leu. mesenteroides* subsp. *cremoris* for diacetyl and CO_2, can be used as direct vat set frozen concentrate.

[*Lac. lactis* subsp. *lactis* biovar diacetilactis is generally not used because it may produce too much acetaldehyde, causing green or yogurt flavor defect]

4. Growth

At 72°C, there is balanced growth of the two species, and balanced production of acid, diacetyl, and CO_2. Above 72°C, the growth of *Lactococcus* species is favored, with more acid and less flavor. Below 72°C, the growth of *Leuconostoc* species is favored, with less acid and more flavor.

5. Biochemistry

Lactose hydrolysis by P-β-galactosidase in *Lactococcus* sp.:

$$\text{Lactose} \begin{cases} \text{Glucose} \longrightarrow \text{Pyruvate} \longrightarrow \text{L(+)-Lactate} \\ \qquad\qquad\qquad\quad \uparrow \\ \text{P-galactose} \rightarrow \text{Tagatose} \end{cases}$$

Lactose hydrolysis by β-galactosidase in *Leuconostoc* sp.:

$$\text{Lactose} \begin{cases} \rightarrow \text{Glucose} \longrightarrow \text{D(-)-lactate} + CO_2 + \text{Acetate (ethanol)} \\ \qquad\qquad \uparrow \\ \rightarrow \text{Galactose} - \end{cases}$$

Citrate metabolism by *Leuconostoc* sp.:

$$\text{Citrate} \longrightarrow CO_2 + \text{Diacetyl} + \text{Acetaldehyde}$$
$$-O_2 \downarrow \quad \uparrow + O_2$$
$$\text{Acetoin (no flavor)}$$

(For a desirable flavor, diacetyl:acetaldehyde should be at >3:1 to <4.5:1.)

6. Genetics

Lac. lactis strains should hydrolyze lactose (Lac^+), metabolize P-galactose by tagatose pathway and galactose by Leloir pathway, should not produce slime, and should not be very proteolytic; also, they should be resistant to phages.

Leuconostoc species should be able to utilize citrate to produce more diacetyl and less acetaldehyde and ferment lactose; they should not produce slime and should be phage resistant.

Strains should not produce inhibitory compounds (such as bacteriocins) against each other, but can have antimicrobial activity against undesirable organisms.

7. Microbial Problems (also see Chapter 18)

Green (yogurt flavor): due to too much acetaldehyde production (especially if biovar diacetylactis is used). A slimy texture implies contamination with bacteria that produce slime (*Alcaligenes viscolactis*) or that starter cultures are slime formers. A yeast flavor implies contamination with lactose-fermentating yeasts, and a cheesy flavor alludes to contamination with proteolytic psychrotrophs (during storage).

D. Microbiology of Yogurt Fermentation

1. Characteristics

Plain yogurt has a semisolid mass due to coagulation of milk (skim, low, or full fat) by starter culture bacteria. It has a sharp acid taste with flavor similar to walnuts and a smooth mouth feel. The flavor is due to the combined effects of lactate, acetaldehyde, diacetyl, and acetate, but 90% of the flavor is due to acetaldehyde.

There are many types of yogurt available in the market, some of which are: plain yogurt, fruit yogurt, flavored and colored yogurt, sweetened yogurt, heated yogurt, frozen yogurt, dried yogurt, low lactose yogurt, and carbonated yogurt.[3,4]

2. Processing

Yogurt is generally fermented in batches, but a continuous method has also been developed. The batch process for a low-fat (2%) plain yogurt is outlined below.

A stabilizer is added to give desired gel structure. Heating helps to destroy vegetative microbes and slightly destabilize casein for good gel formation. Starter

Homogenized milk (12% TS), stabilizer (1%)	Heated to 185°F/30 min, cooled to 110°F	Starter added, incubated at 110°F to pH 4.8 (about 6 h), acidity ≃ 0.9%	Cooled to 85°F, agitated, pumped to filler machine	Packaged in containers, cooled by forced air to 40°F	Held for 24 h, pH drops to 4.3

used as either direct vat set (frozen) or bulk culture (2 to 3%). Then, it is quickly cooled to 85°F in about 30 min to slow down further starter growth and acid production, especially by *Lactobacillus* species. Final cooling to 40°F by forced air results in a rapid drop in temperature to stop the growth of starters.

3. Starters (Controlled Fermentation)

Frozen concentrate or direct vat set starters can be used. Normally, two species are used: *Lab delbrueckii* subsp. *bulgaricus* and *Str. thermophilus*. Some processors also combine these two with *Lab. acidophilus* and *Bifidobacterium* sp. However, in general, the latter two strains do not compete well in growth with the other two. They also do not survive well when present in yogurt with the regular yogurt starter cultures.

For a good product, the two species should be added in a 1:1 cell ratio; in the final product, it should not exceed 3:2 (*Streptococcus:Lactobacillus*). However, *Lactobacillus* cells are more susceptible to freezing and freeze-drying. In a frozen concentrate for use as DVS, the survivors may not be present in the desired ratio unless a proven preservation method is used.

4. Growth

For balanced growth of the two species, the fermentation is conducted at about 110°F (43.3°C). At this temperature, both acid and flavor compounds are produced at the desired level. If the temperature is raised above 110°F, the *Lactobacillus* sp. will predominate, which causes more acid and less flavor production. Conversely, at temperatures below 110°F, growth of *Streptococcus* sp. is favored with a product containing less acid and more flavor.

The two species show symbiotic growth while growing together in milk. Initially, *Streptococcus* sp. grows rapidly in the presence of dissolved oxygen and produces formic acid and CO_2. The anaerobic condition, formic acid, and CO_2 stimulate growth of *Lactobacillus* sp., which has good proteinase and peptidase systems and produces peptides and amino acids from milk proteins. Some of the amino acids, such as glycine, valine, histidine, leucine, and methionine, are necessary for good growth of the *Streptococcus* sp., which lacks proteinase enzymes. *Streptococcus* sp., then grows rapidly until the pH drops to about 5.5, at which time the growth of *Streptococcus* sp. slows down. However, growth of *Lactobacillus* sp. continues fairly rapidly until the temperature is reduced to 85°F, following a drop in pH to 4.8. At 85°F, both grow slowly, but *Streptococcus* sp. has the edge. At 40°F with about pH 4.3, growth of both species stops.

The two species also have a synergistic effect on growth rate, rate of acid production, and amounts of acetaldehyde formation when growing together as compared to when growing individually. The species growing separately in milk produce about 8 to 10 ppm acetaldehyde; when grown together, acetaldehyde production increased to a desirable level of 25 ppm or higher.

5. Biochemistry

a. Lactose Metabolism

Both species have a β-galactosidase system, and lactose is hydrolyzed to glucose and galactose. Both species are homofermentative and produce lactate from glucose by the EMP pathway. *Lab. delbrueckii* subsp. *bulgaricus* strains have enzymes for the Leloir pathway to metabolize galactose, but while actively metabolizing glucose do not utilize galactose well. Most *Str. thermophilus* strains do not have the enzymes of the Leloir pathway (or have a very weak system) and thus do not metabolize galactose. As a result, galactose is excreted outside, causing its accumulation in yogurt.

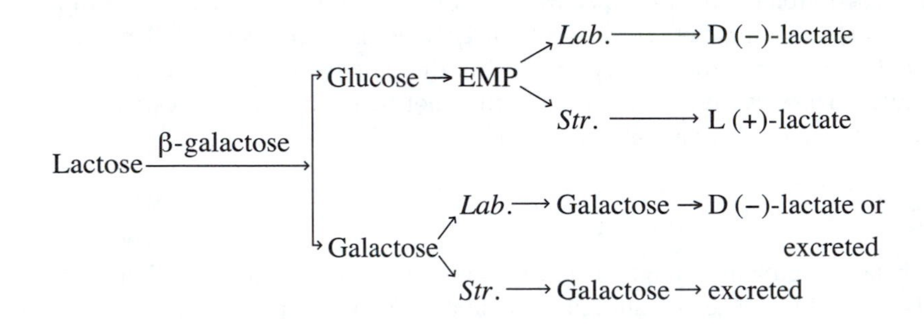

b. Flavor Production

The major flavor compound in yogurt is acetaldehyde (25 ppm), with some diacetyl (0.5 ppm) and acetate. Acetaldehyde is produced in two ways: from glucose via pyruvate by *Streptococcus* sp. and from threonine (supplied or produced through proteolysis in milk) by *Lactobacillus* sp.

CO_2 (stimulates *Lab.*)

Glucose → *Str.* → Pyruvate → Acetaldehyde (flavor) ↓ Diacetyl (flavor)

Threonine → *Lab.* → Glycine (stimulates *Str.*) ↗ Acetaldehyde

c. Formate Production

Formate is produced by *Str. thermophilus* from pyruvate by the action of formate lyase.

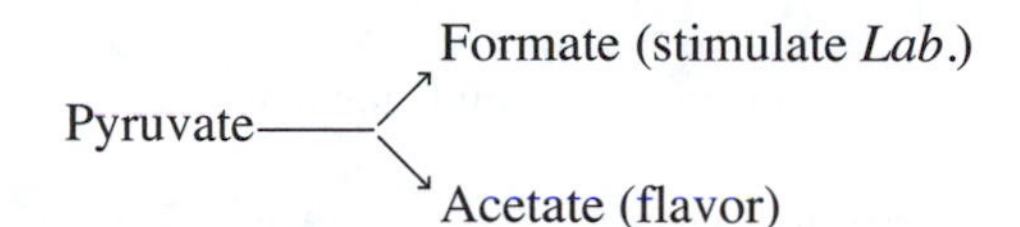

d. Slime Formation (glycan)

β-Galactosidase in some strains of *Str. thermophilus* polymerizes glucose to produce oligosaccharides and glycan, which may give a viscous texture to yogurt

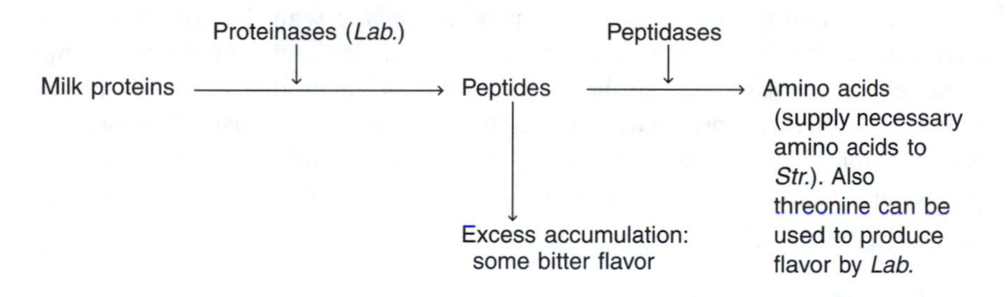

6. Genetics

a. Lac⁺ Phenotype

In both species, the trait (β-galactosidase/permease) is chromosomally linked, constitutive, and quite stable. Some strains can have strong β-galactosidase activity.

b. Gal⁺ Phenotype

Lab. delbrueckii subsp. *bulgaricus* is Gal^+, but *Str. thermophilus* usually is Gal^- phenotype. Strains with good Gal^+ phenotype can be developed to reduce galactose concentrations in the product.

c. Pro⁺

Lab. delbrueckii subsp. *bulgaricus* strains differ in protein hydrolysis ability. Strains with desirable proteolytic activity should be used.

d. Phage Resistance

Both species have phages; resistant strains need to be developed and used.

e. Symbiotic and Synergistic Relationship

Strain selection for best combinations is necessary.

f. Antagonistic Effect

Strains should not produce inhibitory compounds against each other.

g. Good Survival to Freezing and Drying

Possible genetic basis needs to be studied to develop resistant strains.

7. Microbial Problems (also see Chapter 18)

In plain yogurt, flavor problems can be associated with the concentrations of acetaldehyde. While a low concentration gives a chalky and sour flavor, too much acetaldehyde can give a green flavor. Similarly, too much diacetyl gives a buttery aroma. Too much acid production during storage causes a sour taste. Proteolysis and accumulation of bitter peptides during storage are associated with bitter flavor. Growth of yeasts during storage can also produce a fruity flavor, especially in yogurt containing fruits and nuts.

E. Cheeses

Cheeses are made by coagulating the casein in milk with lactic acid produced by lactic acid bacteria and without or with the enzyme rennin, followed by collecting the casein for further processing, which may include ripening.[5,6] The story goes that the process was accidentally discovered around 7000 B.C. from the coagulation of milk stored in a calf stomach by lactic acid bacteria (also probably rennin). At present, many varieties of cheeses are made worldwide and probably use over 20% of total milk produced. In the United States, the total production of different varieties of cheese in 1982 was 4.3 billion pounds and, in 1987, that increased to 5.3 billion pounds. Due to the worldwide increase in cheese consumption, cheese production will continue to increase not only in the United States, but also in other countries, especially in dairy-rich countries in Europe and in New Zealand.

The cheese varieties have been grouped in several different ways. Examples of several varieties based on starter cultures used and some important secondary flora are listed.

1. Unripened

- Soft

Cottage cheese with starters *Lac. lactis* subsp. and *Leuconostoc* sp.
Mozzarella cheese with starters *Str. thermophilus* and *Lab. delbrueckii* subsp. *bulgaricus*.

2. *Ripened*

- Soft
  Brie cheese with starter *Lac. lactis* subsp.; *Penicillium* sp. and yeasts are secondary flora.
- Semi-hard
  Gouda cheese with starters *Lac. lactis* subsp. and *Leuconostoc* sp.; dairy *Propionibacterium* may be secondary flora.
  Blue cheese with starter *Lac. lactis* subsp., *Leuconostoc* sp.; *Penicillium roquefortii*, yeasts and micrococci are secondary flora.
- Hard
  Cheddar cheese with starters *Lac. lactis* subsp.; lactobacilli and pediococci are secondary flora.
  Swiss cheese with starters *Str. thermophilus, Lab. helveticus*, and dairy *Propionibacterium* sp.; enterococci can be secondary flora.

From these, cottage, cheddar, Swiss, and blue cheeses are further discussed.

F. Microbiology of Cottage Cheese

1. *Characteristics*

Cottage cheese is made from low-fat or skim milk, has a soft texture with about 80% moisture. It is unripened and has a buttery aroma due to diacetyl (along with lactic acid and little acetaldehyde).

2. *Processing (from Skim Milk)*

Pasteurized, cooled to 70°F, starter added and incubated for 12 h	Firm curd set, cut in cubes, cooked at 125°F for 50 min or more	Whey drained off, stirred to remove more to get dry curd	Salted, creamed, preservative added	Packaged, refrigerated

3. *Starters (Controlled Fermentation)*

Frozen concentrate for direct vat set can be used.

Mixed strains of *Lac. lactis* subsp. *cremoris* and *lactis* are predominantly used for acid.

Leu. mesenteroides subsp. *cremoris* can be added initially (in that case, citrate is added to milk), mainly for diacetyl.

Lac. lactis subsp. *lactis* biovar diacetilactis can be used for diacetyl, but not inoculated in milk due to formation of too much CO_2, which causes curd particles to float. Instead, they are grown separately in cream, which is then used to cream the dry curd.

4. Growth, Biochemistry, and Genetics

The growth, biochemistry, and genetics are as described with the microbiology of buttermilk (see Section III.C).

5. Microbial Problems (also see Chapter 18)

1. Slow growth: Weak or loss of Lac^+ phenotype, phage attack or less viable cells in frozen concentrate, antagonistic effect among starter strains.
2. Flotation of curd: Due to too much CO_2 production by flavor producers.
3. Harsh flavor: Due to more acetaldehyde and less diacetyl production.
4. Low yield: Due to partial proteolysis of casein by heat-stable proteinases produced in refrigerated raw milk by psychrotrophs like *Pseudomonas* spp.
5. Flavor loss: Due to reduction of diacetyl to acetoin (also due to growth of undesirable bacteria during storage, such as *Pseudomonas* spp).
6. Spoilage: Due to high moisture and low acid content, spoilage psychrotrophic bacteria (such as *Pseudomonas* spp.), and yeasts and molds can grow. Preservatives such as sorbates can be used to extend shelf life under refrigerated storage.

G. Microbiology of Cheddar Cheese

1. Characteristics

Cheddar cheese is made from whole milk, contains less than 39% moisture, 48% fat, is generally orange-yellow in color (due to added color), and ripened. The smoothness of the texture and the intensity of characteristic flavor vary with the period of ripening. The typical flavor is the result of a delicate balance between flavor components produced during ripening through enzymatic breakdown of carbohydrates, proteins, and lipids in the unripened cheddar cheese.

2. Process

Pasteurized, color (annato) and starter added	Incubated at 86°F for acidity to increase by 0.2%, rennet added	Incubated for coagulation (about 30 min), cut in cubes	Cooked at 100°F, whey drained, cheddaring to lose whey and curd to mat	Milled, salted, put in form, pressed 16 h to drain whey, removed	Dried for 5 d at 50°F, waxed, cured at 40°F for 2 to 12 months

3. Starters (Controlled Fermentation)

Starters include mixed strains of *Lac. lactis* subsp. *cremoris* or *lactis*. *Leuconostoc* may be added for flavor. Can be used as frozen concentrate (direct vat set).

4. Growth

Growth is accomplished by incubation at 86°F for mesophilic starters to start early growth and produce lactic acid to the 0.2% level by 60 min. This is important for the coagulation of casein to occur rapidly following the addition of rennin. During curing, the cells of starter (also of secondary flora) slowly die and release intracellular enzymes in the cheese.

5. Biochemistry

A large number of biochemical reactions occur during the different stages of processing, from the initial incubation to the end of ripening. Many of these are not yet properly understood. Some are described here.

Initially, lactose metabolism will produce lactic acid as well as some diacetyl, acetate, ethanol, and acetaldehyde, especially under aerobic conditions. Rennin destabilizes casein to produce paracasein, which coagulates at low acidity and low temperature. In addition, the proteinases and peptidases of the starter cultures metabolize milk proteins to peptides and amino acids and transport in the cells. Normally very little, if any, lipolysis occurs due to microbial activity.

During curing, breakdown of remaining lactose in the curd will continue as indicated before. However, a large change will occur in the proteins and other nitrogenous compounds. By the action of rennin (retained in curd) and cellular exo- and endoproteinases and peptidases, peptides of different sizes and amino acids will be released. Further breakdown of amino acids will produce hydrogen sulfide, methanethiol and other sulfur compounds, amines, and other products. Lipids will also undergo lipolysis, causing release of fatty acids, including the C_4 to C_8 fatty acids. Other reactions will produce lactones, ketones, and thioesters. Some of the reactions are nonenzymatic. The typical cheddar cheese flavor is the result of a delicate balance between the products produced from the carbohydrate, proteins, and lipid breakdown during processing and curing. The concentrations of these components change with curing time.

Some secondary microflora that survive heating or gain later entrance in the milk and curd during processing have definite roles in the flavor of cheddar cheese. These include some enterococci, lactobacilli, pediococci, and micrococci and some Gram-negative rods. They probably contribute to the typical intense flavor that could be missing in cheese made with only defined starter strains. Some of these flora are known to produce several flavor compounds rather rapidly and at higher concentrations (e.g., volatile fatty acids and H_2S).

6. Genetics

Characteristics as described before for these species should be considered. Strains with the capability to produce lactic acid rapidly at the initial stage will be preferred. Also, strains with weak proteinase (Pro^+) activity are desirable as they do not cause the appearance of bitter flavor in the products.

7. *Microbial Problems (also see Chapter 18)*

Bitter flavor results from the accumulation of bitter peptides that are about 1000 to 12,000 Da and rich in hydrophobic amino acids. Starters capable of hydrolyzing proteins rapidly (fast starter) tend to produce bitter peptides more than the slow starters. Their enzymes hydrolyze proteins quickly, releasing large amounts of peptides that are slowly hydrolyzed to small peptides and amino acids by peptidases, resulting in their accumulation. Due to their hydrophobic nature, these peptides are also hydrolyzed slowly. The use of slow starters and/or treatment of cheese with peptidase are effective in reducing bitterness.

Mold growth on the surface can occur after removing packing material from cheese. It is not possible to determine from the colonial morphology if they are mycotoxin producers or not. In the case of heavy growth, it is better not to consume such cheese.

Staphylococcus aureus, following contamination of milk after heating, can grow during processing of cheddar cheese and produce enterotoxins. They will remain in the cheese, even after the death of cells during curing. Food poisoning can occur from the consumption of such cheese (even if they are heated in some preparations).

Biological amines (histamine and tyramine) can form from the decarboxylation of some amino acids, especially in cheese cured for a long time. They can cause allergic reactions following consumption of the cheese. Secondary flora may have an important role in such amine formation.

H. Microbiology of Swiss Cheese

1. *Characteristics*

Swiss cheese is made from partially skimmed milk (cow's) and coagulated with acid and rennin; it is hard and contains about 41% moisture and 43% fat. This cheese should have medium-sized eyes (openings) uniformly distributed. It has a sweet taste due to proline and a nutty flavor.

2. *Processing*

Pasteurized, starter added, incubated (90°F) for acidity to increase by 0.2%	Rennin added, incubated for firm coagulation, cut 1/8" cubes, cooked 1 h at 125°F	Whey removed, curd pressed 16 h, cut in blocks, exposed in brine (55°F) 1 to 3 d	Surface dried, sealed in bags, store 7 d at 55°F, transferred at 75°F, 1 to 4 wk	Cured at 37°F 3 to 9 months

3. *Starters (Controlled Fermentation)*

Starters include *Str. thermophilus* and *Lab. helveticus* as primary for acid and *Propionibacterium* sp. as secondary for eye formation, taste, and flavor.

4. Growth

Primary starters grow during fermentation, cooking, and processing; growth of *Str. thermophilus* is favored. Propionibacteria grow well during storage at 75°F. During curing, none of the starters grow. In fact, the cells will slowly die and release intracellular enzymes.

5. Biochemistry

Lactose is hydrolyzed by β-galactosidase of both lactic acid bacteria; while *Str. thermophilus* produces L(+)-lactic acid, *Lab. helveticus* produces D(–)-lactic acid. Some acetate also forms. Production of large amounts of lactate is very important to facilitate the growth of propionibacteria. During storage at 75°F, propionibacteria convert lactate (by lactate dehydrogenase) to pyruvate, which is then converted to propionic acid, acetate, and CO_2. Eye formation occurs from the production of CO_2, and the size and distribution of eyes is dependent upon the rate of CO_2 production. Proteins are hydrolyzed by rennin, intracellular proteinases, and peptidases of starters, especially propionibacteria, resulting in the production of small peptides and amino acids that are responsible for the nutty flavor and sweet taste (sweet taste is attributed to production of proline by propionibacteria). Very little lipolysis occurs during curing.

6. Genetics

Rapid production of lactate in large amounts from lactose by the lactic acid bacteria and their resistance to phages (some *Lab. helveticus* strains have temperate phages that get activated at high processing temperatures). *Propionibacterium* strains should produce CO_2 at proper rates for desirable eye formation.

7. Microbial Problems (also see Chapter 18)

Spores of *Clostridium tyrobutyricum* can germinate, grow, and cause rancidity and gas blowing in this low-acid cheese. Nisin has been used as a biopreservative to control this problem.

I. Microbiology of Blue Cheese

1. Characteristics

Blue cheese is a semi-hard (46% moisture, 50% fat), mold-ripened cheese made from whole milk (cow's). It has a crumbly body, mottled blue color, and sharp lipolytic flavor.

2. Processing

Homogenized, pasteurized, starter added, incubated at 90°F for 0.2% acidity increase	Rennet added, incubated for firm set, cut, cooked at 100°F, whey drained	Curd collected in hoop, drained 16 h, salted in brine for 7 d	Spiked to let air get inside. Mold spores added, stored at 50°F in high humidity for 4 wk	Stored at 40°F for curing for 3 months

3. Starters and Growth (Controlled Fermentation)

Lac. lactis subsp. *cremoris* or *lactis* and *Leuconostoc* sp. serve as primary starters. *Penicillium roquefortii* spores serve as secondary starters. The lactic starters grow until curing and, from lactose, produce lactate, diacetyl, acetate, CO_2, and acetaldehyde. Mold spores, during storage at 50°F in high humidity germinate quickly, produce mycelia, and spread inside to give the mottled green appearance. Their growth continues during curing. Puncturing helps to remove CO_2 and lets the air in to help the growth of molds.

4. Biochemistry, Genetics, and Problems

Lactococcus species produce mainly lactic acid from lactose, while *Leuconostoc* species produce lactic acid, diacetyl, CO_2, and acetate. Proteolysis is quite limited for the lactic starters. Molds produce extracellular lipases and proteinases and cause lipolysis and proteolysis during curing. Fatty acids are both oxidized and reduced to produce methyl ketone and δ-lactone, respectively. These, along with volatile fatty acids, contribute to the sharp flavor of Blue cheese.

The desired genotypes of the lactic acid bacteria have been previously described. A white variant of the mold has been isolated and used to produce this cheese without the blue mottled color. *Penicillium roquefortii* strains that produce mycotoxins have been identified. Strains need to be selected that do not produce mycotoxins.

J. Accelerated Cheese Ripening

During curing of hard and semi-hard cheeses, milk components are degraded through enzymatic (from starters and secondary flora) and nonenzymatic reactions. This is necessary for the desirable flavor development of these cheeses. However, due to long storage time, it is not quite economical. Thus methods are being studied that will speed up the ripening process. Some of these methods are briefly described.

1. Curing at High Temperature

Since enzymatic (and nonenzymatic) reactions increase as the temperature is increased, studies were performed to cure cheese above 5 to 6°C (40°F). Ripening some cheeses at 13 to 16°C (55 to 61°F) reduced the curing time by 50% or more.

However, growth of spoilage bacteria has been a problem in some cheeses. There is some concern as to the possibility of growth of foodborne pathogens.

2. *Addition of Enzymes*

As intracellular enzymes have an important role in curing, enzymes obtained from cell lysates of starter culture bacteria have been added to increase the rate of curing. In cheddar cheese, the curing time has been substantially reduced, but resulted in a bitter flavor defect.

3. *Slurry Method*

Cheddar cheese slurry has been prepared by mixing water to cheese to 40% solids (in place of the usual 60%). The slurry is incubated at 30°C for 4 to 5 d with agitation. This method increased the flavor greatly and the product could be used to make processed cheeses. The major disadvantages are in properly controlling the enzymatic actions and possible growth of spoilage and pathogenic bacteria.

IV. FERMENTED MEAT PRODUCTS

A. Types

Fermented meat products[8,9] are produced by first mixing meat, fat, salt, sugar, curing agents, and spices, then filling the mixture in a casing and fermenting it either naturally or by adding (during mixing) selected starter culture bacteria. The acids produced by the starters during fermentation and curing agents help control the growth of pathogenic and spoilage bacteria that might be present in the meat. Depending upon type, the fermented products may be dried to reduce the A_w or smoked or heated to ensure the safety and shelf life of the products.

Meat fermentation probably originated in the Mediterranean countries and later spread to European countries and North America. In the United States, the semi-dry sausages are most popular, although some dry sausages are also produced. Following fermentation, the semi-dry sausages are heated (also sometimes smoked) prior to consumption. For dry sausages, following cooking, the products are dried to reduce the A_w. Even now, a large portion of fermented sausages in the United States are produced by natural fermentation, especially those produced by small processors. However, many processors are now using selected starter cultures and controlled fermentation. Starter cultures are available as both frozen and freeze-dried concentrates for direct inoculation in the meat mixture.

Semi-dry and dry sausages include many types, such as pepperoni, Genoa salami, hard salami, summer sausage, beef sticks, beef logs, thuringer, cervelat, Italian salami, and others. Most are made with beef and pork but in recent years, some are being made with meat from chicken and turkey. The microbiology of semi-dry sausage is described here.

B. Microbiology of Semi-Dry Sausage

1. Characteristics

Semi-dry sausages include summer sausage, thuringer, and semi-dry salami. The average composition is about 30% fat, 20% protein, 3% minerals (salts), and 47% water. They have a tangy taste with a desirable flavor from the combined effect of lactate, acetate, and diacetyl, and some breakdown components from proteolysis and lipolysis. Also, the use of spices contributes to the flavor. Those containing nitrite have a pinkish color instead of a grayish color in products without it.

2. Processing

Meat, salts, glucose, cure, spices, starter mixed uniformly	Stuffed in casings, fermented at 85 to 110°F with 80 to 90% relative humidity	Incubated until pH drops to about 5.2 to 4.6, cooked to 140°F internal temperature, cooled to 50°F	Stored at 40 to 50°F for 3–4 d, vacuum-packaged, consumed directly

Cures contain nitrite to give about 100 ppm final concentration. Fermentation can be carried out in a smokehouse. Fermentation time is usually 8 to 12 h during which the pH is dropped to the desired level.

3. Starters (Controlled or Natural Fermentation)

In controlled fermentation, frozen or dried concentrates are used directly at a level of 10^{6-7} cells per gram mix. Starter should not be mixed with salt, cure, or spices as it can kill injured cells. Instead, it should be thawed and immediately put into the meat. Starters vary depending upon the fermentation temperature and final pH of the product desired. For high temperature and low pH, *Pediococcus acidilactici* are preferred; and for low temperature and high pH, *Lab. plantarum* strains are preferred. *Ped. pentosaceus* strains can be used under both conditions. Some starters can have both *Pediococcus* and *Lactobacillus* species. In addition, selected *Micrococcus* spp. or *Staphylococcus carnosus* strains are added as secondary flora for their beneficial effects on desired product color.

In naturally fermented sausages, *Lab. sake, Lab. curvatus,* and *Leuconostoc* spp. present in raw materials are important starter bacteria, especially when fermentation is set at lower temperatures (60 to 70°F) for several days and the final pH reached is not below 5.0.

4. Growth

Because raw meat is used that may contain pathogens and spoilage bacteria, it is extremely important that the starter cultures grow rapidly and produce acid in order to reduce the pH from the original 5.7 to about 5.3 very quickly. This can be achieved by adding large numbers of active starter cells, adding dextrose to the mix, and setting the temperature of fermentation optimum for the starter being used. The optimum growth temperatures for *Ped. acidilactici, Ped. pentosaceus*, and *Lab. plantarum* are around 40, 35, and 30°C (104, 95, and 86°F), respectively. *Micrococcus* spp. and *Sta. carnosus* grow well around 32.2°C (90°F). Cooking to an internal temperature of 60°C (140°F) will kill *Lab. plantarum* and probably *Ped. pentosaceus*, but probably not *Ped. acidilactici*, *Micrococcus*, or *Sta. carnosus*. However, low pH and low A_w will prevent their growth in the finished products.

5. Biochemistry

Both pediococci are homolactic fermenters and metabolize glucose to mainly lactic acid (DL), with small amounts of acetate and diacetyl. *Lab. plantarum*, being facultatively homofermentative, metabolizes glucose to principally lactic acid (DL). However, it can also produce substantial amounts of acetate, ethanol, and diacetyl. Strains of all three species can produce H_2O_2, which can discolor the product due to oxidation of myoglobin during fermentation. *Micrococcus* spp. or *Sta. carnosus* can produce catalase to destroy H_2O_2. *Micrococcus* sp. or *Sta. carnosus* and some strains of *Lab. plantarum* can also reduce nitrate to nitrite. If nitrate is used in place of nitrite in cure, these bacteria can produce nitrite and help to develop the agreeable pinkish color of the product. If the products are cured or stored for long periods of time, some of the intracellular enzymes of the starters are available to cause proteolysis, lipolysis, and the production of biologically active amines (such as histamine).

6. Genetics

Rapid acid-producing lactic acid bacterial strains at temperatures of fermentation and non-H_2O_2 producers are desired. Strain selection can also be done for nonproducers of biogenic amines. Strains producing bacteriocins can be used to control pathogens and spoilage bacteria. *Ped. acidilactici* strains with the inability to hydrolyze sucrose (Suc$^-$) can be used to produce sweet and sour products by supplementing the meat mixture with both glucose and sucrose.

7. Microbial Problems (also see Chapter 18)

Slow acid production can be a serious problem if the starters used are not active or have lower numbers of viable cells or other factors; for example, low glucose and high salts in the mix. Sour and no flavor can happen if the starter, especially *Ped. acidilactici*, is growing very rapidly and reduces the pH below 4.5. Gas formation can occur due to growth of *Leuconostoc* spp. during fermentation and during storage in vacuum packages. *Leuconostoc* spp. are also present in raw meat. Pathogens present in meat can grow if acid production is slow during fermentation. These products can cause health hazards. During long storage or curing, biogenic amines can form. Also, mycotoxin-producing molds can grow in products during curing.

V. FERMENTED VEGETABLE PRODUCTS

Almost all vegetables can be fermented through natural processes since they harbor many types of lactic acid bacteria. Throughout the world, many types of vegetables are fermented, mostly in small volumes.[10] However, some are produced commercially. As stated before, vegetable fermentation originated in the early years of human civilization and was used by many cultures. Examples of some fermented products and vegetables used now for fermentation are: sauerkraut (from cabbage), olives, cucumbers, carrots, celery, beans, peas, corn, okra, green tomatoes, cauliflower, peppers, onions, citron, beets, turnips, radishes, chard, brussels sprouts, and their blends. Most are produced by natural fermentation; however, some are now being produced by controlled fermentation, such as cucumbers. Production of sauerkraut by natural fermentation is described here as an example.[10]

A. Microbiology of Sauerkraut

1. Characteristics

Sauerkraut is produced by fermenting shredded cabbage. The product has a sour taste with clean acid flavor.

2. Processing

Cabbage cleaned, trimmed, and shredded fine and uniform

Packaged tight to exclude air in vat, layered with salt (2.25%)

Top covered to exclude air, fermented at 18°C for 2 months

Fine shredding helps the sugars (3 to 6%) come out of cabbage cells. Tight packaging helps create an anaerobic condition, thus preventing growth of aerobes. Salt stimulates the growth of some lactic acid bacteria, and discourages the growth of some undesirable bacteria and pectinase (in cabbage) action. The top is covered to exclude air and prevent growth of aerobes. Fermentation at 18°C (65°F) discourages rapid growth of some undesirable bacteria (facultative anaerobic or anaerobic), but encourages the growth of desirable lactic acid bacteria. Natural inhibitors in cabbage also discourage the growth of undesirable Gram-negative and Gram-positive bacteria.

3. Starters (Natural) and Growth

The raw material has a large number of undesirable organisms and a small population of lactic acid bacteria (<1%). Among the lactic acid bacteria, most are *Lactococcus* spp. and *Leuconostoc* spp., and a small fraction are *Lactobacillus* spp. and *Pediococcus* spp. During fermentation, sequential growth of these lactic acid bacteria occurs. The presence of 2.25% salt, large amounts of fermentable sugars (sucroses, hexoses, pentoses), the absence of oxygen, and low fermentation temperature facilitate *Leuconostoc* spp., primarily *Leu. mesenteroides* to start rapid growth.

When the acidity has reached about 1% (as lactic acid), growth of *Leu. mesenteroides* slows down. Then *Lab. brevis* starts growing rapidly until the acid production reaches about 1.5%. Then *Ped. pentosaceus* takes over and increases the acidity to about 1.8%. Finally, *Lab. plantarum* starts growing and brings the acid level to about 2%.

4. Biochemistry

Leuconostoc spp. metabolize hexoses and pentoses to lactate, acetate, ethanol, CO_2, and diacetyl. *Lab. brevis* (obligatory heterofermentative, like *Leuconostoc* spp.) ferments hexoses and pentoses to products similar to *Leuconostoc* spp. *Ped. pentosaceus* metabolizes hexoses to form mainly lactic acid and pentoses to lactic acid, acetate, and ethanol. *Lab. plantarum* also produces products from hexoses and pentoses similar to *Ped. pentosaceus*. While *Leuconostoc* spp. produces D(–)-lactate, the other three species produce DL-lactate.

The characteristic flavor of sauerkraut is the result of the combined effects of lactate, acetate, ethanol, CO_2, and diacetyl.

5. Genetics

If starters are developed for controlled fermentation, some of the characteristics that will be important are: rapid acid production, good flavor production, low CO_2 production (to reduce gassy defect), and the ability to produce antimicrobial compounds.

6. Microbial Problems (also see Chapter 18)

Off-flavor, soft texture, and discoloration of sauerkraut can occur due to growth of molds and yeasts when air is not completely excluded. A slimy texture in sauerkraut can occur due to overgrowth of *Leuconostoc* spp. in the presence of sucrose; they metabolize fructose but synthesize dextrans from glucose.[10]

REFERENCES

1. Ray, B., History of food preservation, in *Food Biopreservatives of Microbial Origin*, Ray, B. and Daeschel, M. A., Eds., CRC Press, Boca Raton, FL, 1992, 2–22.
2. Orman, H., Fermented milks, *Microbiology of Fermented Foods*, Vol. 1, Wood, B. J. B., Eds., Elsevier, New York, 1985, 167–195.
3. Mitchell, L. and Sandine, W. E., Associative growth and differential enumeration of *Streptococcus thermophilus* and *Lactobacillus bulgaricus*: a review, *J. Food Prot.*, 47, 245, 1984.
4. Tamime, A. Y. and Deeth, H. C., Yogurt: technology and biochemistry, *J. Food Prot.*, 43, 939, 1980.
5. Law, B. A., Cheeses, in *Fermented Foods*, Rose, A. H., Ed., Academic Press, New York, 1982, 147–198.
6. Kosikowski, F. V., *Cheese and Fermented Milk Products*, 2nd ed., Edwards Bros., Ann Arbor, MI, 1972.

7. El Soda, M. A., The role of lactic acid bacteria in accelerated cheese ripening, *FEMS Microbiol. Rev.*, 12, 239, 1983.
8. Smith, J. L. and Palumbo, S. A., Use of starter cultures in meats, *J. Food Prot.*, 46, 997, 1983.
9. Lücke, F.-K., Fermented sausages, in *Microbiology of Fermented Foods*, Vol. 2, Wood, B. J. B., Ed., Elsevier, New York, 1985, 41–83.
10. Vaughn, R. H., The microbiology of vegetable fermentation, in *Microbiology of Fermented Foods*, Vol. 1, Wood, B. J. B., Ed., Elsevier, New York, 1985, 49–109.

CHAPTER 13 QUESTIONS

1. Define and discuss the advantages of different methods used to produce fermented foods.

2. Briefly discuss the factors to be considered in selecting milk to produce fermented dairy products.

3. List the characteristics of buttermilk and describe how they help to select the specific starter cultures.

4. Describe the symbiotic growth of the starter cultures used in the production of yogurt. What genetic improvement will be important in these strains? Why may the addition of *Lab. acidophilus* and *Bifidobacterium* spp. along with the normal yogurt starters, not be effective in achieving all four species at desirable levels in the final product?

5. In creamed cottage cheese, cream ripened with diacetyl-producing lactic starter is often used instead of using this starter during fermentation. Explain the reason.

6. Discuss the biochemical basis of the typical cheddar cheese flavor. What is the basis of bitter flavor formation? How can it be reduced by using genetic improvement of the starters.

7. Discuss the role of starter cultures in the development of the desired characteristics in Swiss cheese.

8. Describe the role of the mold used to produce Blue cheese. What precautions should be used in the selection of these mold strains?

9. List the primary and secondary starter cultures used in controlled fermentation of semi-dry sausages and discuss their specific roles.

10. Describe the sequential growth of lactic acid bacteria during the natural fermentation of sauerkraut.

CHAPTER 14

Food Ingredients and Enzymes of Microbial Origin

CONTENTS

I. INTRODUCTION

Many microbial metabolites can be used as food additives to improve nutritional value, flavor, color, and texture. Some of these include proteins, essential amino acids, vitamins, aroma compounds, flavor enhancers, salty peptides, peptide sweeteners, colors, stabilizers, and organic acids. Because they are used as ingredients, they need not come only from microorganisms used in food fermentation. They can be produced by many types of microorganisms (also algae) but must have regulatory approval prior to use in foods. Many enzymes from bacteria, yeasts, molds, as well as from plant and mammalian sources, are currently used for processing foods and food ingredients. Some examples are production of high fructose corn syrups, extraction of juice from fruits and vegetables, and enhancing flavor in cheese.

Recombinant DNA technology (or biotechnology) has opened up the possibilities of identifying and isolating gene(s) or synthesizing a gene encoding a desirable trait from plant and animal sources, or from microorganisms that are difficult to grow normally, clone it in a suitable vector (DNA carrier), and incorporate the recombinant DNA in a suitable microbial host that will express the trait and produce the specific additive or enzyme economically. It can then be purified and used as a food additive and in food processing, provided it is safe to use (GRAS-listed compound). The microbiology of the production and uses of some of these additives and enzymes are discussed here.

II. MICROBIAL PROTEINS (SINGLE-CELL PROTEIN, SCP)

Molds, yeasts, bacteria, and algae are rich in proteins, and the digestibility of these proteins ranges from 65 to 96%.[1,2] Proteins from yeasts in general have high digestibility as well as biological values. In commercial production, yeasts are preferred. Some of the species used are from genera *Candida, Saccharomyces,* and *Torulopsis*. Some bacterial species have been used, especially from genus *Methylophilus*.

The use of microbial proteins as food has several advantages over animal proteins. There may not be enough animal proteins to feed the growing human population in the future, especially in many developing countries. Also microbial proteins can be produced under laboratory settings. Thus land shortage and environmental calamities (such as drought or flood) can be overcome. They can be produced on many agricultural and industrial wastes. This will help alleviate waste disposal problems and also reduce the cost of production. Microbial proteins can be a good source of B vitamins, carotene, and carbohydrates.

There are some disadvantages of using microbial proteins as human food. They are poor in some essential amino acids, such as methionine. However, this can be corrected by supplementing microbial proteins with the needed essential amino acids. The other problem is that proteins from microbial sources can have high nucleic acid content (RNA and DNA; 6 to 8%), which can be metabolized in the human body to uric acid. A high serum acid level can lead to kidney stone formation and gout. However, through genetic manipulations, the nucleic acid content in microbial proteins has been reduced.

Even though, at present, the use of microbial proteins as a protein source in human food is limited, they are being used as protein sources in animal feed. An increase of microbial proteins will automatically reduce the use of grains, (such as corn and wheat) as animal feed, which then can be used as human food.[1,2]

III. AMINO ACIDS

Proteins of most cereal grains are deficient in one or more of the essential amino acids, particularly methionine, lysine, and tryptophan.[3] To improve the biological values, the cereals are supplemented with essential amino acids. Supplementation of vegetable proteins with the essential amino acids has been suggested to improve the protein quality for people who either do not consume animal proteins (people on vegetarian diets) or don't have enough animal proteins (such as in some developing countries, especially important for children). To meet this demand as well as for use as nutrient supplements, a large amount of several essential acids are being produced. At present, due to economic reasons, they are mostly produced for the hydrolysis of animal proteins followed by purification.

In recent years, bacterial strains have been isolated, some of which are lactic acid bacteria, which produce and excrete large amounts of lysine in the environment. Isolation of high producing strains of other amino acids, and developing strains by recombinant DNA technology that will produce these amino acids in large amounts, can be important for economical production of essential amino acids.[3]

IV. VITAMINS

Many vitamins are added to foods and are also used by many as supplements regularly.[2] Thus there is a large market for vitamins, especially some B vitamins and vitamins C, D, and E. Some of these are obtained from plant sources, others are synthesized, while several are produced by microorganisms. Vitamin C is now produced by yeast using cheese whey. Microorganisms have also been a source of vitamin D. Many are capable of producing B vitamins.

The possibility of using recombinant DNA technology to improve production of vitamins by microorganisms may not be very practical or economical. Vitamins are produced through multienzyme systems, and it may not be possible to clone the necessary genes.

V. FLAVOR COMPOUNDS AND FLAVOR ENHANCERS

Flavor compounds and enhancers include those that are associated directly with the desirable aroma and taste of foods and indirectly strengthen some flavors.[2-4] Many microorganisms produce different types of flavor compounds such as diacetyl (butter flavor), acetaldehyde (yogurt flavor), some nitrogenous and sulfur-containing compounds (sharp cheese flavor), propionic acid (nutty flavor), pyrazines (roasted nutty flavor) by strains of *Bacillus subtilis* and *Lactococcus lactis*, and terpens (fruity or flowery flavors) produced by some yeasts and molds. Some natural flavors from plant sources are very costly as only limited amounts are available and the extraction process is very elaborate. Using biotechnology, they can be produced economically by suitable microorganisms. Natural vanilla flavor (now obtained from plants), if produced by microorganisms, may cost only one tenth or less. Natural fruit flavors are extracted from fruits. Not only is it costly, but large amounts of fruits are also wasted. Possible production of many of these flavors by microorganisms through recombinant technology is being studied.

Several flavor enhancers are now being used to strengthen the basic flavors of foods. Monosodium glutamate (MSG; enhances meat flavor) is produced by several bacterial species, such as *Corynebacterium glutamicum*. 5′-Nucleotides, such as inosine monophosphate and guanosine monophosphate, give an illusion of greater viscosity and mouth feel in foods such as soups. They can be produced from *Bacillus subtilis*.

Several small peptides such as lysylglycine have strong salty tastes. They can be produced by recombinant DNA technology by microorganisms and used to replace NaCl. Sweet peptides, such as monellin and thaumatins from plant sources, can also be produced by microorganisms through gene cloning. At present, the dipeptide sweetener aspartame is produced synthetically, but a method to produce it by microorganisms has been developed.[2-4]

VI. COLORS

Many bacteria, yeasts, and molds produce different color pigments. The possibility of using some of them, especially from those that are currently consumed by humans, is being studied. This includes the red color pigment astaxanthine of a yeast species (*Phaffia* sp.). This pigment gives the red color to salmon, trout, lobster, and crabs. Another red pigment, produced by the yeast *Monascus* sp., has been used for a long time in the Orient to make red rice wine. Because pigment production may involve multistep reactions, recombinant DNA techniques to produce some fruit colors by microorganisms may not be economical. However, they can be produced by the plant cell culture technique.

VII. STABILIZERS

Different polysaccharides are used in food systems as stabilizers and texturizers.[5] Although many of them are of plant origin, some are obtained from microbial sources. Dextran, produced by *Leuconostoc mesenteroides* while growing in sucrose, is used as a stabilizer in ice cream and confectioneries. Xanthan gum, produced by *Xanthomonas campestris*, is also used as a stabilizer. By introducing lactose-hydrolyzing genes into the *Xanthomonas* species, it can now be grown in whey to produce the stabilizer economically.

VIII. ORGANIC ACIDS

Production of lactic (by lactic acid bacteria), propionic (by propionic acid bacteria), and acetic (by acetic acid bacteria) acids and their different uses in foods are discussed in Chapters 10, 15, and 39. Several other organic acids and their salts are used in foods to improve taste (flavor and texture) and "keeping" quality. Production of ascorbic acid by yeasts and its use as a vitamin supplement have also been discussed. Ascorbic acid is also used in some foods as a reducing agent to maintain color (to prevent loss by oxidation). It also has antibacterial action. Citric acid is used in many foods to improve taste and texture (in beverages) and stabilize color (in fruits). It also has some antibacterial property. Citric acid is produced by the mold *Aspergillus niger*.

IX. MICROBIAL ENZYMES IN FOOD PROCESSING

Many enzymes are used in the processing of food and food additives.[6,7] About 80% of the total enzymes produced, on a dollar basis, are used by the food industries. Use of specific enzymes instead of microorganisms has several advantages. A specific substrate can be converted into a specific product by an enzyme through a single-step reaction. Thus production of different metabolites by live cells from the same substrate can be avoided. In addition, a reaction step can be controlled and enhanced more easily by using purified enzymes. Finally, by using recombinant DNA technology, the efficiency of enzymes can be improved and, by immobilization they can be recycled. The main disadvantage of using enzymes is that if a substrate is converted to a product through many steps (such as glucose to lactic acid), microbial cells must be used for their efficient and economical production.

A. Enzymes Used

Among the five classes of enzymes, three are predominatly used in food processing. They are hydrolases (hydrolyze C–C, C–O, C–N, etc. bonds), isomerases (isomerization and racemization), and oxido-reductases (oxygenation or hydrogenation). Some of these are listed in Table 14.1 and their uses are discussed here.[6]

1. α-Amylase, Glucoamylase, and Glucoseisomerase

Together, these three enzymes are used to produce high fructose corn syrup from starch. α-Amylase hydrolyzes starch at α-1 position randomly and produces oligosaccharides (three hexose units or more, dextrins). Glucoamylase hydrolyzes the dextrins to glucose units, which are then converted to fructose by glucose isomerase.

Table 14.1 Some Microbial Enzymes Used in Food Processing

Enzyme	Class[a]	Source	Substrate	Function; use
α-Amylase	H	Bacteria, molds	Starch	Production of dextrins; brewing and baking
Catalase	OR	Molds	H_2O_2	Removal of H_2O_2; milk, liquid eggs
Cellulase	H	Molds	Cellulose	Hydrolyze cellulose; ethanol production; juice extraction
Glucoamylase	H	Molds	Dextrins	Dextrins to glucose
D-glucose isomerase	I	Bacteria	Glucose	Glucose to fructose; high fructose corn syrup
D-glucose oxidase	OR	Molds	D-glucose, oxygen	Flavor and color of liquid egg, juice
Hemicellulase	H	Bacteria, molds	Hemicellulose	Juice clarification
Invertase	H	Yeasts	Sucrose	Production of invert sugar
Lactase	H	Molds, yeasts	Lactose	Glucose/galactose from whey; low lactose milk
Lipases	H	Bacteria, molds	Lipids	Cheese ripening
Pectinases	H	Molds	Pectin	Clarification of wine, fruit juice, juice extraction
Proteinases	H	Bacteria, molds	Proteins	Meat tenderization; cheese making and ripening

[a] H, hydrolases; I, isomerases; OR, oxido-reductases.

α-Amylase is also used in bread-making to slow down staling (starch crystallization due to loss of water). Partial hydrolysis of starch by α-amylase can help reduce the water loss and extend the shelf-life of bread.

2. Catalase

Raw milk and liquid eggs can be preserved with H_2O_2 prior to pasteurization. However, the H_2O_2 needs to be hydrolyzed by adding catalase before heat processing of the products.

3. Cellulase, Hemicellulase, and Pectinase

By their ability to hydrolyze their respective substrates, the use of these compounds in citrus juice extraction has increased juice yield. Normally, these insoluble polysaccharides trap juice during pressing. Also, they get into the juice and increase viscosity, causing problems during juice concentration. They also cloud the juice. By using these hydrolyzing enzymes, such problems can be reduced.

4. Invertase

Invertase can be used to hydrolyze sucrose to invert sugars (mixture of glucose and fructose) and increase sweetness. It is used in chocolate processing.

5. *Lactase*

Whey contains high amounts of lactose. Lactose can be concentrated from whey and treated with lactase to produce glucose and galactose. It can then be used to produce alcohol.

6. *Lipases*

Lipases may be used to accelerate cheese flavor along with some proteases.

7. *Proteases*

Different proteases are used in many food processing facilities. They are used to tenderize meat, extract fish proteins, separate and hydrolyze casein in cheese making (rennet), concentrate cheese flavor (ripening), and reduce bitter peptides in cheese (specific peptidases).

B. Enzyme Production by Recombinant DNA Technology

The enzymes currently used in food processing are obtained from bacteria, yeasts, molds, plants, and mammalian sources. They have been approved by the regulatory agencies and their sources have been included in the GRAS list. There are some disadvantages in obtaining enzymes from plant and animal sources. The supply of these enzymes could be limited and thus can be costly. Also, molds grow relatively slowly, as compared to bacteria or yeast, and some strains can produce mycotoxins. It would be more convenient and cost effective if the enzymes now obtained from nonbacterial sources (including yeasts, as their genetic system is more complicated than bacteria) could be produced in bacteria. This can be hypothetically achieved through recombinant DNA technology.[7] However, in trying to do so, one has to recognize that the bacterial host strains need to be approved by the regulatory agencies if they are not in the GRAS list. Also, regulatory approval will be necessary for the source, if it is not currently in the GRAS list.

The technique is going rapidly through many improvements. In brief, it involves the separation of specific mRNA (while growing on a substrate) and uses the mRNA to synthesize cDNA by employing the reverse transcriptase enzyme. The cDNA (double-stranded) is cloned in a suitable plasmid vector, which is then introduced by transformation in the cells of a suitable bacterial strain (say *Escherichia coli*). The transformants are then examined to determine the expression and efficiency of production of the enzyme.

This method has been successfully used to produce rennin (from calf) and cellulase (from molds) by bacteria. Rennin thus produced is being used to make cheese.

X. IMMOBILIZED ENZYMES

Enzymes are biocatalysts and can be recycled. By adding an enzyme to a substrate in liquid or solid food, it is used only one time. In contrast, if the molecules of an enzyme are attached to a solid surface (immobilized), it can be repeatedly exposed to a specific substrate.[8] The major advantage is the economical use of an enzyme, especially if the enzyme is very costly.

Enzymes can be immobilized by several physical, chemical, or mechanical means. The techniques can be divided into four major categories (Figure 14.1).

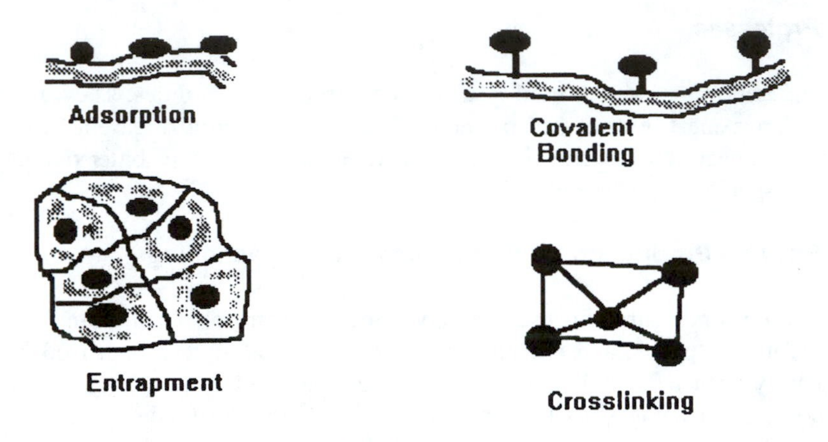

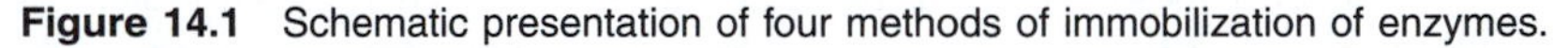

Figure 14.1 Schematic presentation of four methods of immobilization of enzymes.

A. Adsorption on a Solid Support

This technique relies on the affinity of the support for the enzyme molecules. It is achieved by adding an enzyme solution to the support (such as ion exchange resins) and washing away the unattached molecules. The association is very weak, and the molecules can be desorbed and removed.

B. Covalent Bonding

The enzyme molecules are covalently bound to a solid surface (such as porous ceramics) by a chemical agent. The enzyme molecules are accessible to the substrate molecules. The enzymes are more stable.

C. Entrapping

The enzyme molecules are enclosed in a polymeric gel (e.g., alginate) that has an opening for the substrate molecules to come into contact with the catalytic sites. The enzymes are added to the monomer before polymerization.

D. Crosslinking

Crosslinking is achieved by making chemical connections between the enzyme molecules to form large aggregates that are insoluble. This is a very stable system.

There are several disadvantages in enzyme immobilization. Immobilization can reduce the activity of an enzyme. Substrate molecules may not be freely accessible to the immobilized enzymes. The method may not be applicable if the substrate molecules are large. α-Amylase may not be a good candidate for immobilization because its substrate starch molecules are fairly large. However, glucose isomerase can be immobilized, as its substrate is small glucose molecules. The supporting materials can be contaminated with microorganisms that are difficult to remove and can be a source of contamination in food. The materials to be used as support should not be made of substances that are unsafe and should be regulatory approved. Some of the immobilized enzymes currently used are glucose isomerase, β-galactosidase, and amino-acylase.

Microbial cells can also be immobilized by the methods listed above and have been studied in the production of some food ingredients and beverages. Examples include *Aspergillus niger* (for citric and gluconic acids), *Saccharomyces cerevisiae* (for alcoholic beverages), and *Lactobacillus* species (for lactic acid).

XI. THERMOSTABLE ENZYMES

The term *thermostable enzymes* is generally used for those enzymes that can catalyze reactions above 60°C.[9] There are several advantages in using thermostable enzymes in a process. The rate of an enzyme reaction doubles for every 10°C increase in temperature; thus the production rate can be increased or the amount of enzyme used can be reduced. At high temperatures, when an enzyme is being used for a long time (as in the case of immobilized enzymes), the problems of microbial growth and contamination can be reduced.

At high temperature, enzymes denature due to unfolding of their three-dimensional structures. The stability of the three-dimensional structure of an enzyme is influenced by the ionic charges, hydrogen bonding, and hydrophobic interaction among the amino acids. Thus the linear sequences of amino acids in an enzyme greatly influences its three-dimensional structure and stability. Studies have revealed that increases in both ion pairing and H-bonding on the surface of an enzyme (on three-dimensional structure) and increases in internal hydrophobicity will increase the thermostability of an enzyme. Thus the enzyme tyrosinase from a thermolabile strain of *Neurospora* species denatures in 4 min at 60°C, but from a thermostable strain of the same species, it denatures at 60°C in 70 min. An analysis of their amino acid sequences revealed that at position 96, the tyrosinase has an aspergine (uncharged) in the thermolabile strain, but aspartic acid (charged) in the thermostable strain. Thus an extra ionic charge (on the surface) has increased the thermostability of this enzyme.

Several methods can be used to increase the thermostability of an enzyme that include chemical and recombinant DNA techniques. Recombinant DNA technology can be used in two ways. If the enzyme is present in thermostable form in a microorganism that is not on the GRAS list, the gene can be cloned in a suitable vector, which then can be introduced into a GRAS-listed microorganism and examined for expression and economical production. The other method is more complicated and involves the determination of amino acid sequence of the enzyme and its three-dimensional structure (by computer modeling) to recognize the amino acids on the surface (or inside). The next step involves changing one or more amino acids to ionic or H-bonding among those that are on the surface. This can be achieved by site-specific mutagenesis of base sequence of cDNA for the specific amino acid. The synthesized DNA can be incorporated in a vector and introduced in a desired microbial strain for expression of the enzyme and testing for its thermostability.

Several thermostable enzymes obtained from microorganisms on the GRAS list are currently being used. It is expected that in the future their production by different methods and use in food will increase.

XII. ENZYMES IN FOOD WASTE TREATMENT

Food industries generate large volumes of both solid and liquid wastes. The waste disposal methods have utilized different physical, chemical, and some biological methods.[10] The biological methods include anaerobic digestion and production of single-cell proteins. Due to increases in regulatory restrictions in waste disposal, effective and economical alternative methods are being researched. The possibility of using enzymes to reduce wastes and convert the wastes to value-added products are being developed. The availability of specific enzymes at low cost has been a major incentive in their use for waste disposal.

Some of the enzymes used in food waste treatments are polysaccharidases (cellulase, pectinase, hemicellulase, chitinase, and amylase), lactase, and proteinases. Treatment of fruits with cellulase and pectinase increases juice yield and improves separation of solids from the juice. The solids can be used as animal feed. Chitinases are used to depolymerize the shells of shellfish, and the product used to produce single-cell proteins. Amylases are used to treat starch-containing wastewater to produce glucose syrup for use in alcohol production by yeasts. Lactose in whey is treated with lactase (β-galactosidase) to produce glucose and galactose, which is then used in alcohol production by yeast or to produce bakers yeasts. Proteases are used to treat wastewater from fish and meat processing operations. Some of these products are used as fish food.

In the future, development of better and low-cost enzymes through recombinant DNA technology will increase their uses in food waste treatment.

REFERENCES

1. Lipinsky, E. S. and Litchfield, J. H., Single-cell protein in perspective, *Food Technol.*, 28(5), 16, 1974.
2. Hass, M. J., Methods and application of genetic engineering, *Food Technol.*, 38(2), 69, 1984.
3. Best, D., Conference unveils new ingredient technology, *Prepared Foods,* 3, 187, 1987.
4. Trivedi, N., Use of microorganisms in the production of unique ingredients, in *Biotechnology in Food Processing,* Harlander, S. K. and Labuza, T. P., Eds., Noyes Publishing, Park Ridge, NJ, 1986, 115–132.
5. Sinskey, A., Jamas, S., Easson, D., and Rha, C., Biopolymers and modified polysaccharides, in *Biotechnology in Food Processing,* Harlander, S. K. and Labuza, T. P., Eds., Noyes Publishing, Park Ridge, NJ, 1986, 73–114.
6. Neidleman, S., Enzymology and food processing, in *Biotechnology in Food Processing,* Harlander, S. K. and Labuza, T. P., Eds., Noyes Publishers, Park Ridge, NJ, 1986, 37–56.
7. While, T. J., Meade, J. H., Shoemaker, S. P., Koths, K. E., and Innis, M., Enzyme cloning for the food fermentation technology, *Food Technol.,* 38(2), 90, 1984.
8. Maugh, T. H., A renewed interest in immobilized enzymes, *Science*, 223, 474, 1984.
9. Wasserman, B. P., Thermostable enzyme production, *Food Technol.,* 38(2), 78, 1984.
10. Shoemaker, S., The use of enzymes for waste management in the food industry, in *Biotechnology in Food Processing,* Harlander, S. K. and Labuza, T. P., Eds., Noyes Publishers, Park Ridge, NJ, 1986, 259–278.

CHAPTER 14 QUESTIONS

1. What are the advantages and disadvantages of using microbial proteins as human food?

2. List one each of the following from microbial origin that are used in food: amino acids, flavor compounds, flavor enhancers, color, and stabilizers.

3. List the advantages and disadvantages of using microbial enzymes and microbial cells in food processing.

4. List five enzymes of microbial origin and discuss their specific uses in the food industry.

5. Discuss the advantages of using immobilized enzymes in the production of food ingredients. List three methods of immobilization.

6. What are the advantages of using a thermostable enzyme in the production of food ingredients (or a food)? How will recombinant DNA technology be of help in producing thermostable enzymes?

7. Discuss how microbial enzymes can be helpful in the treatment of wastes from food production.

CHAPTER 15

Food Biopreservatives of Microbial Origin

CONTENTS

I. INTRODUCTION

Even in the early days of food fermentation, our ancestors recognized that fermented foods not only have delicate and refreshing tastes, but also have longer shelf life, and the chances of becoming sick from foodborne diseases are less. This method helped them preserve some foods in fermented form longer than the raw materials, which may be a major reason why food fermentation was so popular among the early civilizations located in high-temperature zones. This knowledge of safety and stability of fermented foods has been transferred through the centuries and helped us understand their scientific basis.[1]

It is now known that food-grade bacteria associated with food fermentation are capable of producing different types of metabolites that have antimicrobial properties (Table 15.1). At present, there is an increased interest in the use of these antimicrobials in nonfermented foods to increase their stability and safety.[1] Some of them,

Table 15.1 Antimicrobial Compounds of Food-grade Bacteria

Metabolites	Effectiveness
Organic acids: lactic, acetic, propionic	Against bacteria and fungi
Aldehydes, ketones, and alcohols: acetaldehyde, diacetyl, ethanol	Against bacteria
Hydrogen peroxide	Against bacteria, fungi, phages
Reuterine	Against bacteria and fungi
Bacteriocins	Against Gram-positive bacteria, normally

such as lactate and acetate (vinegar), have been used in many foods for a long time, while others have generated much interest as potential food biopreservatives in place of some currently used nonfood preservatives, such as nitrite, sulfite, parabens, diacetate, ethylformate, and others. Their effectiveness as food biopreservatives is discussed in this chapter. In addition, some yeasts are currently found to inhibit the growth of molds in fruits and vegetables. This aspect is also briefly discussed.

II. USE OF VIABLE CELLS OF STARTER CULTURES TO PRESERVE REFRIGERATED FOODS

The process involves the addition of viable cells of mesophilic *Lactococcus lactis*, some *Lactobacillus* species, and *Pediococcus* species in high numbers to control spoilage and pathogenic bacteria during refrigeration storage of a food at or below 5°C.[2] In the presence of the mesophilic lactic acid bacteria, the growth of psychrotrophic spoilage and pathogenic bacteria is reported to be controlled. Growth of some of these spoilage and pathogenic bacteria at slightly higher temperatures (≤10 to 12°C) is also reduced. Studies were conducted by adding lactic acid bacteria to fresh meat, seafoods, liquid egg, and some processed meat products, such as bacon, against *Clostridium botulinum, Salmonella* spp. and *Staphylococcus aureus*. In refrigerated raw milk, meat, egg, and seafoods, cells of *Lactobacillus, Lactococcus,* and *Leuconostoc* species were added to control the growth of psychrotrophic spoilage bacteria, such as *Pseudomonas* spp. In some studies, the growth of the psychrotrophs was inhibited by 90% or more during 4 to 10 d refrigerated storage. Addition of cells of lactic acid bacteria in refrigerated raw milk also increased the yield of cheese and extended the shelf-life of cottage cheese.

The inhibitory property could be due to the release of antimicrobial compounds from the cells by the nonmetabolizing lactic acid bacteria. This could be organic acids, bacteriocins, or hydrogen peroxide. The antibacterial role of hydrogen peroxide in raw milk is discussed later.

III. USE OF ORGANIC ACIDS AS BIOPRESERVATIVES

In Chapter 10, the ability of starter culture bacteria to produce lactic, acetic, and propionic acids was discussed. Commercially, lactic acid is produced by some *Lactobacillus* spp., capable of producing L(+)lactate (or DL-lactate), acetic acid by

Acetobacter aceti, and propionic acid by dairy *Propionibacterium* spp. They are on the GRAS (Generally Regarded as Safe) list and are used in many foods as additives to enhance flavor and shelf-life and as safety precaution against undesirable microorganisms. These acids and their salts are used in foods at about a 1 to 2% level (also see Chapter 37).

Acetic acid, its salts, and vinegar (which contains 5 to 40% acetic acid and many other compounds that give the characteristic aroma) are used in different foods for inhibiting growth and reducing the viability of Gram-positive and Gram-negative bacteria, yeasts, and molds. It is generally bacteriostatic at 0.2% but bactericidal above 0.3% concentrations and more effective against Gram-negative bacteria. However, this effect is pH dependent and the bactericidal effect is more pronounced at lower pH (below pH 4.5). It is added to salad dressings and mayonnaise as an antimicrobial agent. It has been permitted to be used as a carcass wash.

Propionic acid and its salts are used in food as a fungistatic agent, but it is also effective in controlling growth and reducing the viability of both Gram-positive and Gram-negative bacteria. Gram-negative bacteria seemed to be more sensitive at pH 5.0 and below, even at 0.1 to 0.2% levels. It is used to control molds in cheeses, butter, and bakery products, and to prevent growth of bacteria and yeasts in syrup, apple sauce, and some fresh fruits.

Lactic acid and its salts are used in food more for flavor enhancement than for their antibacterial effect, especially when used above pH 5.0. However, recent studies have shown that they have definite antibacterial effect when used in foods at 1 to 2% levels, even at pH 5.0 and above. Growth of both Gram-positive and Gram-negative bacteria is reduced, indicating increased bacteriostatic action. Below pH 5.0, it can have a bactericidal effect, especially against Gram-negative bacteria. It may not have any fungistatic effect in the food environment. It is used in many processed meat products and has also been recommended as a carcass wash.

The antimicrobial effect of these three acids is considered to be due to their undissociated molecules. The dissociation constants (pKa) are 4.8 for acetic, 4.9 for propionic, and 3.8 for lactic acid. Thus at most food pH values (5.0 and above), the undissociated fractions of the three acids could be quite low, with the lowest being for lactic acid. The lower antimicrobial effectiveness of lactic acid is thought to be due to its low pKa. The antimicrobial action of the undissociated molecules is produced by the dissociation of the molecules in the cytoplasm following their entry through the membrane. H^+ released following dissociation initially reduces the transmembrane proton gradient and neutralizes the proton motive force; then it reduces the internal pH, causing denaturation of proteins and viability loss. However, other studies have suggested that these weak acids produce antimicrobial action through the combined effects of the undissociated molecules and dissociated ions. The acids also can induce sublethal injury of the cells and increase their chance of viability loss. The undissociated molecules as well as the dissociated ions are able to induce cellular injury.

Some pathogenic bacterial strains, such as some strains of *Salmonella typhimurium,* have been found to be relatively resistant to low pH due to their ability to overproduce some protein induced by the acid environment. These proteins enable the cells to withstand lower internal pH. The presence of such a strain in food, which is currently acidified at 1 to 2% level to control microorganisms, may pose a problem.

IV. USE OF DIACETYL AS BIOPRESERVATIVE

Diacetyl is produced by several species of lactic acid bacteria in large amounts, particularly through the metabolism of citrate; this is discussed in Chapter 10.[5] Several studies have shown that it is antibacterial against many Gram-positive and Gram-negative bacteria. Gram-negative bacteria are particularly sensitive at pH 5.0 or below. It is effective at about 0.1 to 0.25% levels. Recent studies have shown that in combination with heat, diacetyl is more bactericidal than when used alone. Diacetyl has intense aroma and thus its use is probably limited to some dairy-based products where the diacetyl flavor is not unexpected. Also, it is quite volatile and thus may lose its effectiveness in foods that are expected to have long storage life. Under reduced conditions, it is converted to acetoin, which may have reduced antibacterial effects. This will pose difficulties in its use in vacuum-packaged products. The antibacterial action is probably produced by deactivating some important enzymes. The dicarbonyl group (–CO–CO–) reacts with arginine in the enzymes and modifies their catalytic sites.[5]

V. HYDROGEN PEROXIDE AS A FOOD PRESERVATIVE

[See Chapter 37 and Reference 6.]

Some lactic acid bacteria produce H_2O_2 under aerobic conditions of growth and, due to the lack of cellular catalase, pseudocatalase, or peroxidase, they release it into the environment to protect themselves from its antimicrobial action. Some strains can produce, under proper growth conditions, enough H_2O_2 to induce bacteriostatic (6 to 8 μg/ml) but rarely bactericidal action (30 to 40 μg/ml). It is a strong oxidizing agent and can be antimicrobial against bacteria, fungi, and viruses (bacteriophages). Under anaerobic condition very little H_2O_2 is expected to be produced by these strains. In refrigerated raw milk, the antibacterial action produced from the addition of nongrowing cells of mesophilic lactic acid bacteria is thought to be due to its ability to activate the lactoperoxidase-thiocyanate system in raw milk. Raw milk contains lactoperoxidase enzyme and thiocyanate (SCN^-). In the presence of H_2O_2, lactoperoxidase generates hypothiocyanate anion ($OSCN^-$), which at milk pH can be in equilibrium with hypothiocyanous acid (HOSCN). Both $OSCN^-$ and HOSCN are strong oxidizing agents and can oxidize the –SH group of proteins, such as membrane proteins of Gram-negative bacteria that are especially susceptible. This system is inactivated by pasteurization.

Hydrogen peroxide is permitted in refrigerated raw milk and raw liquid eggs (about 25 ppm) to control spoilage and pathogenic bacteria. Prior to pasteurization, catalase (0.1 to 0.5 g/1000 lb) is added to remove the residual H_2O_2. It produces antibacterial action by its strong oxidizing property and damage to cellular components, especially the membrane. Due to its oxidizing property, it can produce undesirable effects in food quality such as discoloration in processed meat, and thus has limited use in food preservation. However, its application in some food processing and equipment sanitation is being studied.

VI. REUTERINE AS A FOOD PRESERVATIVE

Some strains of *Lactobacillus reuteri*, found in the gastrointestinal tract of humans and animals, produce a small molecule, reuterine (β-hydroxypropionaldehyde; $CHO–CH_2–CH_2OH$) that is antimicrobial against Gram-positive and Gram-negative bacteria.[6] It produces antibacterial action by inactivating some important enzymes such as ribonucleotide reductase. However, reuterine is produced by the strains only when glycerol is supplied in the environment, which reduces its use in food preservation. In limited studies, food supplemented with glycerol and inoculated with reuterine-producing *Lac. reuteri* effectively controlled the growth of undesirable bacteria. Addition of reuterine to certain foods also effectively controlled growth of undesirable bacteria.[6]

VII. BACTERIOCINS OF STARTER CULTURES AS FOOD BIOPRESERVATIVES

[Also see Chapter 37 and References 8–10.]

Many strains of lactic acid bacteria and dairy *Propionibacterium* produce proteinaceous compounds, designated as bacteriocins, that are bactericidal to cells of other strain(s) of Gram-positive bacteria.[8-10] Some are relatively large and heat labile, whereas others are small molecules and heat stable. The heat-stable small peptides are discussed here.

Many strains of species from genera *Lactococcus, Streptococcus, Leuconostoc, Pediococcus, Bifidobacterium,* and *Propionibacterium* used in food fermentation have been reported to produce different bacteriocins. Many of these bacteriocins have not been studied for the different characteristics that might be necessary before they can be used as food biopreservatives. Those that have been thoroughly studied and others with limited information provide an understanding about the general characteristics of these molecules (Table 15.2). They are relatively small, consisting of 50 or fewer amino acids. The most studied bacteriocin, nisin, produced by some strains of *Lactococcus lactis*, has 35 amino acids. Amino acid contents of several other bacteriocins are: 44 for pediocin AcH (same as pediocin PA-1) in *Pediococcus acidilactici* strains, 43 in sakacin A in *Lactobacillus sake* LB 706, 37 in leucocin A in *Leuconostoc gelidum* UAL 187, 57 in lactacin F in *Lactobacillus acidophilus* 11988, and 55 in lactococcin A in *Lactococcus lactis* subsp. *cremoris* 9B4. The genetic determinants of several bacteriocins have been identified. Some are linked to plasmids of different size (such as pediocin AcH or pediocin PA-1 in a 8.9-kb plasmid, lactococcin A in *Lactococcus lactis* in a 60-kb plasmid), some others are linked to chromosome (plantaracin A and sakacin 674), while some are linked to transposons (such as nisin). Location of the structural gene (one that encodes for a

Table 15.2 Characteristics of Small Molecular Peptide Bacteriocins of Starter Culture Bacteria

Characteristic	Explanation
Chemical nature	Cationic; hydrophobic; tendency to aggregate; some have modified amino acids (lanthionine) and a disulfide bridge
Genetic determinants	Linked to plasmids, chromosome, and transposons; structural gene is present in a cluster with other genes involved in the production and immunity
Synthesis	Ribosomally translated as prepeptides; processed to active peptides; translocated through the membrane by transport protein(s)
Mode of action	Bactericidal; destabilize functions of cytoplasmic membrane (by affecting energy synthesis and permeability); producer cells are immune to their own bacteriocin, adsorb on cell surface of Gram-positive bacteria, including producer cells; may cause cell lysis
Host range	Wide to narrow; against Gram-positive bacteria; some have been found to act on stressed cells of Gram-negative bacteria
Toxicity	Two (nisin and pediocin AcH) have been found not to be toxic to laboratory animals
Stability (activity)	Stable to heat, low pH, refrigeration and freezing, many organic solvents, salts, and enzymes, but sensitive to proteolytic enzymes
Effectiveness (in food)	Nisin has been tested effective in many foods; pediocin AcH has been tested effective in fresh and processed meats and some dairy foods against many spoilage and pathogenic Gram-positive and Gram-negative bacteria; pediocin PA-1 (similar to pediocin AcH) is effective in salad dressings against spoilage bacteria

bacteriocin) for several bacteriocins have been identified. Generally, the structural gene for a bacteriocin is located with several other genes in a cluster. The other genes encode for proteins necessary for immunity, processing, and membrane translocation of the molecules. The bacteriocin molecules are translated as inactive prepeptides and contain an N-terminal leader peptide and a C-terminal propeptide. The leader peptide is removed by a specific peptidase. Also, some modifications are necessary before the propeptides (such as lanthionine formation in nisin) is converted to an active molecule. The active molecules are excreted in the environment. At pH 5.0 and above, many bacteriocin molecules remain adsorbed on the surface of the producer cells. The molecules have a positive charge and have a tendency to form aggregates. They are adsorbed on the cell surface of Gram-positive bacteria. Some bacteriocins, such as nisin, contain the modified amino acid lanthionine, which is formed after translation of the molecule

The bactericidal action of a bacteriocin against sensitive cells is produced through the destabilization of membrane functions as a permeability barrier and energy generator. Some strains also are lysed. The producer cells of a bacteriocin are immune to their own bacteriocin due to the specific immunity protein.

These bacteriocins are normally effective against cells of other Gram-positive bacteria. However, some have a narrow host range and are bactericidal only against a few related strains. In contrast, some can be effective against many strains and species of Gram-positive bacteria from many genera. Nisin and pediocin AcH (pediocin PA-1) are examples of bacteriocins with a wide host range (Figure 15.1). The sensitive strains differ in their sensitivity toward a bacteriocin; thus among two strains of *Listeria monocytogenes*, strain Scott A can be more sensitive than strain CA to the same concentration of nisin. Even in the most sensitive strain, there can be cells that are resistant to a particular bacteriocin. Similarly a strain sensitive to one bacteriocin can be resistant to another bacteriocin. The spores of a sensitive bacterial strain is resistant to the same bacteriocin; however, following germination and outgrowth, they are killed by it. Normally, the Gram-negative bacteria are resistant to these bacteriocins. But following a sublethal stress, such as freezing, drying, EDTA treatment, or acid treatment, they become susceptible to bacteriocins, such as nisin and pediocin AcH. The sublethally stressed cells, with their impaired outer membrane, probably allow the bacteriocin molecules to enter inside and come in contact with the cytoplasmic membrane to destabilize its functions.

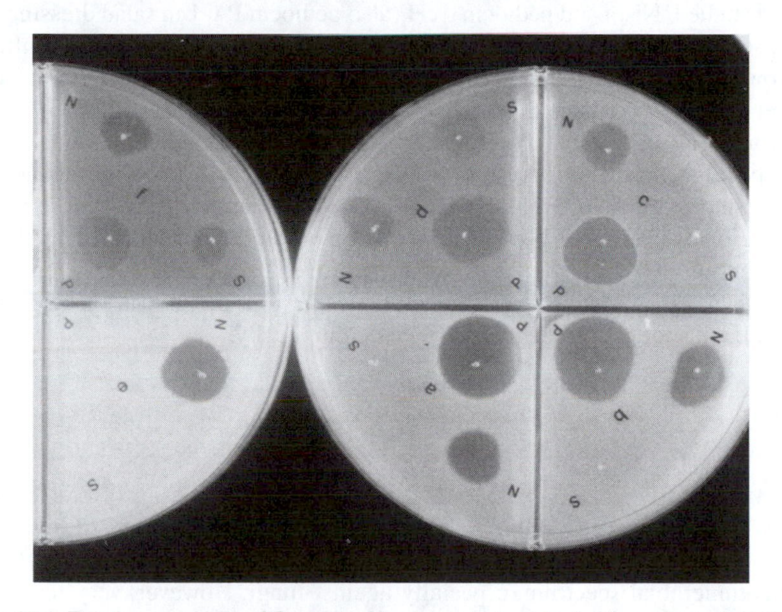

Figure 15.1 Relative sensitivity of Gram-positive bacteria to nisin A (N), pediocin AcH (P), and sakacin A (S): (a) *Lactobacillus plantarum* NCDO 955, (b) *Pediococcus acidilactici* LB42, (c) *Leuconostoc mesenteroides* Ly, (d) *Enterococcus faecalis* MB1, (e) *Micrococcus luteus* T1, (f) *Listeria monocytogenes* Scott A. Nisin A is effective against all six, pediocin AcH is effective against five, but sakacin A is effective against only two strains tested.

Feeding trials with nisin have demonstrated that it is nontoxic to animals and humans. It has been approved for use in the United States and many other countries as a preservative in several foods. Limited studies with pediocin AcH have also shown that it is nontoxic to laboratory animals and cultured cells. Bacteriocins produced by bacteria used in food fermentation are probably nontoxic to humans, as many of them have been consumed by humans for a long time with no known deleterious effects.

The bactericidal effect of bacteriocins is relatively stable at high temperatures. Following heating at 100°C for 10 min, nisin and pediocin AcH lost about 30 and

10% activity, respectively. Heating to 121°C for 15 min reduced the activity by about 25 to 50% for some bacteriocins. The activity is very stable at acidic pH but can be destroyed at pH 9.0 or above, especially by heating. Many proteolytic enzymes can destroy the biological activity.

Depending upon the bacteriocin, many have been found to be bactericidal to different Gram-positive pathogenic and spoilage bacteria important in foods. However, only a few have been tested in food systems. In some studies, a bacteriocin-producing strain was inoculated with a pathogenic or spoilage bacteria, and the ability of the bacteriocin producer to control growth of the pathogens and spoilage bacteria determined. Strains producing sakacin A, pediocin PA-1, pediocin ACH, pediocin JD, and nisin were studied this way. In other studies, purified bacteriocins was added in food systems and their ability to control spoilage and pathogenic bacteria studied. Nisin and pediocin AcH (also pediocin PA-1 in salad dressing only) were also studied in this way. In general, they are found to be effective in controlling the growth of the pathogenic and spoilage bacteria, especially in refrigerated and vacuum-packaged products that are expected to have a long shelf-life.

Bacteriocins of starter culture bacteria, particularly those with a wide antibacterial spectrum, have the potential of being used as food biopreservatives. They might be more effective in some types of food, such as refrigerated and vacuum-packaged foods, and when used in combination with other preservation methods or with other hurdles. However, they must be approved by the regulatory agencies and for that their different properties characteristics have to be known. Finally, for use in a food, a bacteriocin must be produced economically.

VIII. YEASTS

Certain yeasts, including strains of *Saccharomyces cerevisiae*, produce several proteins that have limited antimicrobial properties.[11] These proteins (designated as killer toxins or zymocins) can, through genetic manipulation, be altered to have wider antimicrobial spectrum, especially against fungi. However, very few studies are currently being conducted in this area.

Several yeast isolates normally present on the surface of fruits and vegetables were reported to prevent spoilage of the products by molds. Some of the inhibitory compounds are small proteins, while some others are enzymes. Cells of one such yeast isolate was found to adhere tightly with the mold mycelia and produce β-gluconase that degrade the cell wall of the molds and kill them. As many of these yeasts are normally present in fruits and vegetables that are eaten raw, they are not considered pathogenic and thus can be used in place of fungicides to enhance the preservation of fruits and vegetables.

REFERENCES

1. Daeschel, M. A., Antimicrobial substances from lactic acid bacteria for use as food preservatives, *Food Technol.,* 43(1), 164, 1989.

2. Ray, B., Cells of lactic acid bacteria as food biopreservatives, *Food Biopreservatives of Microbial Origin,* Ray, B. and Daeschel, M. A., Eds., CRC Press, Boca Raton, FL, 1992, 81–101.
3. Ray, B. and Sandine, W. E., Acetic, propionic and lactic acids of starter culture bacteria as biopreservatives, in *Food Biopreservatives of Microbial Origin,* Ray, B. and Daeschel, M. A., Eds., CRC Press, Boca Raton, FL, 1992, 103.
4. Baird-Parker, A. C., Organic acids, in *Microbial Ecology of Foods,* Vol. 1, Silliker, J. H., Ed., Academic Press, New York, 1980, 126.
5. Ray, B., Diacetyl of lactic acid bacteria as a food biopreservative, in *Food Biopreservatives of Microbial Origin,* Ray, B. and Daeschel, M. A., Eds., CRC Press, Boca Raton, FL, 1992, 137.
6. Daeschel, M. A. and Penner, M. H., Hydrogen peroxide, lactoperoxide systems, and reuterine, in *Food Biopreservatives of Microbial Origin,* Ray, B. and Daeschel, M. A., Eds., CRC Press, Boca Raton, FL, 1992, 155.
7. Wolfson, L. M. and Summer, S. S., Antibacterial activity of the lactoperoxidase system: a review, *J. Food Prot.,* 56, 887, 1993.
8. Ray, B., Bacteriocins of starter culture bacteria as food biopreservatives, in *Food Biopreservatives of Microbial Origin,* Ray, B. and Daeschel, M. A., Eds., CRC Press, Boca Raton, FL, 1992, 177.
9. Klaenhammer, T. R., Genetics of bacteriocins produced by lactic acid bacteria, *FEMS Microbiol. Rev.,* 12, 39, 1993.
10. Jack, R., Tagg, J., and Ray, B., Bacteriocins of Gram-positive bacteria, *Microbiol. Rev.,* 59, 171, 1995.
11. Bakalinsky, A. T., Metabolites of yeasts as biopreservatives, in *Food Biopreservatives of Microbial Origin,* Ray, B. and Daeschel, M. A., Eds., CRC Press, Boca Raton, FL, 1992, chap. 12.
12. Wilson, C. L., Wisniewski, M. E., Biles, C. L., McLaughlin, R., Chalutz, E., and Dorby, S., Biological control of post-harvest diseases of fruits and vegetables: alternatives to synthetic fungicides, *Crop Prot.,* 10, 172, 1991.

CHAPTER 15 QUESTIONS

1. List the antimicrobial compounds produced by different starter culture bacteria.

2. Discuss the mode of antibacterial action of organic acids produced by starter culture bacteria. At pH 6.0, why are propionic and acetic acids more antibacterial than lactic acid? List the advantages of using these acids in food preservation.

3. Describe the mode of antibacterial action of diacetyl and discuss why it has limited use as a food biopreservative.

4. Under what conditions can some lactic acid bacteria generate sufficient quantities of H_2O_2 to inhibit other microorganisms in a food system? What are some disadvantages of using H_2O_2 in food?

5. Discuss the advantages of using small amounts of H_2O_2 in raw milk to inhibit psychrotrophic bacteria like *Pseudomonas* species. What needs to be done prior to pasteurization of this milk?

6. Define bacteriocins of starter culture bacteria and list their characteristics.

7. Explain the host range and mode of antibacterial action of a bacteriocin. What are its advantages and disadvantages as a food biopreservative?

8. What are the requirements of a bacteriocin to be considered a food biopreservative?

CHAPTER 16

Health Benefits of Beneficial Bacteria

CONTENTS

I. INTRODUCTION

Since the discovery of food fermentation, our ancestors recognized that the technique yielded products that not only had better shelf life and desirable qualities, but also had some health benefits, especially to combat some intestinal ailments. The belief in the health benefit of fermented foods continued throughout civilization and even today remains an interest among many consumers and researchers.[1] There are differences between the early beliefs and the current interest; while the old belief probably emerged from associated effect (benefit from consuming) without knowing the scientific basis, current interest is based on understanding the microbiology and biochemistry of the fermented foods, the microbial ecology of the human gastrointestinal tract, and their roles in human health. The researchers, however, are divided in their opinions. There are some who advocate different health benefits that consumers can have from the consumption of fermented foods, especially fermented dairy products; but there are others who have doubts about those attributes.[1]

In Western countries, the beneficial role of fermented milk (yogurt) in the prolongation of life was first advocated by Metchnikoff in 1907. He suggested that the bacteria and their metabolites in yogurt neutralized the harmful products yielded from foods in the gastrointestinal (GI) tract and provided protection to human health. However, later studies produced controversial results.

In recent years, especially since the 1970s, consumers in many developed countries are particularly interested in the health benefits of foods. There is an increased interest in foods that are not harshly processed and preserved, but are natural, and fermented foods are considered natural and healthy. Consumers' interest in and demand for fermented foods has resulted in a large production increase of many products that had very small markets before. Many ethnic products have earned commercial status. In addition, consumer interest in the health benefits of some bacterial species has stimulated production of new products containing these bacteria.

This trend will continue not only in the developed countries, but also in many developing countries, particularly where fermented foods have been consumed for a long period of time. However, it will be of extreme importance to resolve the present controversies on their health benefits, which can only be possible through the understanding of the current problems and then conducting well-designed experiments. In this chapter, the current status of our knowledge on intestinal microbial ecology, beneficial microorganisms of the GI tract, beneficial effects of fermented milk products, and the probable causes of some controversies are briefly presented.[1]

II. MICROBIOLOGY OF THE HUMAN GI TRACT

The GI tract of humans contains about 10^{16} microorganisms, many more than the total number of our body cells. They are metabolically diverse and active; thus it is quite likely that they have great influence on our well-being. It is estimated that there are about 500 bacterial species in the human GI tract, but only 30 to 40 species constitute 95% of the population.[2] Normally, the microbial level in the small intestine (particularly in the jejunum and ileum) is about 10^{6-7} per gram, and in the large intestine (colon) about 10^{10-11} per gram of the content. The most predominant types in the small intestine are several species of *Lactobacillus* and *Enterococcus,* and in the large intestine are several genera of *Enterobacteriaceae*, different species of *Bacteroides, Fusobacterium, Clostridium, Eubacterium, Enterococcus, Bifidobacterium,* and *Lactobacillus*.[2]

The intestine of a fetus in the uterus is sterile. At birth, it is inoculated with vaginal and fecal flora from the mother. Subsequently, a large variety of microorganisms enters into the digestive tract of infants from the environment. From these, the normal flora of the GI tract are established. In both breast-fed and formula-fed babies during the first couple of days, *Escherichia coli* and *Enterococcus* appear in large numbers in the feces. Then, in the breast-fed babies, large numbers of *Bifidobacterium* species and a lower level of both *Esc. coli* and *Enterococcus* species appear. In formula-fed babies, in contrast, *Esc. coli* and *Enterococcus*, together with *Clostridium* and *Bacteroides,* predominate, with *Bifidobacterium* being almost absent; this presents a situation that may result in the incidence of diarrhea. As breast-fed babies are introduced to other foods, the levels of *Esc. coli, Enterococcus, Bacteroides, Clostridium,* and others increase, but *Bifidobacterium* still remains high. When breast-feeding is completely stopped, *Bacteroides*, *Bifidobacterium,* and *Lactobacillus* species become predominant, along with some *Esc. coli, Enterococcus, Clostridium,*and others. By the second year of life, the different microflora establish themselves at their specific ecological niche in the GI tract, and the population resembles that of adult GI tracts.

The intestinal microflora are divided into indigenous (autochthonous) and transient (allochthonous) types. Many indigenous species are capable of adhering to the intestinal cells that help maintain them in their specific niche. While the indigenous types are permanent inhabitants, the transient types are either passing through or temporarily colonizing a site from where the specific indigenous type(s) has been removed due to some inherent or environmental factor (such as antibiotic intake).

Among the indigenous microbial flora, several species of *Lactobacillus* in the jejunum and ileum, and *Bifidobacterium* in the large intestine, are thought to have beneficial effects on the health of the GI tracts of the hosts. From the intestines and intestinal content of humans, *Lab. acidophilus, Lab. fermentum, Lab. reuteri, Lab. casei, Lab. lactis, Lab. leichmannii, Lab. plantarum, Bif. bifidum, Bif. longum, Bif. adolescentis, Bif. infantis,* and others have been isolated. However, age, food habits, and health conditions greatly influence the species and their levels. There is some belief that a portion of the intestinal *Lactobacillus* species is transient. At present, it is generally considered that *Lab. acidophilus, Lab. reuteri,* and some *Bifidobacterium* are indigenous species. The presence of high numbers of the indigenous *Lactobacillus* species in the feces (and content of the large intestine) probably results from their constant removal from the small intestine.[2]

III. IMPORTANT CHARACTERISTICS OF BENEFICIAL BACTERIA

Some relevant characteristics of *Lab. acidophilus, Lab. reuteri*, and *Bifidobacterium* species are briefly discussed here.[3-5] All three are found in the GI tract of humans as well as in animals and birds. They are Gram-positive rods and are able to grow under anaerobic conditions. While *Lab. acidophilus* is an obligatory homolactic fermentator, *Lab. reuteri* is a heterolactic fermentator and produces lactic acid, ethanol, and CO_2, and *Bifidobacterium* species produce lactic and acetic acids (in a

2:3 ratio). They are less sensitive to stomach acid than many other bacteria under a given condition and highly resistant to bile and lysozyme present in the GI tract. While the two *Lactobacillus* species are present in low numbers in the jejunum, but in relatively high numbers in the ileum, especially toward the distal part, *Bifidobacterium* are present in the proximal part of the colon (near the ileum). All three are able to colonize in their respective niche, but studies with *Lab. acidophilus* revealed that all strains do not adhere to the GI mucosa of the host. Other studies have shown that the ability to adhere to the intestinal epithelial cells by *Lab. acidophilus* strains could be species specific, i.e., a specific strain can be adherent to a particular species only.

Under normal conditions, these three species are thought to help in maintaining the ecological balance of GI tract microflora by controlling the growth rate of undesirable microflora. This effect is produced through their ability to metabolize relatively large amounts of lactic and acetic acids. In addition, they are capable of producing specific inhibitory substances, several of which have been recognized. Thus many strains of *Lab. acidophilus* are known to produce bacteriocins, although they are most effective against closely related Gram-positive bacteria. Also, due to the sensitivity of bacteriocins to proteolytic enzymes of the GI tract, their actual role in controlling undesirable Gram-positive bacteria in the GI tract can be disputed. Some strains also produce compounds that are not well characterized, but have been reported to have an antibacterial effect against both Gram-positive and Gram-negative bacteria. Some *Lab. reuteri* strains, while growing in glycerol, produce reuterine, which is inhibitory to both Gram-positive and Gram-negative bacteria. Antibacterial metabolite production by a few *Bifidobacterium* species was also reported. Some strains of *Lab. acidophilus* are able to deconjugate bile acids to produce compounds that are more inhibitory. Some *Lactobacillus* strains are also capable of producing H_2O_2, but probably not under anaerobic conditions in the GI tract. Identification of specific antibacterial compounds other than acids (such as reuterine) in these beneficial bacteria will help us understand their role in maintaining intestinal health.

Several studies have indicated that beneficial effects of these bacteria are produced when they are present in relatively high numbers in the intestinal tract ($10^{6\text{-}7}$ or more per gram of intestinal content). Diets rich in foods from plant sources, as opposed to those rich in foods from animal sources, seem to favor bacterial presence in higher numbers. Many other conditions in a host also can reduce bacterial numbers in the GI tract, such as antibiotic intake, mental stress, starvation, improper dietary habits, alcohol abuse, and sickness and surgery of the GI tract. This in turn can allow the undesirable indigenous or transient bacteria grow to high levels and produce enteric disturbances, including diarrhea, flatulence, and infection by enteric pathogens.

IV. BENEFICIAL EFFECTS OF PROBIOTICS

In the last 30 years, studies were conducted to determine specific health benefits from the consumption of live cells of beneficial bacteria.[1,4-7] The live cells were consumed from three principal sources: (1) as fermented milk products, such as yogurt containing live cells of *Lab. delbrueckii* subsp. *bulgaricus* and *Str. thermophilus*, and acidophilus milk containing *Lab. acidophilus*, (2) as supplementation of foods and drinks with live cells of one, two, or more types of beneficial intestinal bacteria, *Lab. acidophilus, Lab. reuteri, Lab. casei*, and *Bifidobacterium* species, and (3) as pharmaceutical products of live cells in the form of tablets, capsules, and granules. The beneficial effects from consumption of these live cells were attributed to their ability to provide protection against enteric pathogens, supplying enzymes to help in metabolizing some food nutrients (such as lactase to hydrolyze lactose) and detoxifying some harmful food components and metabolites in the intestine, stimulating intestinal immune systems, and improving intestinal peristaltic activity. Some of these are briefly discussed here.

A. Lactose Hydrolysis

Lactose-intolerant individuals are unable to produce lactase in the small intestine due to a genetic disorder. When they consume milk, lactose molecules are not hydrolyzed in or absorbed from the small intestine and passed to the colon. They are then hydrolyzed in the colon by lactase of different bacteria to glucose and galactose and then further metabolized to produce acids and gas resulting in fluid accumulation, diarrhea, and flatulence. Consumption of yogurt, acidophilus milk, live cells of *Lactobacillus*, especially *Lab. acidophilus* in fresh milk and pharmaceutical products, was found to reduce the symptoms in lactose-intolerant individuals. This benefit was attributed to the ability of the beneficial bacteria to supply the needed lactase in the small intestine. However, as *Lab. delbrueckii* subsp. *bulgaricus* and *Str. thermophilus* do not survive stomach acidity well and are not normal intestinal bacteria, the benefit of consuming yogurt is considered to be due to the reduced amounts of lactose in yogurt, as compared to milk, and to the supply of lactase from the dead cells. In contrast, the intestinal bacteria, especially some *Lactobacillus* species, could, under proper conditions, colonize the small intestine and subsequently supply lactase. Different studies, however, did not unequivocally produce desired benefits. This could be due to differences in study methods, including the use of strains that lack β-galactosidase, the use of strains that are not host specific, the use of species that are not an intestinal type, the use of preparations with low viable cells, and the lack of expertise in microbiology and gastroenterology research.

B. Reducing Serum Cholesterol Level

Consumption of fermented dairy products (some containing unknown microorganisms) and high numbers of live cells of beneficial intestinal bacteria have been associated with the lower levels of serum cholesterol in humans. This was attributed to two possible factors. One is the ability of some intestinal lactobacilli to metabolize dietary cholesterol and thus reduce the amounts absorbed in blood. The other possibility is that some lactobacilli can deconjugate bile salts and prevent their reabsorption in the liver. The liver in turn uses more serum cholesterol to synthesize bile salts and indirectly helps to reduce the cholesterol level in serum. However, the results of several studies by different researchers were not always in favor of this hypothesis. Again, one reason for the differences could be due to differences in experimental design, such as use of strains that do not metabolize cholesterol or deconjugate bile acid and other reasons described before.

C. Reducing Colon Cancer

Many of the undesirable bacteria in the colon have enzymes that can activate procarcinogens, either present in food or produced through metabolism of undesirable bacteria, to active carcinogens that in turn can cause colon cancer. Beneficial intestinal bacteria, both *Lactobacillus* and *Bifidobacterium* species, by their ability to control the growth of undesirable bacteria in the colon, can reduce the production of these enzymes. Also, the beneficial bacteria, by increasing intestinal peristaltic activity, aid in the removal of fecal materials. This, in turn, lowers the concentrations of the enzymes and carcinogens in the colon and reduces the incidence of colon cancer. Several studies have shown that oral consumption of large numbers of live cells of the beneficial bacteria does reduce fecal concentrations of enzymes such as β-glucuronidase, azoreductase, and nitroreductase of undesirable colon bacteria. However, the relationship between reduction of these enzymes to reduction in colon cancer from the consumption of beneficial intestinal bacteria has not been studied and thus contributions of these bacteria in controlling colon cancer is not clearly known. Other studies have shown that formation of aberrant crypts, that are considered to be putative precancerous lesions, is reduced by the consumption of live cells of beneficial GI tract bacteria, especially bifodobacteria. Again, the results are not consistent.

D. Reducing Intestinal Disorders

Under certain conditions, indicated before, the intestinal population of beneficial bacteria can be reduced. The undesirable bacteria in the intestine and some transient bacteria (such as enteric pathogens) from the environment can then cause enteric disorders, including infection. Ingestion of large numbers of live cells of beneficial intestinal bacteria over a period of time was reported to reduce these problems. Both infants and adults on oral antibiotic therapy can develop diarrhea, due to a loss of desirable bacteria in the intestine and an increase in undesirable bacteria such as pathogenic *Esc. coli*. It was suggested that the beneficial bacteria, when consumed in large numbers, establish in the intestine and produce antibacterial compounds (acids, bacteriocins, reuterine, and others unknown), which in turn control the undesirable bacteria. Deconjugation of biles by the beneficial species also produces compounds that are more antibacterial than the bile acids; this has also been suggested as a mechanism to control the growth of undesirable enteric bacteria. In many studies, the results were not always positive. This again could be due to the variation in study methods, as described before, including the use of strains without the specific trait.

E. Miscellaneous Benefits

Many other health benefits of fermented dairy products and beneficial intestinal bacteria have been attributed, and some of them have been tested. But the results were not always either in favor or against. Some of these are: stimulation of immune systems, prophylaxis against urogenital infections (yeast infection), increased calcium absorption from the intestine, stimulation of endocrine systems, and growth promotions.

V. SOME ASPECTS TO CONSIDER

The health benefit theory of fermented foods and beneficial intestinal bacteria is controversial. Although an association effect, i.e., some benefits from their consumption can not be denied, many studies were not able to prove those benefits without doubts. As suggested before, the differences in study methods could be one reason for the differences in results. In designing these studies, several aspects have to be recognized. There are definitely differences in human responses, but there are also differences in bacterial response. In selecting bacterial strains, the following aspects are important.[7-10]

A. Strain Variation

Beneficial strains differ in adherence ability and specificity (Figure 16.1). An adherent strain probably should be favored over a nonadherent strain. Also the strains adherent to humans should be preferred over strains adherent to other species. The selected strain(s) should have a strong, in place of weak adherence property. The adherent property could be lost during long maintenance under laboratory conditions. Many studies were conducted without even knowing the source and identity of the strains. In selecting a strain for a study, these factors need to be considered.

B. Sensitivity to Stomach Acids

Survivability of strains to low stomach pH varies greatly. This effect can be either reduced by reducing stomach acidity with food or using strains that are resistant to acid environment.

C. Viability and Injury of Cells

Cells of beneficial bacteria, when frozen, dried, exposed to low pH, high salts, and many chemicals, can die. Among the survivors, many can be injured and killed by stomach acid, and bile salts and lysozyme in the intestine. In studies, it may be

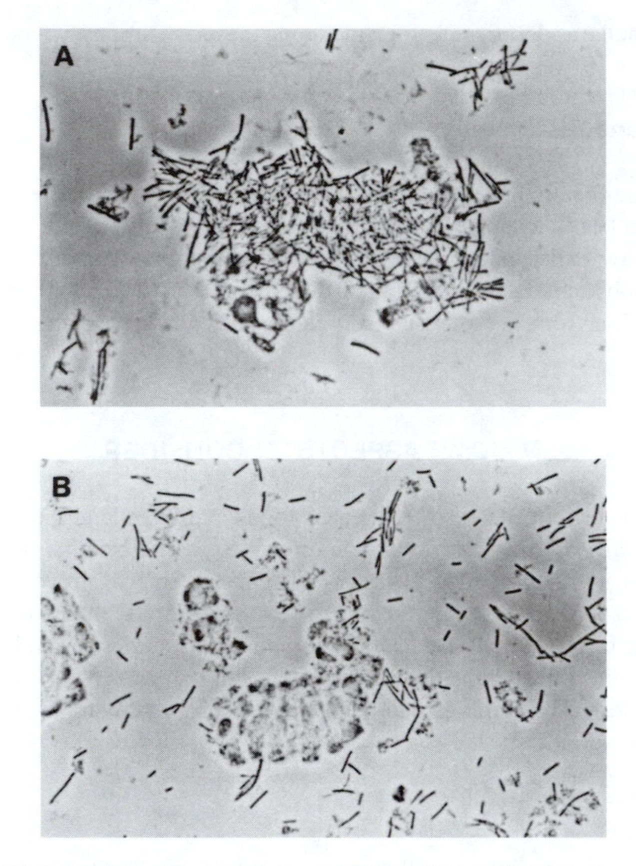

Figure 16.1 (A) Cells of an adherent strain of *Lactobacillus acidophilus* are associated with the calf intestinal epithelial cells. (B) Cells of a nonadherent strain of *Lactobacillus acidophilus* remained uniformly distributed and did not show adherence to the epithelial cells from the same source.

better to use cells that are grown for 16 to 18 h and maintained prior to feeding under conditions that retain their maximum viability.

D. Dose Level and Duration

Consumption of large numbers of live cells that are not stressed over a period of time is advocated to obtain benefit. Using preparations that have low levels of viable cells, many of which could be stressed, are not believed to provide expected results.

E. Induced Lactase Trait

In *Lab. acidophilus,* lactase is an induced enzyme. To study the lactase effect, the strains should be grown in lactose-containing media. In commercial preparations, a strain may be grown in glucose and thus will not have lactase when consumed.

F. Antibacterial Substances

Many studies have reported that strains of beneficial intestinal bacteria produce metabolites that are active against many Gram-positive and Gram-negative bacteria. Some of these were identified, such as several bacteriocins and reuterine. Other substances need to be identified and examined in purified form for their antibacterial effectiveness.

G. True Species/Strain

Many species previously regarded as *Lab. acidophilus* were found to be different species, and many were not of intestinal origin. Before selecting a strain for a study, one needs to be sure, through different methods, that the strain being used is what it is supposed to be.

H. Expertise in Microbiology Research

Lack of an understanding in microbiological research, particularly of the GI system, as well as differences in response by humans and animals in feeding trials can lead to poor experimental design. Such a study is not expected to produce valid data.

Research conducted by considering these factors will help reduce bacterial variability. This in turn will help compare results of different studies and determine if the health benefits of these bacteria are real or imaginary.

REFERENCES

1. Gilliland, S. E., Health and nutritional benefits from lactic acid bacteria, *FEMS Microbiol. Rev.,* 87, 175, 1990.
2. Darsar, B. S. and Barrow, P. A., *Intestinal Microbiology,* Am. Soc. Microbiol., Washington, DC, 1985.
3. Sandine, W. E., Roles of bifidobacteria and lactobacilli in human health, *Contemporary Nutr.,* 15, 1, 1990.
4. Hoover, D. G., Bifidobacteria: activity and potential benefits, *Food Technol.,* 47(6), 120, 1993.
5. Speck, M. L., Dobrogosz, W. J., and Casas, I. A., *Lactobacillus reuteri* in food supplementation, *Food Technol.,* 47(6), 90, 1993.
6. Marteau, P. and Rambaud, J.-E., Potential of using lactic acid bacteria for therapy and immunomodulation in man, *FEMS Microbiol. Rev.,* 12, 207, 1993.
7. Brennan, M., Wanismail, B., and Ray, B., Prevalence of viable *Lactobacillus acidophilus* in dried commercial products, *J. Food Prot.,* 46, 887, 1983.
8. Johnson, M. C., Ray, B., and Bhowmik, T., Selection of *Lactobacillus acidophilus* for use in acidophilus products, *Antonie van Leeuwenhock,* 53, 215, 1987.
9. Brennan, M., Wanismail, B., Johnson, M. C., and Ray, B., Cellular damage in dried *Lactobacillus acidophilus*., *J. Food Prot.,* 49, 47, 1986.

10. Bhowmik, T., Johnson, M. C., and Ray, B., Factors influencing synthesis and activity of β-galactosidase in *Lactobacillus acidophilus*, *J. Indust. Microbiol.*, 2, 1, 1987.

CHAPTER 16 QUESTIONS

1. Discuss the sequential establishment of GI flora in humans from birth to 2 years of age. How does the predominant flora differ between breast-fed and formula-fed babies?

2. Discuss the terms with examples: Indigenous bacteria, transient bacteria, beneficial bacteria, and undesirable bacteria in the GI tract of humans.

3. List the important characteristics of beneficial bacteria present in the GI tract. Where are they present? What are some of the antibacterial substances they produce?

4. List the factors that could adversely affect the presence of beneficial bacteria in the human GI tract.

5. What is lactose intolerance? Explain the possible mechanisms by which consumption of some fermented dairy products or live cells of beneficial intestinal bacteria can help overcome this problem.

6. How could beneficial intestinal bacteria or some fermented foods possibly reduce serum cholesterol levels?

7. "Beneficial GI tract bacteria may be effective in reducing colon cancer and enteric diseases." What is the basis for these suggestions and observations?

8. Experimental data did not always support the health benefits from the consumption of fermented foods and live cells of beneficial intestinal bacteria. Some of this could be due to wide differences in experimental methods used. If you are planning to conduct such a study, what are some of the microbiological factors you should consider?

SECTION IV

MICROBIAL FOOD SPOILAGE

A food is considered spoiled when it loses its acceptance qualities. The factors considered in judging the acceptance qualities of a food include color, texture, flavor (smell and taste), shape, and absence of abnormalities. Loss of one or more normal characteristics in a food is considered to be due to spoilage.

Food spoilage not only causes economic loss, but also causes loss of consumable foods. In the United States and some other countries where foods are produced much more than the need, spoilage up to a certain level is not considered serious. However, in countries where food production is not efficient, food spoilage can adversely affect the availability of food. With the increase in population in the world, serious consideration needs to be given not only to increase food production, but also to reduce food spoilage, which for certain produce in some countries could reach 25% or more.

The acceptance qualities of a food can be lost due to infestation with insects and rodents, undesirable physical and chemical actions, and growth of microorganisms. An example of physical spoilage is dehydration of vegetables (wilting). Chemical spoilage includes oxidation of fat, browning of fruits and vegetables, autolytic degradation of some vegetables (by pectinases) and fishes (by proteinases). Microbial spoilage results either as a consequence of microbial growth in a food or is due to the action of some microbial enzymes present in a food. In this section, food spoilage due to microbial growth and microbial enzymes is discussed under the following topics.

SECTION IV

CHAPTER 17

Important Factors in Microbial Food Spoilage

CONTENTS

I. INTRODUCTION

Microbial food spoilage occurs as a consequence either of microbial growth in a food or the release of extracellular and intracellular (following cell lysis) enzymes in the food environment. Some of the detectable parameters associated with spoilage of different types of foods are changes in color, odor, and texture, formation of slime, accumulation of gas (or foam), and release of liquid (exudate, purge). Comparatively, spoilage due to microbial growth occurs much faster than spoilage caused by microbial extra- or intracellular enzymes in the absence of the viable microbial cells. Between initial production (such as harvesting of plant foods, slaughter of food animals) and final consumption, different methods are used to preserve the acceptance qualities of foods that include the reduction of microbial numbers and growth. Yet microorganisms grow and cause food spoilage, which for some food could be relatively high. It is important to understand the factors associated with microbial food spoilage, both for recognizing the cause of an incidence and developing an effective means of control.

II. SEQUENCE OF EVENTS

Generally, for microbial food spoilage to occur, several events need to take place in sequence. The microorganisms have to get into the food from one or more sources, the food environment (pH, A_w, O–R potential, nutrients, and inhibitory agents) should favor growth of one or more types of these contaminating microorganisms, the food must be stored (or abused) at a temperature that enables one or more types to multiply, and finally, the food must be stored under conditions of growth for a sufficient length of time for the multiplying microbial type(s) to attain the high numbers necessary to cause the detectable change(s) in a food. In a heat-treated food, the microorganisms associated with spoilage either survive the specific heat treatment (thermodurics) or get into the food following heating (as post-heat contaminants). Spoilage of a heated food from microbial enzymes, in the absence of viable microbial cells, can result from some heat-stable enzymes produced by microorganisms in the foods prior to heat treatment. In addition, the foods need to be stored at a temperature for a sufficient length of time for the catalytic activities of the enzymes to occur to produce the detectable changes.

III. SIGNIFICANCE OF MICROBIAL TYPES

Raw and most processed foods normally contain many types of molds, yeasts, and bacteria capable of multiplying and causing spoilage (viruses do not multiply in foods). As multiplication is an important component in spoilage, bacteria (due to a shorter generation time), followed by yeasts, are in favorable positions over molds to cause rapid spoilage of foods. However, in foods where bacteria or yeasts do not grow favorably and the foods are stored for a relatively longer period of time, such as breads, hard cheeses, fermented dry sausages, and acidic fruits and vegetables, spoilage due to mold growth is more prevalent. Recent advances in anaerobic packaging of foods has also greatly reduced the spoilage of food by molds and to some extent by yeasts, but not by anaerobic and facultative anaerobic bacteria. Thus, among the three microbial groups, the highest incidence of spoilage, especially rapid spoilage, of processed foods is caused by bacteria, followed by yeasts and molds.

IV. SIGNIFICANCE OF MICROBIAL NUMBERS

To produce detectable changes in color, odor, and texture of a food accompanied with slime formation and/or gas and liquid accumulation, microorganisms (mainly bacteria and yeasts) must multiply and attain certain levels, often referred to as the spoilage detection level. Although it varies with the type of foods and microorganisms, bacteria and yeasts need to reach to about 10^7 cells per gram, per milliliter, or per square centimeter of a food from the level present normally in a food. Depending upon the specific nature of spoilage and microbial types, the spoilage detection level can range from 10^6 to 10^8 cells per gram, milliliter, or square centimeter. Spoilage associated with H_2S, some amines, and H_2O_2 formation can be detected at a lower microbial load, while formation of lactic acid may be detected at a higher microbial load. Slime formation, associated with the accumulation of microbial cells, is generally detected at 10^8 cells or higher per gram, milliliter, or square centimeter of a food. It appears then that a food with higher initial loads of spoilage bacteria (or yeasts), and a storage condition that favors rapid growth (shorter generation time), will spoil more rapidly than a food with a low initial load and longer generation time (Figure 17.1). In this hypothetical example, the population reached the spoilage detection level within 7 d with high initial load (about 5×10^5/g) as opposed to 20 d with low initial load (about 5×10^2/g) during storage at 12°C. However, when the product with a low initial load was stored at 4°C (to increase the generation time), it took about 55 d for the spoilage bacteria to reach the spoilage detection level. To reduce microbial spoilage of a food, one needs to aim at achieving both the low initial load and longer generation time of spoilage microorganisms during storage. Just the mere presence of 10^7 cells per gram or square centimeter without growth (e.g., from a massive initial contamination) will not immediately cause a food to lose its acceptance quality; but such a food will spoil very rapidly following growth of the contaminants. Bioprocessed foods, in general, contain very high numbers of microorganisms (10^8 to 10^9 cells per gram or milliliter). However, under normal conditions, they are desirable types and the fermented foods are not considered spoiled. Spoilage of these foods can occur due to growth of undesirable bacteria, such as slime formation and off flavor in cottage cheese by *Alcaligenes* and *Pseudomonas* spp. In such products, selective methods should be used to determine the populations of undesirable bacteria or yeasts.

V. SIGNIFICANCE OF PREDOMINANT MICROORGANISMS

The microbiological profile of a food is quite different from that of a pure culture growing in a laboratory medium. An unspoiled, nonsterile food generally contains many types of microorganisms consisting of bacteria, yeasts, and molds (also viruses) from different genera and may be more than one species from the same genus. The population level of each type can vary greatly. However, when the same food is spoiled, it is found to contain predominantly one or two types, and they may not even be present initially in the highest numbers in the unspoiled product. Among the different species initially present and capable of growing in a particular food,

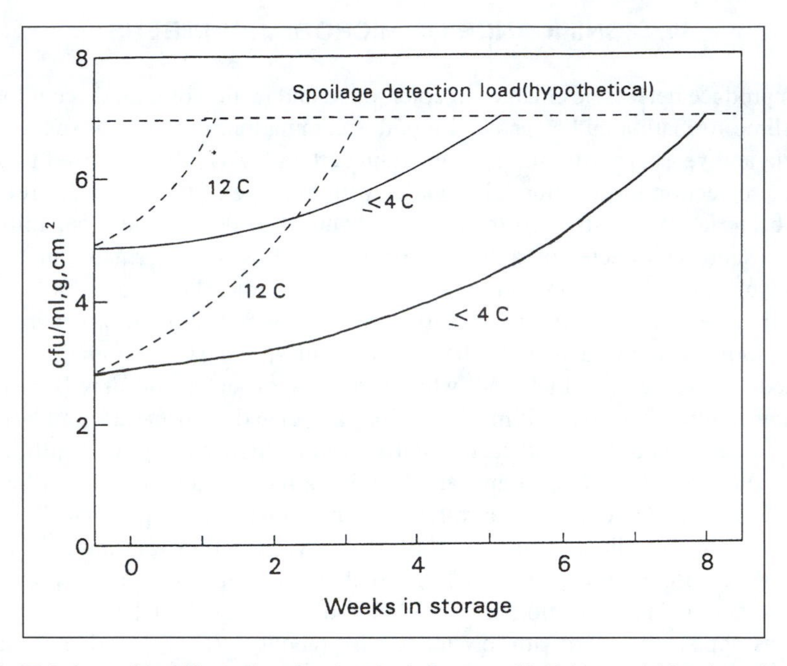

Figure 17.1 A graphical illustration showing the influence of initial bacterial levels and storage temperatures on the shelf-life of a refrigerated product.

only those with the shortest generation time under the conditions of storage will attain the numbers rapidly and cause spoilage. In a study, a beef sample (pH 6.0) was found to initially contain about 10^3 bacterial cells per gram, with relative levels of *Pseudomonas* spp. 1%, *Acinetobacter* and *Moraxella* 11%, *Brochothrix thermosphacta* 13%, and others (*Micrococcus, Staphylococcus, Enterobacteriaceae*, lactic acid bacteria, etc.) 75%. Following aerobic storage at 2°C for 12 d, the population reached to 6×10^7 cells per gram, with the relative levels of *Pseudomonas* spp. 99% and all others 1%. Many of the bacterial species present initially were capable of growing at the storage condition of the meat, but *Pseudomonas* spp. had the shortest generation time. As a result, initially they constituted only 1% of the total population; after 12 d, they became predominant (99%). If the same meat sample was stored at 2°C anaerobically (such as in vacuum-package) until the population had reached 10^7 per gram, the predominant bacteria would have been, with most probability, *Lactobacillus* and/or *Leuconostoc* due to their growth advantages.

In this context, it is important to recognize that the generation time of a microbial species, even under optimum conditions of growth, is much longer in a food than in a microbiological broth. Also, under the same storage conditions, the growth behavior of a microbial mixed population could be quite different in a food as compared to in a broth. Because of this, the predominant type(s) obtained following growing an initial mixed microbial population from a food in a broth and in the same food under identical conditions could be different. Although growing in a broth is convenient and, if properly designed, can provide valuable initial information, it is always better to include studies with the specific food(s).

VI. SOME IMPORTANT FOOD SPOILAGE BACTERIA

[See also Chapter 2.]

Theoretically, any microorganism that can multiply in a food to reach the high level (spoilage detection level) is capable of causing it to spoil. Yet, in reality, bacterial species from only several genera have been implicated with spoilage of most foods. This is dictated by the bacterial characteristics, food characteristics, and storage conditions. The influence of some of these factors in determining predominant spoilage bacteria in foods are briefly discussed.

A. Psychrotrophic Bacteria

As indicated before, psychrotrophic bacteria constitute the bacterial species capable of growing at 5°C and below, but multiply quite rapidly at 10 to 25°C. Many foods are stored on ice (chilling) and in refrigerators, and some are expected to have a long shelf-life (50 d or more). Between processing and consumption, they can be temperature abused to 10°C and higher. The psychrotrophic bacteria (also many yeasts and molds that are psychrotrophic) can cause spoilage in these foods. If the food is stored under aerobic conditions, psychrotrophic aerobes will be the predominant spoilage bacteria. In foods stored under anaerobic conditions, anaerobic and facultative anaerobic bacteria will be the predominant bacteria. If the food is given low heat treatment and not exposed to post-heat contamination during storage at low temperature, psychrotrophic thermoduric bacteria can cause it to spoil.

1. Some Important Psychrotrophic Aerobic Spoilage Bacteria

Pseudomonas fluorescens, Pseudomonas fragi, other *Pseudomonas* species, *Acinetobacter, Moraxella,* and *Flavobacterium*. (Some molds and yeasts are included in this group.)

2. Some Important Psychrotrophic Facultative Anaerobic Spoilage Bacteria

Brochothrix thermosphacta, Lactobacillus viridescens, Lactobacillus sake, Lactobacillus curvatus, unidentified *Lactobacillus* spp., *Leuconostoc carnosum, Leuconostoc gelidum, Leuconostoc mesenteroides*, some *Enterococcus* spp., *Alcaligenes* spp., *Enterobacter* spp., *Serratia liquifaciens*, some *Hafnia* and *Proteus* spp., and *Shewanella* (previously *Alteromonas*) *putrefaciens*. (Some yeasts are also included in this group.)

3. Some Important Thermoduric Psychrotrophs

Facultative anaerobes: spores of *Bacillus coagulans* and *Bacillus megaterium*, some strains of *Lactobacillus viridescens*. Anaerobes: spores of *Clostridium laramie, Clostridium esterthetícum, Clostridium algidicarnis, Clostridium putrefaciens*, and unidentified *Clostridium* spp.

When a food is temperature abused above 5°C, some true mesophiles (growth temperature range 15 to 45°C, optimum 25 to 40°C) can also grow. However, at 10 to 15°C storage temperature psychrotrophs will generally grow much faster than these mesophiles.

B. Thermophilic Bacteria

By definition, the bacteria in this group grow between 40 and 90°C, with optimum growth at 55 to 65°C. Some high heat-processed foods are kept warm at temperatures between 50 and 60°C for a long period of time (at delis, fast food establishments, and restaurants). Spores of some thermophilic *Bacillus* and *Clostridium* spp. can be present in these heat-treated foods, which at warm temperatures germinate and multiply to cause spoilage. In addition, some thermoduric vegetative bacteria surviving low heat processing (such as pasteurization) or thermophiles getting in food as post-heat contamination can also multiply in these warm foods, especially if the temperature is close to 50°C. These include some lactic acid bacteria, such as *Pediococcus acidilactici* and *Streptococcus thermophilus,* as well as some *Bacillus* and *Clostridium* spp. They can also cause spoilage of foods that are cooked at low heat (60 to 65°C, as for some processed meats) or kept warm for a long period of time.

C. Aciduric Bacteria

Bacteria capable of growing relatively rapidly in food at pH 4.6 or below are generally regarded as aciduric (or acidophilic). They are usually associated with spoilage of acidic food products such as fruit juices, pickles, salsa, salad dressings, mayonnaise, and fermented sausages. Heterofermentative lactic acid bacteria (such as *Lactobacillus fructivorans, Lactobacillus fermentum,* and *Leuconostoc mesenteroides*) and homofermentative lactic acid bacteria (such as *Lactobacillus plantarum* and *Pediococcus acidilactici*) have been associated with such spoilage. (Yeasts and molds are aciduric and thus are associated with spoilage of such foods.)

VII. FOOD TYPES

Foods differ greatly in their susceptibility to spoilage by microorganisms. This is mainly because of their differences in intrinsic factors (A_w, pH, O–R potential, nutrient content, antimicrobial substances, and protective structures). A food with a lower A_w (say 0.90) or a lower pH (say 5.3) is less susceptible to bacterial spoilage than one with A_w = 0.98 or pH 6.4. However, molds and yeasts will probably grow equally well under both conditions. The influence of each of the intrinsic parameters on microbial growth has previously been described (Chapter 6). On the basis of ease of spoilage, foods can be grouped as perishable (spoil quickly, in days), semiperishable (have a relatively long shelf-life, few weeks or months), and nonperishable (have a very long shelf-life, many months or years). In addition to the intrinsic parameters, extrinsic parameters (storage conditions) play important roles in determining the ease of microbial spoilage of many foods.

VIII. METABOLISM OF FOOD NUTRIENTS

Microbial growth in a food is associated with the metabolism of some food carbohydrates, proteinaceous and nonprotein nitrogenous (NPN) compounds, and lipids. Major types of carbohydrates (polysaccharides, trisaccharides, disaccharides, monosaccharides, and sugar alcohols), proteinaceous compounds (proteins, peptides), NPN compounds (amino acids, urea, creatine, trimethylamine oxide), and other lipids (triglycerides, phospholipids, fatty acids, sterols) present in foods are briefly discussed here. The metabolic pathways of some of these compounds by microorganisms were discussed in Chapters 8 and 10. It is evident from these discussions that microorganisms differ greatly in their ability to metabolize different food nutrients (such as ability or inability to utilize cellulose and lactose as carbon sources, casein as a nitrogen source, and oxidation of oleic acid). Similarly, the same nutrient (substrate) can be utilized by different microorganisms by different metabolic pathways to produce different end-products (such as glucose metabolized by homolactic and heterolactic acid bacteria). The same nutrient (substrate) can be degraded to produce different end-products under aerobic and anaerobic metabolism (respiration and fermentation, respectively). Thus glucose is metabolized (catabolism) by *Micrococcus* spp. aerobically to produce CO_2 and H_2O, and by *Lactobacillus acidophilus* anaerobically to produce mainly lactic acid. *Saccharomyces cerevisiae* metabolizes glucose aerobically to CO_2 and H_2O, but anaerobically to ethanol and CO_2. Under specific conditions, some microorganisms can also synthesize (anabolism) polymeric compounds as end-products such as dextran (polymer of glucose) production by *Leuconostoc mesenteroides* while metabolizing sucrose. Microorganisms can also secrete extracellular enzymes to break down large molecular nutrients (polymers) in a food (such as the breakdown of starch by amylase produced by some molds). Finally, some microorganisms can synthesize pigments while growing in a food (such as *Micrococcus luteus* producing yellow pigment).

Thus the metabolism of food nutrients during growth of microorganisms in a food can adversely change its acceptance quality. Some of the changes are: odor (due to production of volatile end-products), color (pigment production or oxidation of natural color compounds, such as oxidation of meat myoglobin), texture (breakdown of pectin by pectinases in vegetables, or softening of the tissues in meat by proteinases), accumulation of gas (due to production of CO_2, H_2, or H_2S_2), formation of slime (due to production of dextran or too many microbial cells resulting in confluent growth), and loss of liquid (purge accumulation; in meat, due to the breakdown of structures holding the water of hydration). Some of these changes also occur from the effect of microbial metabolites on food pH. Production of organic acid by microorganisms, causing lowering of food pH, can reduce the water-holding ability of food (such as growth of some lactic acid bacteria in low fat-high pH processed meat products). Similarly, production of basic compounds by microorganisms in a food can shift its pH to the alkaline side and reduce its acceptance quality (such as decarboxylation of amino acids in some low heat-processed meat products with the production of amines, shifting the pH to basic, and changing product color from light brown to pink in some processed meats).

Some of the end-products of microbial metabolism of food nutrients that are attributed to food spoilage are listed in Table 17.1. The end-products vary with the nature of metabolism (i.e., aerobic respiration, anaerobic respiration, or fermentation; see Chapters 8 and 10). It is evident that many of these metabolites are able to produce the changes associated with microbial food spoilage (change in odor, gas formation, or slime formation).

Table 17.1 Some End-products from Microbial Metabolism of Food Nutrients

Food nutrients		
Carbohydrates	**Proteinaceous and NPN compounds**	**Lipids**
CO_2, H_2, H_2O_2, lactate, acetate, formate succinate, butyrate, isobutyrate, isovalerate, ethanol, propanol, butanol, isobutanol, diacetyl, acetoin, butanediol, dextran, levans	CO_2, H_2, NH_3, H_2S, amines, keto acids, mercaptans, organic disulfides, putrescine, cadaverine, skatole	Fatty acids, glycerol, hydroperoxides, carbonyl compounds (aldehydes and ketones), nitrogenous bases

IX. PREFERENCE FOR UTILIZATION OF FOOD NUTRIENTS

Almost all foods contain some amount of carbohydrates, proteinaceous and NPN compounds, and lipids that are available for microbial use during growth. However, the characteristics of food spoilage differ greatly. This is mainly due to differences in the nature and amount of a specific nutrient present in a food, the type of microorganism growing in the food, and the nature of metabolism (respiration or fermentation). In general, microorganisms prefer to use carbohydrates first, followed by NPN and proteinaceous compounds, and then lipids. However, it again depends on whether or not a particular species has the ability to use a specific carbohydrate (such as ability or inability to utilize lactose). Also, small nutrient molecules are used before the large molecules (polymers).

If a food has a carbohydrate that can be fermented by the contaminating microorganisms, then it will usually be metabolized first. If the metabolizable carbohydrates are present in sufficient quantities, then the metabolic pathway remains unchanged. However, if the carbohydrates are present in limiting concentrations, then after the carbohydrates are used up, the microorganisms usually start using NPN and proteinaceous compounds. For example, yeasts growing in a fruit juice containing relatively high amounts of metabolizable carbohydrates (fructose, glucose, sucrose) will produce either CO_2 and H_2O (aerobically), or alcohol and CO_2 (anaerobically). However, *Pseudomonas fluorescens* growing aerobically in fresh meat with limiting amounts of glucose will first metabolize it and then start metabolizing free amino acids and other NPN compounds. If it is allowed to grow for a long time, it will produce extracellular proteinases to break down meat proteins to produce small peptides and amino acids for further metabolism. With time, it may even be able to produce lipase to break down meat lipids and use up some fatty acids. In a food (such as milk) containing large amounts of both carbohydrates (lactose) and proteins, a lactose-metabolizing microorganism will preferentially utilize the lactose and produce acid or acid and gas (*Lactococcus lactis* subsp. *lactis* or a *Leuconostoc* sp.), but a microorganism unable to utilize lactose will use the NPN and proteinaceous compounds to multiply (*Pseudomonas* spp.). The spoilage patterns will be quite different.

In a mixed microbial population, as normally present in a food, availability and amount of metabolizable carbohydrates greatly affect the spoilage pattern. Fresh meats, due to a low level of glucose, are susceptible to spoilage through microbial degradation of NPN and proteinaceous compounds. However, if a metabolizable carbohydrate (such as glucose, sucrose, or lactose) is added to meat, metabolism of carbohydrates will predominate. If lactic acid bacteria are present as natural microflora, they will produce enough acids to arrest the growth of many normal microflora that preferentially metabolize NPN and proteinaceous compounds. This is commonly known as the protein sparing effect (proteins are not metabolized), and carbohydrate supplementation is used for this to produce many processed meat products.

X. MICROBIAL GROWTH IN SUCCESSION

The intrinsic and extrinsic factors or environments of a food dictate which, among the mixed microbial species normally present, will multiply rapidly and become predominant to cause spoilage. However, as the predominant types grow, they produce metabolites and change the food environment. In the changed environment, some other species, originally present but not able to compete, may be in a favorable position to grow rapidly and again change the food environment to enable a third type to grow rapidly. If sufficient time is given, the predominant microbial types and the nature of spoilage of a food can change. Sequential growth of *Lactococcus* spp., yeasts, and Gram-negative rods (such as *Pseudomonas* spp.) in a milk sample can be used as a hypothetical example. Initially, rapid growth of *Lactococcus* spp. (able to metabolize lactose) under a favorable growth condition will reduce the pH from 6.5 to 4.5 and prevent growth of many other microbial species present. As the pH dropped to 4.5, the generation time of *Lactococcus* spp. becomes longer. However, yeasts, due to their aciduric nature, can then start multiplication and increase the pH (say 5.8). At high pH, the *Pseudomonas* spp. present can grow initially by metabolizing NPN and proteinaceous compounds, and can increase the pH further by producing basic metabolites (amines, NH_3). This way, the predominant spoilage microorganisms and the metabolites associated with spoilage (i.e., nature of spoilage) of a food can change if a food is stored for a long time.

REFERENCES

1. Gill, C. O., The control of microbial spoilage in fresh meat, in *Advances in Meat Research: Meat and Poultry Microbiology,* Vol. 2, Pearson, A. M. and Dutson, T. R., Eds., AVI Publishing, Westport, CN, 1986, 49.
2. Kraft, A. A., Health hazard vs. food spoilage, *Psychrotrophic Bacteria in Foods,* CRC Press, Boca Raton, FL, 1992, 113.
3. Sinell, H. J., Interacting factors affecting mixed populations, in *Microbial Ecology of Foods,* Vol. 1, Silliker, J. H., Ed., Academic Press, New York, 1980, 215.
4. Ray, B., Foods and microorganisms of concern, *Food Biopreservatives of Microbial Origin,* Ray, B. and Daeschel, M. A., Eds., CRC Press, Boca Raton, FL, 1992, 25.

CHAPTER 17 QUESTIONS

1. Describe when a food is considered spoiled by microorganisms. You put a packet of lunch meat in the refrigerator several weeks back and forgot about it. When you opened it, you thought it had a slightly sour odor. You asked your friend to smell it, but he (she) said it smelled fine. What could be the reason(s) for this difference?

2. You opened a 2-l bottle of soft drink and kept it in the refrigerator and forgot about it. After 2 weeks, you found some molds growing on the surface of the liquid. Describe the sequence of events that resulted in this spoilage.

3. Discuss the significance of microbial types and numbers in food spoilage.

4. A food is usually spoiled by limited (usually one or two) types of microorganisms among the many types initially present. Describe how this predominance is established.

5. At present, psychrotrophic bacteria seem to be the most important spoilage bacteria. Suggest the possible reasons.

6. How can psychrotrophic bacteria be arbitrarily grouped? Give examples of two bacteria from each group. Also, give two examples each of thermoduric, thermophilic, and aciduric spoilage bacteria.

7. Define aerobic respiration, anaerobic respiration, and fermentation. List four metabolites associated with (a) odor and (b) gas formation from food nutrients.

8. Discuss the importance of preferential nutrient metabolism by microorganisms and the nature of food spoilage.

CHAPTER 18

Spoilage of Specific Food Groups

CONTENTS

I. INTRODUCTION

The predominant microbial types normally expected to be present in different food groups were listed previously (Chapters 2 and 4). Initially, a food produced under proper sanitary conditions generally contains microorganisms at a level (per g, ml, or cm^2) much lower than that at which spoilage is detected. Subsequently, growth of some of the microbial species among those present initially enable the microorganisms to reach the spoilage detection level. Many factors dictate which species will multiply relatively rapidly to become the predominant spoilage microorganisms. Along with microbial types, food types and food environments (both intrinsic and extrinsic factors) have important roles in determining the predominant spoilage microflora in a food. These aspects have been discussed before (see Chapters 4 and 6). In this chapter, predominant microorganisms associated with the spoilage of different food groups are described.

II. FRESH AND READY-TO-EAT MEAT PRODUCTS

A. Raw Meat

Fresh meats from food animals and birds contain a large group of potential spoilage bacteria that include species of *Pseudomonas, Acinetobacter, Moraxella, Shewanella, Alcaligenes, Aeromonas, Escherichia, Enterobacter, Serratia, Hafnia, Proteus, Brochothrix, Micrococcus, Enterococcus, Lactobacillus, Leuconostoc, Carnobacterium,* and *Clostridium,* as well as yeasts and molds.[1-6] The predominant spoilage flora in a meat is determined by the nutrient availability, oxygen availability, storage temperature, pH, the storage time of the product, and the generation time of the spoilage microorganisms under a given environment. Post-rigor meats are rich in nonprotein nitrogenous compounds (about 13 mg/g; amino acids and creatine), peptides, and proteins, but contain low concentrations of carbohydrates (about 1.3

mg/g; glycogen, glucose, glucose-6-phosphate), with a pH of about 5.5 and $A_w > 0.97$. DFD (dark, firm, dry) meats have almost no carbohydrates and a pH 6.0 or above.

To delay microbial spoilage, fresh meats are stored at refrigerated temperature ($\leq 5°C$), unless the facilities are not available. Thus normally psychrotrophic bacteria will be the most predominant types in raw meat spoilage. Under aerobic storage at low temperature, growth of psychrotrophic aerobes and facultative anaerobes is favored. In retail cut meats, due to shorter generation time, *Pseudomonas* spp. will grow rapidly, using glucose first and then amino acids; the metabolism of amino acids is accompanied by the production of malodorous methyl sulfides, esters, and acids. In meats with high pH and/or low glucose content, *Acinetobacter* and *Moraxella*, which preferentially metabolize amino acids instead of glucose, can grow rapidly and produce undesirable odors. Spoilage by these strict aerobes in the form of off-odor is detected at a population of about 10^8 cells per square centimeter and slime at about 10^9 cells per square centimeter. The oxygenated red color of myoglobin will undergo oxidation to produce gray or brown metmyoglobin. DFD meats are spoiled more rapidly, inasmuch as the bacteria utilize amino acids immediately due to the absence of carbohydrates.

Refrigerated meat in a modified atmosphere, such as in a mixture of CO_2 and O_2, will favor the growth of facultative anaerobic *Brochothrix thermosphacta*, especially in meat with pH 6.0 or higher (DFD meat). It metabolizes glucose to acetic acid and acetoin, and leucine and valine to isovaleric and isobutyric acids to produce off-odor (cheesy odor). Under anaerobic conditions, they metabolize glucose to produce small amounts of lactic acid (which is not considered to be a cause of spoilage).

Psychrotrophic facultative anaerobes and anaerobes can grow in vacuum-packaged meats to produce different types of spoilage. *Lactobacillus curvatus* and *Lab. sake* metabolize glucose to produce lactic acid and the amino acids leucine and valine to isovaleric and isobutyric acids. These volatile fatty acids impart a cheesy flavor in meat at a population level of over 10^8 cells per square centimeter. This spoilage is not considered highly undesirable, inasmuch as after opening the package, the flavor disappears. However, when they metabolize cysteine and produce H_2S, the product will have undesirable odor and color. Heterofermentative *Leuconostoc carnosum* and *Leuconostoc gelidum* produce CO_2 and lactic acid, causing accumulation of gas and liquid in the package. *Shewanella putrefaciens*, which can grow under both aerobic and anaerobic conditions, metabolizes amino acids (particularly cysteine) to produce methylsulfides and H_2S in large quantities. Along with offensive odors they adversely affect the normal color of meats. H_2S oxidizes myoglobin to a form of metmyoglobin, causing a green discoloration. Facultative anaerobic *Enterobacter, Serratia, Proteus,* and *Hafnia* species metabolize amino acids while growing in meat to produce amines, ammonia, methylsulfides, and mercaptans, and cause putrefaction. Some strains also produce H_2S in small amounts to cause greening of the meat. Due to production of amines and ammonia, the pH of the meat usually changes to the alkaline range, and meat can have a pinkish to red color. Psychrotrophic *Clostridium* spp., such as *Clo. laramie*, have been found to cause spoilage associated with proteolysis and loss of texture of meat, accumulation of liquid in the bag, and offensive odor, with an H_2S smell predominating. The color of the meat becomes unusually red initially and then changes to green (due to oxidation by H_2S). Some *Clostridium* spp., and probably *Enterococcus,* can cause spoilage of beef rounds and ham deep near the bone, designated as bone sour or bone taint.

Comminuted meats spoil more rapidly than the retail cuts since they have more surface area. Under aerobic storage, growth of aerobic bacteria (predominantly *Pseudomonas* spp.) will cause the changes in odor, texture, color, and sliminess. Inside, which is initially microaerophilic (due to dissolved or trapped air) and then changed to anaerobic, growth of facultative bacteria will predominate. In vacuum-packaged products, growth of lactic acid bacteria will predominate at the initial stage. Heterofermentative lactic acid bacteria can cause accumulation of gas in the package. When the glucose is used up, Gram-negative facultative anaerobes will degrade amino acids and produce putrid odor.

To reduce spoilage of fresh meats, storage at low temperatures (close to 0 to –1°C), modified atmosphere packaging, and vacuum packaging are extensively used. Several other methods to reduce initial microbial load and slow down growth of Gram-negative rods are being either used or tested. These include the addition of small amounts of organic acids to lower the pH of meat (slightly above pH 5.0), drying of meat surfaces (to reduce A_w), and a combination of the above factors including lower storage temperature.

B. Ready-to-Eat Meat Products

This group includes high heat-processed and low heat-processed uncured and cured meat products. High heat-processed cured and uncured meats are given heat treatment to make them commercially sterile. Thus they may only have some thermophilic spores surviving, which will not germinate unless the products are temperature abused. This aspect has been explained, along with microbial growth in the canned products (see Chapter 4 and p. 229).

Low heat-processed uncured meats, such as roasts, are given heat treatment ranging from 140 to 150°F internal temperature (60 to 65°C). Generally, the surface of the meats (and thus the microorganisms) is exposed to the final temperature for 1 h or more depending upon the size of the meat (which could be over 10 lb). Under this condition, only the spores of *Bacillus* and *Clostridium* spp. and some extremely thermoduric vegetative species (*Lactobacillus viridescens,* some *Enterococcus, Micrococcus*) can survive. However, the products, even when cooked in bags, are opened and handled prior to final vacuum packaging and refrigeration storage. Many types of microorganisms can enter as post-heat contaminants into the products from equipment, personnel, and air. In some situations, spices and other ingredients are added to the products after heating, which in turn can add to the microbial contamination of the products. Some products are sliced before vacuum packaging, which increases the surface area of the product as well as chances of contamination from the equipment and environment. Psychrotrophic facultative anaerobic and anaerobic bacteria have been implicated in the spoilage of these products. In roast beef, heterofermentative *Lactobacillus* spp. and *Leuconostoc* spp. have been involved in the accumulation of large quantities of gas (CO_2) and liquid (due to acid production) inside the bag, without causing much flavor, color, or texture changes. Gas production and purge accumulation by psychrotrophic *Clostridium* spp., along with off-flavor and color changing from brown to pink to red (after 4 weeks), have been detected. In spoiled sliced roast beef, the brown color of roast beef changed to pink (in 1 week) and the beef had a putrid odor (after 6 weeks). *Proteus* and *Hafnia* spp. were involved in spoilage of this product (see Chapter 20).

Low heat-processed cured meats include a wide variety of products such as franks, bologna, ham, and luncheon meats made from beef, pork, and poultry. Meats are mixed with different types of additives to improve color, texture, flavor, shelf-life, and safety. Some of these additives are nitrite, salt, dextrose, phosphate, sorbate, erythorbate, nonfat dry milk, soy proteins, carrageenan, and different types of spices. Some of the products have very low fat (≤2%) as compared to products with ≥30% fat (some franks). Some of the products, especially low-fat products, have pH values as high as 6.8 (compared to pH < 6.0 in other products). They are cooked at 150 to 160°F (65 to 71°C) internal temperature. Depending upon the size of the products, the surface can be exposed to the final temperature for a long time as compared to the center. As the products are made from ground or chopped meats, microorganisms are distributed throughout the products (as compared to only on the surface of a roast or retail cuts). Thus the thermodurics surviving cooking will be present throughout the products. The products, following cooking, are extensively handled before they are again vacuum packaged or packed with modified atmosphere of CO_2 or $CO_2 + N_2$ or sold unpackaged. Depending upon the post-heat treatment processing steps (slicing, portioning, and skinning) involved, the products can be contaminated with microorganisms from equipment, personnel, air, and water. Some of these microorganisms establish themselves in the processing environment, especially in places difficult to sanitize. In products like franks, they contaminate only the surface; but in sliced products, they are distributed over the slices during cutting.

The vacuum-packaged and gas-packaged products, during storage, can be spoiled by psychrotrophic *Lactobacillus* spp. (including homofermentative *Lab. sake, Lab. curvatus,* heterofermentative *Lab. viridescens*) and *Leuconostoc* spp. (such as *Leu. carnosum, Leu. gelidum, Leu. mesenteroides*). The products, depending upon bacterial types, show cloudy appearance, large accumulation of gas (CO_2) and/or liquid, slime formation due to bacterial cells and dextran production by *Leuconostoc* spp. in products containing sucrose (honey), and slight acidic flavor. In some products, growth of *Serratia* spp. (*Serratia liquifaciens*) causes amino acid breakdown, which then causes an ammonia-like flavor (diaper smell). Low fat-containing, vacuum-packaged turkey rolls (portions or sliced) develop a pink color after about 5 weeks, probably as a result of growth of some lactic acid bacteria. Under aerobic conditions of storage (unpackaged or permeable film wrapped), some *Lactobacillus* spp. rapidly produce H_2O_2, which can oxidize nitrosohemochrome (formed by heating nitrosomyoglobin) to brown metmyoglobin or to oxidized porphyrins, some of which are green in color. Refrigerated luncheon meat can develop brown to yellow spots with fluorescens from the growth of *Enterococcus casseliflavus* (see Chapter 20).

Unpackaged cooked products that do not have carbohydrates can become putrid from the growth and protein degradation by the proteolytic Gram-positive bacteria. If the products are stored for a long time, yeasts and molds can also grow, causing off-flavor, discoloration, and sliminess. Due to the growth of H_2O_2-producing lactic acid bacteria, the products may have green to gray discolorations.

III. EGGS AND EGG PRODUCTS

A. Shell Eggs

The pores in the eggshell and inner membrane do not prevent entrance of bacteria and hyphae of molds, especially when the pore size increases during storage. The presence of moisture enhances the entrance of the motile bacteria. The albumen (egg white) and yolk have about 0.5 to 1.0% carbohydrate and are high in protein but low in nonprotein nitrogen. During storage, the pH can shift to the alkaline side (pH 9 to 10). In addition, lysozyme (causes lysis of bacterial cell wall mucopeptide), conalbumin (chelates iron), antivitamin proteins (avidin binds riboflavin), and protease inhibitors in eggs have inhibitory effects on microbial growth. The most predominant spoilage of shell eggs is caused by Gram-negative motile rods from several genera that include *Pseudomonas, Proteus, Alcaligenes, Aeromonas,* and coliform group. The different types of spoilage are designated as rot. Some examples are: "green rot" causing greening of albumen due to growth of *Pseudomonas fluorescens*; "black rot" causing muddy discoloration of yolk due to H_2S production by *Proteus vulgaris*; "red rot" by *Serratia mercescens* due to red pigment production. On some occasions, molds from the genera *Penicillium, Alternaria,* and *Mucor* can grow inside eggs, especially when the eggs are oiled, and produce different types of "fungal rot."

B. Egg Products

Liquid eggs consisting of whole, yolk, and white are generally pasteurized and frozen to prevent microbial growth. If the liquid products are held at room temperature following breaking prior to pasteurization, spoilage bacteria can grow and cause off-flavor (putrid), sourness, or fish flavor (due to formation of trimethylamine). Pasteurized eggs at refrigerated temperature have limited shelf-life, unless additional preservatives are added. The predominant bacteria in pasteurized products can be some Gram-positive bacteria surviving pasteurization, but predominantly, spoilage is caused by the psychrotrophic Gram-negative bacteria getting into products after heat treatment. Dried eggs are not susceptible to microbial spoilage due to low A_w.

IV. FISH, CRUSTACEANS, AND MOLLUSKS

A. Fish

Fish harvested from both fresh and saltwater are susceptible to spoilage through autolytic enzyme actions, oxidation of unsaturated fatty acids, and microbial growth.

Protein hydrolysis by autolytic enzymes (proteinases) is predominant if the fish are not gutted following catch. Oxidation of unsaturated fatty acids is also high in fatty fish. Microbial spoilage is determined by the microbial types, their level, fish environment, fish types, methods used for harvest, and subsequent handling. These aspects have been discussed previously (see Chapter 4). Fish tissues have high levels of nonprotein nitrogenous compounds (free amino acids, trimethylamine oxide, and creatinine), peptides, and proteins, but almost no carbohydrates; the pH is generally above 6.0. Gram-negative aerobic rods, such as *Pseudomonas* spp., *Acinetobacter, Moraxella, Flavobacterium,* and facultative anaerobic rods like *Shewanella, Alcaligenes, Vibrio,* and coliforms constitute the major spoilage bacteria. However, due to the relatively shorter generation time, spoilage due to psychrotrophic *Pseudomonas* spp. predominates under aerobic storage at both refrigerated and slightly higher temperatures. In fish stored under vacuum or CO_2, lactic acid bacteria can become predominant.

Gram-negative rods initially metabolize the nonprotein nitrogenous compounds by decay (oxidation), followed by putrefaction to produce different types of volatile compounds such as NH_3, trimethylamine ($N:CH_3$; from reduction of trimethylamine oxide), histamine (from histidine; cause of Scombroid poisoning), putrescine, cadaverine, indoles, H_2S, mercaptans, dimethyl sulfide (especially by *Shewanella putrefaciens*), and volatile fatty acids (acetic, isobutyric, or isovaleric acids). The proteolytic bacterial species also produce extracellular proteinases that hydrolyze fish proteins and supply peptides and amino acids for further metabolism by the spoilage bacteria. The volatile compounds will produce different types of off-odor, namely stale, fishy (due to trimethylamine), and putrid. Bacterial growth is also associated with slime production, discoloration of gills and eyes (in whole fish), and loss of muscle texture (soft due to proteolysis).

In fish stored by vacuum or CO_2 packaging, growth of aerobic spoilage bacteria is prevented. However, anaerobic and facultative anaerobic bacteria can grow, including lactic acid bacteria. Under refrigeration, the products have a relatively long shelf-life due to the slower growth of the spoilage bacteria. Salted fish, especially lightly salted fish, are susceptible to spoilage by halophilic bacteria, such as *Vibrio* (at lower temperature) and *Micrococcus* (at higher temperature). Smoked fish, especially with lower A_w, will inhibit growth of most bacteria. However, molds can grow on the surface. Minced fish flesh, surimi, and seafood analogs prepared from fish tissue generally have high initial bacterial levels due to extensive processing (about 10^{5-6}/g). The types include those present in fish and those that get in during processing. These products, like fresh fish, can spoil rapidly by Gram-negative rods, unless frozen rapidly or used soon after thawing. Canned fish (tuna, salmon, and sardines) are given heat treatment to produce commercially sterile products. They can be spoiled by thermophilic sporeformers unless proper preservation and storage conditions are used.[1-4]

B. Crustaceans

Microbial spoilage of shrimp is more prevalent than that of crabs and lobsters. While crabs and lobsters remain alive until they are processed, shrimp die during harvest. The flesh of crustaceans is rich in nonprotein nitrogenous compounds (amino acids, especially arginine, and trimethylamine oxide), contains about 0.5% glycogen and has a pH above 6.0. The predominant microflora are *Pseudomonas* and several Gram-negative rods. If other necessary factors are present, the nature of spoilage will be quite similar to that in fresh fish. Microbial spoilage of shrimp is dominated by odor changes due to production of volatile metabolites of nonprotein nitrogenous compounds (from decay and putrefaction), slime production, and loss of texture (soft) and color. If the shrimp are processed and frozen rapidly, the spoilage can be minimized. Lobsters are frozen following processing or sold live and thus are not generally exposed to spoilage conditions. Crabs, lobsters, and shrimp are also cooked to extend their shelf life. However, they are subsequently exposed to conditions that cause post-heat contamination and then stored at low temperature (refrigerated and frozen). Blue crabs are steamed under pressure and the meat is picked and marketed as fresh crab meat. To extend shelf-life (and safety), the meat is also heat processed (85°C for 1 min) and stored at refrigerated temperature.

C. Mollusks

As compared to fish and crustaceans, oyster, clam, and scallop meats are lower in nonprotein nitrogenous compounds but higher in carbohydrates (glycogen, 3.5 to 5.5%) with pH normally above 6.0. The mollusks are kept alive until processed (shucked); thus, microbiological spoilage occurs only after processing. The resident microflora are predominantly *Pseudomonas* and several other Gram-negative rods. During refrigerated storage, microorganisms metabolize both nonprotein nitrogenous compounds and carbohydrates. Carbohydrates can be metabolized to produce organic acids by lactic acid bacteria (*Lactobacillus* spp.), enterococci, and coliforms with the lowering of pH. Breakdown of nitrogenous compounds primarily by *Pseudomonas* and *Vibrio*, especially at refrigerated temperature, results in the production of NH_3, amines, and volatile fatty acids.

V. MILK AND MILK PRODUCTS

A. Raw Milk

Raw milk contains many types of microorganisms coming from different sources. The average composition of cow's milk is 3.2% protein, 4.8% carbohydrates, 3.9% lipids, and 0.9% minerals. Besides casein and lactalbumin, it has free amino acids that provide a good N-source (and some C-source, if necessary). As the main carbohydrate is lactose, those microorganisms with lactose-hydrolyzing enzymes (lactase or β-galactosidase) have an advantage over those unable to metabolize lactose. Milk fat can be hydrolyzed by microbial lipases, with the release of small molecular volatile fatty acids (butyric, capric, and caproic acids).

Microbial spoilage of raw milk can potentially occur from the metabolism of lactose, proteinaceous compound, fatty acids (unsaturated), and the hydrolysis of triglycerides. If the milk is refrigerated immediately following milking and stored for days, the spoilage will be predominantly caused by the Gram-negative psychrotrophic rods, such as *Pseudomonas, Alcaligenes, Flavobacterium* spp., and some coliforms. *Pseudomonas* and related species, being lactose-negative, will metabolize proteinaceous compounds to change the normal flavor of milk to bitter, fruity, or unclean. They also produce heat-stable lipases (producing rancid flavor) and heat-stable proteinases that have important implications; this is discussed separately (Chapter 19). The growth of lactose-positive coliforms will produce lactic, acetic, and formic acids, CO_2, and H_2 (by mixed acid fermentation) and will cause curdling, foaming, and souring of milk. Some *Alcaligenes* spp. (*Alc. viscolactis*) and coliforms can also cause ropiness (sliminess) due to production of viscous polysaccharides. However, if the raw milk is not refrigerated soon, growth of mesophiles predominates. These include species of *Lactococcus, Lactobacillus, Enterococcus, Micrococcus, Bacillus, Clostridium,* and coliforms, along with *Pseudomonas, Proteus,* and others. However, growth of lactose-hydrolyzing species, such as *Lactococcus* spp., will generally predominate, with the production of enough acid to lower the pH considerably and prevent or reduce growth of others. In such situations, curdling of milk and sour flavor will be the predominant spoilage. If other microorganisms also grow, gas formation, proteolysis, and lipolysis will become evident. Yeast and mold growth, under normal conditions, is generally not expected.[1-3,7]

B. Pasteurized Milk

Raw milk is pasteurized before it is sold for consumption as liquid milk. Thermoduric bacteria (*Micrococcus, Enterococcus,* some *Lactobacillus*, *Streptococcus, Corynebacterium,* and spores of *Bacillus* and *Clostridium*) will survive the process. In addition, coliforms *Pseudomonas, Alcaligenes, Flavobacterium,* and similar types can enter as post-pasteurization contaminants. Pasteurized milk, under refrigerated storage, has a limited shelf life, mainly due to growth of these psychrotrophic contaminants. Their spoilage pattern is the same as described for raw milk spoilage. Flavor defects from their growth are detectable when the population reaches $\geq 10^6$ cells/ml. Growth of psychrotrophic *Bacillus* spp., such as *Bac. cereus*, has been implicated in the spoilage of pasteurized refrigerated milk, especially when the levels of post-pasteurization contaminants are low. Spores of psychrotrophic *Bacillus* spp., surviving pasteurization, germinate and multiply to cause a defect known as "bitty." They produce the enzyme lecithinase, which hydrolyzes phospholipids of the fat globule membrane, causing aggregation of fat globules that adhere to the container surfaces. Production of rennin-like enzymes by the psychrotrophs can cause sweet curdling of milk at higher pH than required for acid curdling.

Ultrahigh temperature-treated milk (150°C for a few seconds) is an essentially commercially sterile product that can only contain viable spores of some thermophilic bacteria. The milk is not susceptible to spoilage at ambient storage temperature, but can be spoiled if exposed to high temperatures as such with canned foods.

C. Concentrated Liquid Products

Evaporated milk, condensed milk, and sweetened condensed milk are principal types of concentrated dairy products that are susceptible to limited microbial spoilage during storage. All these products are subjected to sufficient heat treatments to kill vegetative microorganisms as well as spores of molds and some bacteria.

Evaporated milk is condensed whole milk with 7.5% milk fat and 25% total solids. It is packaged in hermetically sealed cans and heated to obtain commercial sterility. Under proper processing conditions, only thermophilic spores of spoilage bacteria can survive, and exposure to high storage temperatures (43°C or higher) can trigger their germination and subsequent growth. Under such conditions, *Bacillus* species, such as *Bac. coagulans,* can cause coagulation of milk (flakes, clots, or a solid curd).

Condensed milk is generally condensed and has about 10 to 12% fat and 36% total solids. The milk is initially given a low heat treatment, close to pasteurization temperature, and then subjected to evaporation under partial vacuum (at about 50°C). Thus it can have thermoduric microorganisms that subsequently can grow and cause spoilage. Other microorganisms can also get into the product during the condensing process. Even at refrigerated temperature, this product has a limited shelf-life, as does pasteurized milk.

Sweetened condensed milk contains about 8.5% fat, 28% total solids, and 42% sucrose. The milk is initially heated to a high temperature (80 to 100°C) and then condensed at about 60°C under vacuum and put into containers. Due to low A_w, it is susceptible to spoilage from the growth of osmophilic yeasts (such as *Torula* spp.), causing gas formation. If the containers have enough head space and oxygen, molds (e.g., *Penicillium* and *Aspergillus*) can grow on the surface.

D. Butter

Butter contains 80% milk fat and can be salted or unsalted. The microbiological quality of butter depends upon the quality of cream and the sanitary conditions used in the processing. Growth of bacteria (*Pseudomonas* spp.), yeasts (*Candida* spp.), and molds (*Geotrichum candidum*) on the surface have been implicated in flavor defects (putrid, rancid, or fishy) and surface discoloration. In unsalted butter, coliforms, *Enterococcus,* and *Pseudomonas* can grow favorably in water-phase and produce flavor defects.

VI. VEGETABLES AND FRUITS

A. Vegetables

Fresh vegetables contain microorganisms coming from soil, water, air, and other environmental sources, and can include some plant pathogens. They are fairly rich in carbohydrates (5% or more), low in proteins (about 1 to 2%), and, except for tomatoes, have high pH. Microorganisms grow more rapidly in damaged or cut vegetables. The presence of air, high humidity, and higher temperature during storage increases the chances of spoilage. The most common spoilage is caused by different types of molds; some of those are from the genera *Penicillium, Phytopthora, Alternaria, Botrytis,* and *Aspergillus*. Among the bacterial genera, species from *Pseudomonas, Erwinia, Bacillus,* and *Clostridium* are important.

Microbial vegetable spoilage is generally described by the common term *rot*, along with the changes in the appearance, such as black rot, gray rot, pink rot, soft rot, stem-end rot. In addition to changes in color, microbial rot causes loss of texture and off-odor.

Refrigeration, vacuum or modified atmosphere packaging, freezing, drying, heat treatment, and chemical preservatives are used to reduce microbial spoilage of vegetables. Spoilage of canned vegetables, vegetable juices, and fermented vegetables are discussed later in this chapter.[1-3]

B. Fruits

Fresh fruits are high in carbohydrates (generally 10% or more), very low in proteins ($\leq$1.0%), but have pH 4.5 or below. Thus microbial spoilage of fruits and fruit products is confined to molds, yeasts, and aciduric bacteria (lactic acid bacteria, *Acetobacter, Gluconobacter)*. Like fresh vegetables, fresh fruits are susceptible to rot by different types of molds from genera *Penicillium, Aspergillus, Alternaria, Botrytis, Rhizopus,* and others. According to the changes in appearance, the mold spoilages are designated as black rot, gray rot, soft rot, brown rot, and others. Yeasts from genera *Saccharomyces, Candida, Torulopsis,* and *Hansenula* have been associated with fermentation of some fruits such as apples, strawberries, citrus fruits, and dates. Bacterial spoilage associated with the souring of berries and figs has been attributed to the growth of lactic acid and acetic acid bacteria.

To reduce spoilage, fruits and fruit products are preserved by refrigeration, freezing, drying, reducing A_w, and heat treatment. Spoilage of fruit juices, jams, and wine is discussed later.[1-3]

VII. SOFT DRINKS, FRUIT JUICES AND PRESERVES, AND VEGETABLE JUICES

Carbonated and noncarbonated soft drinks, fruit juices, and preserved and concentrated fruit juices and drinks are low pH products (2.5 to 4.0). The carbohydrate (sucrose, glucose, and fructose) content ranges from 5 to 15% in juices and drinks, but 40 to 60% in concentrates and preserves. High sugar content reduces the A_w of these products, which in the concentrates and preserves can be about 0.9. Carbonated beverages also have low O–R potentials.

Among the microorganisms that can be present in these products, only aciduric molds, yeasts, and bacteria (*Lactobacillus, Leuconostoc,* and *Acetobacter*) are able to cause spoilage if appropriate preservation methods are not used. In the carbonated beverages, some yeast species from genera *Torulopsis, Candida, Pichia, Hansenula,* and *Saccharomyces* can grow and make the products turbid. Some *Lactobacillus* and *Leuconostoc* species can also grow to cause cloudiness and ropiness (due to production of dextrans) of the products. Noncarbonated beverages can be similarly spoiled by the yeasts, *Lactobacillus* and *Leuconostoc* spp. In addition, if there is enough dissolved oxygen, molds (*Penicillium, Aspergillus, Mucor,* and *Fusarium*) and *Acetobacter* can grow; the latter will produce acetic acid to give vinegar-like flavor. Fruit juices are susceptible to spoilage by molds, yeasts, *Lactobacillus, Leuconostoc,* and *Acetobacter* spp. However, a particular type of juice may be susceptible to spoilage by one or another type of microorganism. Molds and *Acetobacter* can grow if enough dissolved oxygen is available. Yeasts can cause both oxidation (CO_2 and H_2O) and fermentation (alcohol and CO_2) of the products. *Acetobacter* can use alcohol to produce acetic acid. Heterofermentative *Lactobacillus fermentum* and *Leuconostoc mesenteroides* can ferment carbohydrates to lactate, ethanol, acetate, CO_2, diacetyl, and acetoin. In addition, *Leu. mesenteroides* and some strains of *Lab. plantarum* can produce slime due to dextran production. In fruit drinks, *Lactobacillus* and *Leuconostoc* spp. can also convert citric and malic acids (additives) to lactic and acetic acids and reduce the sour taste (flat flavor). In concentrated fruit drinks and fruit preserves, due to low A_w (0.9), only osmophilic yeasts can grow; molds can also grow if oxygen is available.

To prevent growth of these potential spoilage microorganisms, several additional preservation methods are used for these products; these include heat treatment, freezing, refrigeration, and addition of specific chemical preservatives.

Tomato juice has a pH of about 4.3. It is generally given high heat treatment to kill vegetative microorganisms. However, bacterial spores can survive. Flat sour spoilage of tomato juice due to germination and growth of *Bacillus coagulans* has been documented. Most other vegetable juices have pH values between 5.0 and 5.8 and many have growth factors for lactic acid bacteria. These products are susceptible to spoilage due to the growth of many types of microorganisms. Effective preservation methods are used to control their growth.

VIII. CEREALS AND THEIR PRODUCTS

Some of the products susceptible to microbial spoilage include high-moisture cereal grains, refrigerated dough, breads, soft pastas, and pastries.[1,2]

A. Cereal Grains

Grains normally have 10 to 12% moisture, which lowers the A_w to ≤ 0.6 and thus inhibits microbial growth. However, during harvesting, processing, and storage, if the A_w increases above 0.6, some molds can grow. Some species of storage fungi from genera *Aspergillus*, *Penicillium,* and *Rhizopus* can cause spoilage of high-moisture grains.

B. Refrigerated Dough

Refrigerated dough (for biscuits, roles, and pizza) are susceptible to spoilage (gas formation) from the growth of psychrotrophic heterolactic species of *Lactobacillus* and *Leuconostoc*. Rapid CO_2 production can blow the containers, especially when the storage temperature increases to 10°C and above.

C. Breads

Breads normally have low enough A_w (0.75 to 0.9) to prevent growth of bacteria. However, some molds (bread molds: *Rhizopus stolonifer*) can grow, especially if moisture is released due to starch crystallization during storage. Molds are killed during baking; however, spores can get in from air and equipment following baking. When breads are frozen, they may contain ice crystals in the bags. Following thawing, some portions can absorb enough moisture for yeasts and bacteria to grow and cause spoilage (sour taste, off flavor). A specific type of bread spoilage, designated as ropiness and characterized by soft, stringy, brown mass with fruity odor, is caused by the growth of some mucoid variants of *Bacillus subtilis*. The spores, coming from flour or equipment, survive baking and then germinate and grow inside within 1 to 2 d. They also produce extracellular amylases and proteases and break down the bread structure. High moisture inside the bread, slow cooling and pH above 5.0 favor ropiness.

D. Pastas

Pastas can be spoiled by microorganisms prior to drying due to improper manufacturing practices. Dry pastas do not favor microbial growth. However, soft pastas can be spoiled by bacteria, yeasts, and molds. Anaerobic packing and refrigeration storage can prevent mold growth and slow down the growth of yeasts and anaerobic and facultative anaerobic psychrotrophic bacteria. Suitable preservatives can be used to prevent their growth.

E. Pastries

Pastries include cakes and baked shells filled with custard, cream, or sauces. They can be spoiled by microorganisms coming with the ingredients that are added after baking such as icing, nuts, toppings, and cream. Most products, due to low A_w, will allow only molds to grow. However, some materials used as fillings may have high A_w, which allows for bacterial growth.

IX. LIQUID SWEETENERS AND CONFECTIONIERIES

Liquid sweeteners include honey, sugar syrups, maple syrups, corn syrups, and molasses. Confectionery products include soft-centered fondant, cream, jellies, chocolate, and Turkish delight. Most of these products have an A_w of 0.8 or below and are normally not susceptible to bacterial spoilage. Under aerobic conditions, some xerophilic molds can produce visible spoilage. However, osmophilic yeasts from genera *Zygosaccharomyces (Zyg. rouxii), Saccharomyces (Sac. cerevisiae), Torulopsis (Tor. holmii),* and *Candida (Can. valida)* can ferment these products.

To prevent growth of yeasts in some of these products with slightly higher A_w (such as in maple syrup), chemical preservatives are added.

X. MAYONNAISE, SALAD DRESSINGS, AND CONDIMENTS

These products normally contain some molds, yeasts, spores of *Bacillus* and *Clostridium,* and aciduric bacteria, such as *Lactobacillus* and related species. Due to low pH, acid-sensitive bacteria may not survive long. Mayonnaise, with 65% or more edible oil and about 0.5% acetic acid, has A_w about 0.92 and pH 3.6 to 4.0. Salad dressings generally contain 30% or more edible oil, 0.9 to 1.2% acetic acid, A_w 0.92, and pH 3.2 to 3.9. The major factors controlling microbial growth are undissociated acetic acid, low pH, and relatively low A_w. However, some aciduric microorganisms can grow to cause spoilage. Molds can grow only on the surface exposed to air. Some microaerophilic and facultative anaerobic yeasts and heterolactic *Lactobacillus* spp. (especially those that can grow at A_w 0.92) can also multiply. *Lab. fructivorans* hydrolyzes sucrose present in the products and produces gas (CO_2), especially from the rapid metabolism of fructose that is released from sucrose. Strains of *Saccharomyces bailii* have also been implicated in gas spoilage (CO_2 and alcohol) of these products. *Lab. fructivorans* cells usually die rapidly after multiplication and are difficult to isolate unless specific methods are used. This can lead to a wrong assumption as to which microorganism(s) is the causative agent(s) for the spoilage of these products.

Low-calorie salad dressing, in which oil and acetic acid are added in much lower concentrations, have high pH and A_w values. Many microorganisms can grow in these products. To enhance their shelf-life (and safety), refrigerated storage is recommended.

Yeasts and *Lactobacillus* spp. are also capable of causing gassy spoilage of catsup, salsa, sauces, and prepared mustard. Some *Bacillus* spp. have also been implicated with gassy spoilage of mustard preparations. To control growth of spoilage microorganisms, additional preservation methods, especially chemical preservatives, are used.[1,2,9]

XI. FERMENTED FOODS

Desirable microorganisms are used directly or indirectly to produce many types of fermented foods and beverages from meat, fish, milk, vegetables, fruits, cereal grains, and others (Chapter 13). The desirable microorganisms are present in very high numbers and the products contain either high levels of organic acids or alcohol. In addition, the products have low pH and some have low A_w (e.g., dry salami).

Generally, these products have a longer shelf-life. However, under certain conditions, they are susceptible to microbial spoilage. This aspect is discussed briefly here (also see Chapter 13).[1,2]

A. Fermented Meat Products

[See Chapter 13.] Fermented meat products normally have a pH ranging between 4.5 and 5.0 and have A_w between 0.73 and 0.93. During fermentation, if the acid production of homofermentative lactic acid bacteria is slow, undesirable bacteria can grow. *Clostridium, Bacillus,* and other mesophilic bacteria have been reported to cause spoilage in such conditions. Products with pH lower than 5.0 but A_w 0.92 or above and vacuum packaged can be spoiled by heterofermentative *Leuconostoc* and *Lactobacillus* spp. with the accumulation of gas and liquid inside the package and creamy white growth of bacterial cells. If they are not vacuum packaged and have low A_w (0.72 to 0.90), yeasts and molds can grow on the surface, resulting in slime formation, discoloration, and undesirable flavor of the products.

B. Fermented Dairy Products

[See Chapter 13]. Cultured buttermilk, yogurt, and cheeses are a few of the fermented dairy products generally produced by inoculating milk with specific starter culture bacteria. They differ in their acidity, A_w, and storage stability. Buttermilk generally has about 0.8% lactic acid and a pH of 4.8. Yeasts can cause spoilage with production of gas. Some strains of starter cultures can produce polysaccharides to give a slimy texture.

Yogurt generally has a pH 4.5 or lower (with about 1% lactic acid) and is not spoiled by undesirable bacteria. However, the product can develop a bitter flavor due to the production of bitter peptides by some strains of *Lactobacillus delbrueckii* subsp. *bulgaricus* used as a starter culture. During storage, the starter bacteria can continue to produce lactic acid, causing an objectionable sharp acid taste. Yeasts (especially in fruit yogurts) may grow in the acid environment and produce CO_2 as well as yeasty and fruity off-flavors. Some species of molds can grow on the surface if yogurt is stored for a long time.

Microbial cheese spoilage is greatly influenced by A_w and pH. Unripened cottage cheese with high moisture content and low acidity is susceptible to spoilage by Gram-negative psychrotrophic rods, yeasts, and molds. *Alcaligenes* and *Pseudomonas* spp. are frequently involved, causing a slimy texture and an unclean, putrid flavor. Some unripened ethnic soft cheeses (such as Mexican-style) are vacuum packaged and stored at refrigerated temperature for a shelf life of about 60 d (see Chapter 20). They are occasionally spoiled by heterofermentative *Leuconostoc* spp., characterized by gas (CO_2) and liquid accumulation in the bag. Gassiness in some cheeses with high pH, low salt, and relatively high A_w (such as Gouda, Emmentaler, and Provolone) can also occur from the growth of some *Clostridium*, (e.g., *Clo. tyrobutyricum*). Their spores survive pasteurization of milk, germinate, and grow in the anaerobic environment to produce CO_2, H_2, and butyrate from the metabolism of lactate. Hard-ripened cheese, such as cheddar, can have a bitter taste due to rapid production of bitter peptides during ripening. Fast acid-producing strains of *Lactococcus lactis* used as starter are generally associated with this defect. Sharp ripened cheeses can also have large amounts of biologically active amines (e.g., histamine and tyramine), produced from the decarboxylation of the respective amino acids. Decarboxylase enzymes can be present in some starter strains or in secondary microflora of the cheeses (*Enterococcus,* some coliforms). Lysis of the cells releases the enzymes during the ripening process, causing decarboxylation of amino acids and accumulation of these amines. Hard and semi-hard cheeses are generally susceptible to spoilage from mold growth on the surface and produce undesirable color and flavor defects in the products. Anaerobic packaging greatly reduces this problem.

C. Fermented Vegetable and Fruit Products

Many types of vegetables are fermented, among which cucumber and sauerkraut are produced in large volumes. In salt stock pickles containing about 15% salt, yeasts and halophilic bacteria can grow, especially if the acidity is not sufficient. Dill pickles with low salt (<5%) can have a bloating defect from CO_2 production by yeasts, heterofermentative lactic acid bacteria, and coliforms, especially if the desirable bacteria associated with fermentation do not grow properly. *Candida, Torulopsis,* and *Saccharomyces* spp. are often the causative yeasts. Sweet and sour pickles preserved with sugar and vinegar can be spoiled from the growth of yeasts and lactic acid bacteria, especially when the acid level is not sufficient. Sauerkraut can be spoiled from the growth of yeasts and molds if the air is not excluded during fermentation of cabbage (also see Chapter 13). Failure of lactic acid bacteria to grow rapidly and in proper sequence can lead to low acid production. Under this condition, coliforms and other Gram-negative bacteria can multiply to produce undesirable flavor and texture and color defects. Olives are fermented for a long time and are susceptible to many types of spoilage. The most common problem is gassiness (bloating) due to CO_2 production by heterofermentative lactic acid bacteria, coliforms, and yeasts. Softening of texture can be caused by pectinases of yeasts (*Rhodotorula* spp.).

D. Fermented Beverages

Wines can support growth of film yeasts and acetic acid bacteria (*Acetobacter* and *Gluconobacter*) under aerobic conditions. Film yeasts oxidize alcohol and organic acids and form surface pellicle, while acetic acid bacteria oxidize alcohol to acetic acid and CO_2. Under anaerobic conditions, several lactic acid bacteria can grow (e.g., *Lactobacillus, Leuconostoc,* and *Pediococcus* spp.) in wine. Some heterofermentative *Lactobacillus* spp. can ferment glucose and fructose and increase acidity of the wine, producing a defect called "tourne spoilage" (*Lab. brevis, Lab. buchneri*). They can also produce cloudiness and a mousy odor. Some *Leuconostoc* spp. can produce sliminess and cloudiness. *Leuconostoc oenos* can convert malic acid to lactic acid + CO_2 and reduce the acidity of a wine. Sometimes this malolactic fermentation is used advantageously to reduce the sourness of wine.

Beer spoilage can be caused by some lactic acid bacteria and yeasts. Growth of *Pediococcus* spp. causes an increase in acidity and cloudiness. Some *Lactobacillus* spp. can also multiply and cause turbidity. *Acetobacter* and *Gluconobacter*, in the presence of air, can produce cloudiness and sliminess as well as make the beer sour. Wild yeasts (any yeast other than that used in fermentation) can grow in beer, causing off-flavor.

XII. CANNED FOODS

[See Chapter 4 and References 1, 2, and 10.]

Canned foods are heat treated to kill microorganisms present and the extent of heat treatment is predominantly dependent on the pH of a food. High pH (above 4.6; also called low acid) foods are heated to destroy most heat-resistant spores of pathogenic bacteria, *Clostridium botulinum*, to ensure that a product is free of any pathogen. However, spores of some spoilage bacteria, which have greater heat resistance than the spores of *Clo. botulinum*, can survive. Thus these products are called commercially sterile (instead of sterile, which means free of any living system) foods. The spores that survive the heat treatment designed to destroy *Clo. botulinum* spores are thermophilic and can germinate at 43°C and above. However, once germinated, some can outgrow and multiply at temperatures as low as 30°C. The other food group, designated as low pH or high acid food with pH 4.6 and below, is given heat treatment to kill all vegetative cells and some spores. Although low pH will inhibit germination and growth of *Clo. botulinum*, spores of some aciduric thermophilic spoilage bacteria can germinate and grow when the products are stored at higher temperatures, even for a short time, which facilitates germination. Some spores of thermoduric mesophilic spoilage bacteria (including pathogenic) can also survive heating in these products, but they are inhibited by the low pH.

Canned food spoilage is due both to nonmicrobial (chemical and enzymatic reactions) and microbial reasons. Production of hydrogen (hydrogen swell), CO_2, browning, corrosion of cans due to chemical reactions and liquification, gelation, discoloration of products due to enzymatic reactions are some examples of nonmicrobial spoilage. Microbial spoilage is due to three main reasons: (1) inadequate cooling after heating or high-temperature storage, allowing germination and growth of thermophilic sporeformers; (2) inadequate heating, resulting in survival and growth of mesophilic microorganisms (vegetative cells and spores); and (3) leakage (can be microscopic) in the cans, allowing microbial contamination from outside following heat treatment and their growth.

A. Thermophilic Sporeformers

Thermophilic sporeformers can cause three types of spoilage of low-acid (high pH) foods (such as corn, beans, peas) when the cans are temperature abused at 43°C and above, even for short duration.

1. Flat Sour Spoilage

The cans do not swell but the products become acidic due to germination and growth of facultative anaerobic *Bacillus stearothermophilus*. Germination occurs at high temperature (43°C and above) but growth can take place at lower temperature (30°C). The organism ferments carbohydrates to produce acids without gas.

2. Thermophilic Anaerobe (TA) Spoilage

The spoilage is caused by the growth of anaerobic *Clostridium thermosaccharolyticum* with the production of large quantities of H_2 and CO_2 gas and swelling of cans. Following germination in the thermophilic range (43°C or above), the cells can grow at lower temperatures (30°C and above).

3. Sulfide Stinker Spoilage

This spoilage is caused by the Gram-negative anaerobic sporeformer *Desulfotomaculum nigrificans*. The spoilage is characterized by flat container but darkened products with the odor of rotten eggs due to H_2S produced by the bacterium. H_2S produced from the sulfur-containing amino acids dissolves in the liquid and reacts with iron to form the black color of iron sulfide. Both germination and growth occur in the thermophilic range (43°C and above).

B. Spoilage Due to Insufficient Heating

Insufficient heat treatment results in the survival of mainly spores of *Clostridium* and some *Bacillus* spp. Following processing, they can germinate and grow to cause spoilage. The most important concern is the growth of *Clo. botulinum* and production of toxins.

Spoilage can be either from the breakdown of carbohydrates or proteins. Several *Clostridium* spp., *Clo. butyricum* and *Clo. pasteurianum*, ferment carbohydrates to produce volatile acids and H_2 and CO_2 gas, causing swelling of cans. Proteolytic species, *Clo. sporogenes* and *Clo. putrefaciens* (also proteolytic *Clo. botulinum*), metabolize proteins and produce foul-smelling H_2S, mercaptans, indole, skatole, ammonia, as well as CO_2 and H_2 (causing swelling of cans).

Spores of aerobic *Bacillus* spp., surviving inadequate heating, will not grow in cans. However, spores of some facultative anaerobic *Bacillus* spp., such as *Bac. subtilis* and *Bac. coagulans*, will grow with the production of acid and gas (CO_2).

C. Spoilage Due to Container Leakage

Damaged and leaky containers will allow different types of microorganisms to get inside from the environment after heating. They can grow in the food and cause different types of spoilage depending upon the microbial types. Contamination with pathogens will make the product unsafe.

REFERENCES

1. Silliker, J. H., Ed., *Microbial Ecology of Foods,* Vol. 2, Academic Press, New York, 1980.
2. Vanderzant, C. and Splittstoesser, D. F., Eds., *Compendium of Methods for the Microbiological Examination of Foods,* American Public Health Association, Washington, DC, 1992, chap. 44 to 61.
3. Kraft, A. A., *Psychrotrophic Bacteria in Foods: Diseases and Spoilage,* CRC Press, Boca Raton, FL, 1992.
4. Sofos, J. N., Microbial growth and its control in meat, poultry, and fish, in *Advances in Meat Research,* Vol. 9, Pearson, A. M. and Dutson, T. R., Eds., Chapman Hall, New York, 1994, 359.
5. Gill, C. O., The control of microbial spoilage in fresh meats, in *Advances in Meat Research,* Vol. 2, Pearson, A. M. and Dutson, T. R., Eds., AVI Publishing, Westport, CN, 1986, 49.
6. Tompkins, R. B., Microbiology of ready-to-eat meat and poultry products, in *Advances in Meat Research,* Vol. 2, Pearson, A. M. and Dutson, T. R., Eds., AVI Publishing, Westport, CN, 1986, 89.
7. Cousin, M. A., Presence and activity of psychrotrophic microorganisms in milk and dairy products: a review, *J. Food Prot.,* 45, 172, 1982.
8. Meer, R. R., Baker, J., Bodyfelt, F. W., and Griffiths, M. W., Psychrotrophic *Bacillus* spp. in fluid milk products, *J. Food Prot.,* 54, 969, 1991.
9. Smittle, R. B. and Flowers, R. M., Acid tolerant microorganisms involved in the spoilage of salad dressings, *J. Food Prot.,* 45, 977, 1982.
10. Anonymous, Thermophilic organisms involved in food spoilage, *J. Food Prot.,* 44, 144, 1981.

CHAPTER 18 QUESTIONS

1. List the important psychrotrophic spoilage bacteria of raw meats, and describe the metabolic pattern and corresponding spoilage associated with each microbial genus.

2. Briefly discuss the following: (a) sources of microorganisms in low heat-processed meat products; (b) predominant spoilage bacteria in low heat-processed, vacuum-packaged meat products stored at ≤4°C; and (c) factors associated with greening of unpackaged and packaged meat and meat products during refrigerated storage.

3. List four microbial inhibitors in shell eggs and predominant spoilage microflora of eggs and egg products.

4. Briefly explain the following: (a) metabolism of nutrients by predominant spoilage bacteria while growing in fish tissues; (b) post-heat contamination and spoilage of crab meat; and (c) metabolism of nutrients by the spoilage bacteria in shucked oyster.

5. List the predominant microflora of pasteurized milk and their sources. Discuss the reason(s) of limited shelf-life of pasteurized milk.

6. Discuss the predisposing factors associated with microbial spoilage of UHP-treated milk, sweetened condensed milk, and butter.

7. List bacteria, yeasts, and molds (two of each) associated with the spoilage of vegetables and fruits. Discuss the major differences in spoilage by bacteria of these two groups of food.

8. Discuss the influence of intrinsic factors associated with microbial spoilage of fruit drinks, fruit juices, and vegetable juices, and the nature of spoilage of these products produced by the predominant microflora.

9. List the similarities and differences in the microbial spoilage of: (a) breads, (b) pastries, (c) soft pastas, and (d) confectioneries.

10. List the microorganisms involved and the predisposing cause(s) for the following spoilage conditions: (a) gassy defect in mayonnaise; (b) gassy defect of prepared mustard; (c) gassy defect in unripened soft cheese (such as Mexican-style cheese) and Gouda cheese; (d) bitter taste in yogurt and cheddar cheese; (e) tourne spoilage of wine; (f) malolactic fermentation; and (g) sour beer.

11. List the major causes of spoilage of canned foods. Describe the factors, causative bacterial species, and the nature of spoilage of canned foods produced by thermophilic sporeformers.

CHAPTER 19

Food Spoilage by Microbial Enzymes

CONTENTS

I. INTRODUCTION

The metabolism of food nutrients, principally carbohydrates, nitrogenous compounds, and lipids, by spoilage bacteria enables the cells to increase in number as well as produce metabolites that can adversely reduce the acceptance quality of a food. A food is considered spoiled when these changes are detectable and the microbial population has reached about 10^{7-9}/ml, g, or cm^2. These changes are brought about by the catalytic actions of a large number of microbial enzymes. Most of the microbial enzymes are intracellular and act on the nutrients that are transported inside the cells through several transport mechanisms. Thus in a live microbial cell, most of the enzymes act on relatively small nutrient molecules that can be transported inside, although many intracellular enzymes are also capable of acting on intracellular large molecules, such as endonucleases, mucopeptidase, and proteinases. In addition to intracellular enzymes, many microbial cells also produce extracellular enzymes that after synthesis, either remain bound to the cell surface or are excreted in the food environment. Many of the latter group are capable of hydrolyzing large nutrient molecules of food (e.g., polysaccharides, proteins, and lipids) to small molecules, before they are transported into the cells.

Most foods have some amounts of small molecular metabolizable carbohydrates (mono- and disaccharides and their derivatives, such as glucose-6-phosphate), nitrogenous compounds (small peptides, amino acids, nucleosides, nucleotides, urea, creatinine, and trimethylamine oxide), free fatty acids, and some organic acids (lactic, citric, and malic acids). Many spoilage microorganisms, particularly spoilage bacteria, are able to utilize these low-molecular-weight food components to reach a population of $10^{7\text{-}9}$/g, ml, or cm^2 food and cause detectable food spoilage. Thus the supply of extra nutrients from the hydrolysis of the macromolecules of foods by bacterial extracellular enzymes is not necessary for the onset of spoilage in many foods. In fact, studies with *Pseudomonas* spp. and *Bacillus* spp. have shown that in the presence of low-molecular-weight nitrogenous compounds, the synthesis of extracellular proteinases is repressed. When the supply of these small molecules is used up, the mechanism is derepressed, leading to the synthesis and excretion of the extracellular proteinases. When that occurs, these proteinases hydrolyze the large protein molecules of food to produce small peptides and amino acids for transport and endogenous metabolism in the cells, which in turn intensify the spoilage. In general, microbial food spoilage from the metabolism of small molecular weight nutrients occurs at the early stage of microbial growth; spoilage from the breakdown of macromolecules by extracellular enzymes appears late in the sequence of events.

After microbial cells die normally or are killed by nonthermal treatments so that the intracellular and extracellular enzymes are not inactivated or destroyed, the enzymes can cause food spoilage even in the absence of viable cells or growth of microorganisms. Many microbial cells in a food that are subjected to freezing (then thawing), drying (then rehydration), modified atmosphere or vacuum packaging, refrigeration, high hydrostatic pressure, electric pulse field, high-intensity light, or exposure to some preservatives may die, and undergo lysis to release intracellular enzymes. If these foods are stored for a long time under conditions that favor catalysis of one or more intra- and extracellular enzymes, they can undergo spoilage. Ripening (not spoilage) of cheddar cheese (at low temperature and low A_w) is a good example where, following death, bacterial intracellular enzymes are involved in the breakdown of milk nutrients, not to cause spoilage but to impart desirable product characteristics. In most foods, the initial microbial population is generally low and spoilage by their enzymes in the absence of growth may not be of practical significance. However, if a product is heavily contaminated with a high initial microbial load (in the absence of growth) and then subjected to a nonthermal treatment(s) that kills microorganisms but does not inactivate enzymes, spoilage of the food by microbial enzyme(s) can occur. In thermally processed foods, several heat-stable enzymes of the microorganisms retain their activity even after the producer cells are killed. During subsequent storage of the food under favorable conditions, these enzymes can break down the food nutrients to cause spoilage. Among the heat-stable enzymes, some extracellular proteinases, lipases, and phospholipases of several psychrotrophic bacteria found in food are recognized as causing spoilage of thermally processed dairy products.[1,2]

II. CHARACTERISTICS OF HEAT-STABLE ENZYMES OF PSYCHROTROPHIC BACTERIA

Raw milk, between production and pasteurization in commercial operations, is usually stored for 1 to 2 weeks at refrigerated temperatures (<7°C). The psychrotrophic bacteria in the raw milk, coming from water, equipment, and environment, can multiply during storage. Depending upon initial numbers, bacterial species and strains, temperature and time of storage, and extent of temperature abuse, the population can reach a level where it can produce sufficient amounts of extracellular heat-stable enzymes. Even though the raw milk may not be spoiled, the heat-stable enzymes produced have the potential of causing the spoilage of heat-treated dairy products manufactured from this raw milk. Heating, such as pasteurization and ultrahigh heat treatments (UHT), kills the psychrotrophic bacteria but does not inactivate the heat-stable enzymes.

In the raw milk, heat-stable proteinases, lipases, and to some extent phospholipases produced by several psychrotrophic Gram-negative bacteria are considered to be of major economic importance due to the spoilage potential of the products. Species from genera *Pseudomonas (Pse. fluorescens, Pse. fragi), Aeromonas, Flavobacterium, Shewanella (Alteromonas), Serratia,* and *Acinetobacter* are known to produce heat-stable extracellular proteinases, and *Pseudomonas, Alcaligenes, Shewanella, Acinetobacter,* and *Serratia* produce heat-stable lipases. *Pseudomonas* spp. are also known to produce heat-stable phospholipases. Many species from these genera are normally present in raw milk, meat, fish, and other food products. During refrigerated storage, the psychrotrophs are able to grow and produce the heat-stable enzymes in foods.

Production of heat-stable proteinases and lipases in milk by some of these bacteria, especially *Pseudomonas* spp., and the characteristics of these enzymes have been well studied. The results show that different species and strains produce proteinases and lipases that differ in molecular weight and activity. Thus a highly active proteinase or lipase produced by a *Pseudomonas* strain can produce extensive proteolysis or lipolysis, even at a level of $10^{5\text{-}6}$ cells per milliliter milk. But another strain may need to reach 10^9 cells per milliliter to produce similar changes. The enzymes can be produced in raw milk at refrigerated temperature (1 to 7°C) in sufficient quantities to hydrolyze proteins and lipids in detectable levels within 3 to 7 d. The catalytic activity of the proteinases is maximum between pH 6.0 and 7.0, with a range between pH 5.0 and 9.0. Pasteurization of milk (at 63°C for 30 min or 71°C for 15 s) results in a loss of 6 to 36% activity and, even after heating at 121°C for 10 min, some activity of the proteases is retained. UHT treatment (140 to 150°C for 1 to 5 s) failed to completely inactivate proteinases produced by some species and strains of *Pseudomonas* and other psychrotrophs. UHT-treated milk, thus can be spoiled by the remaining activity of the proteinases during storage. Lipases are only partially inactivated by pasteurization or by heating (in cream) at

90°C for 2 min. They are generally inactivated by UHT treatment. Lipases from some *Pseudomonas* strains retain sufficient activity even after heating at 100°C for 10 min.

The heat-stable proteinases of psychrotrophic bacteria differ in their substrate specificity and rate of substrate degradation. Proteinases of *Pseudomonas* spp. preferentially degrade casein of milk by a different mechanism. While proteinases from some species-strains initially degrade β-casein, proteinases from other species-strains initially degrade k-casein. With time, they can also degrade other casein fractions. Proteinases from psychrotrophic species-strains of *Flavobacterium, Aeromonas,* and *Serratia* also showed initial differences in the degradation of β- and k-caseins. α-Casein was degraded last by all strains.

Lipases of psychrotrophic bacteria differ in their specificity toward lipids. A lipase from a strain of *Pse. fragi* specifically hydrolyzed fatty acids from positions 1 and 3 of the triglycerides, while another lipase from a second strain hydrolyzed only in position 1. Similar specificities have been observed with extracellular lipases from *Pse. fluorescens*. Some lipases of psychrotrophic bacteria hydrolyze equally all three fatty acids.

Among the phospholipases produced by psychrotrophic Gram-negative bacteria, phospholipase C, from *Pseudomonas* spp., has relatively high heat stability. It is not destroyed by pasteurization. Phospholipase C from *Pse. fluorescens* is more active against phosphatidylethanolamine than other phospholipids.[1-4]

III. SPOILAGE OF FOODS WITH HEAT-STABLE MICROBIAL ENZYMES

The presence of heat-stable extracellular enzymes of psychrotrophic bacteria in raw milk can cause spoilage of dairy products made from it. In addition, when these dairy products are used as ingredients to make other food products, the action of heat-stable enzymes can also reduce their acceptance qualities. Several examples are used here to emphasize the spoilage potential of these enzymes.[1,2,5-7]

A. Pasteurized Milk

Heat-stable proteinases and lipases of psychrotrophic bacteria are not inactivated by pasteurization and can cause proteolysis of casein and lipolysis of milk lipids to produce flavor defects. However, under normal short-term refrigerated storage, these defects may not be enough to detect. In addition, psychrotrophic bacteria contaminating milk after pasteurization can also multiply during storage and cause spoilage, especially when the milk is either stored for a long time or is temperature abused. In that event, it is difficult to differentiate the role of bacterial growth and heat-stable enzymes in spoilage.

B. Ultrahigh Temperature (UHT)-Treated Milk Products

UHT-treated milks, heated at 140 to 150°C for 1 to 5 s, are considered commercially sterile products with a shelf-life of 3 months at 20°C. Spoilage of these products during storage at 20°C has been observed in the form of bitter flavor and gel formation due to the action of heat-stable proteinases and rancid flavor from the action of heat-stable lipases. Generally, the changes produced by the proteinases are more predominant than the changes associated with the lipases. However, in the presence of heat-stable phospholipases, the lipolysis can be detected. It is speculated that phospholipases degrade the phospholipids in the membrane of the fat globules and increase the susceptibility of fat to lipases.

The time of spoilage of UHT-treated milk by heat-stable proteinases is dependent upon the numbers and strains of *Pseudomonas* species growing in the raw milk. In a controlled study, UHT milk was prepared from raw milk inoculated with a *Pseudomonas fluorescens* strain and grown to 8×10^5 to 5×10^7 cells per milliliter. During storage at 20°C, the products with 5×10^7 cells per milliliter gelled in 10 to 14 d; the products with 8×10^6 cells per milliliter gelled in 8 to 10 weeks; and the products with 8×10^5 cells per milliliter did not gel in 20 weeks but had sediments. Raw milk whose caseins have been highly degraded by the proteolytic enzymes may even be unstable to high heat treatment and thus cannot be used for the manufacture of UHT milk.

C. Cheeses

Proteolytic activity by the extracellular proteinases of psychrotrophic bacteria in raw milk was reported to reduce cheese yield and increase the levels of nitrogenous compounds in whey. Depending upon the proteolysis of caseins, the loss in cheese yield can be as high as 5%. The loss was found to be directly related to the storage time of the raw milk and psychrotrophic counts. In addition, the heat-stable proteolytic enzymes were associated with increased proteolysis of cheeses (especially soft cheeses such as cottage cheese), lower flavor quality, and higher texture problems in cheddar cheese. Lipases have also been implicated in the development of off-flavor in cheese.

D. Cultured Dairy Products

Buttermilk and yogurt, made from raw milk with substantial growth of psychrotrophic bacteria, generally have poor texture and rapidly develop off-flavor during storage, even though they have low pH and are stored at about 10°C. These defects were attributed to bacterial heat-stable proteinases.

E. Cream and Butter

Cream and butter are more susceptible to spoilage by heat-stable lipases than proteinases. Extracellular lipases of psychrotrophic bacteria preferentially partition with the cream phase of milk. The lipases are responsible for off-flavor development in cream. A cream that has undergone lipolysis, foams excessively and takes a longer time to churn (to make butter). The butter prepared from such cream is susceptible to developing rancidity more quickly.

Butter containing residual heat-stable bacterial lipases undergoes rapid lipid hydrolysis even during storage at –10°C. Those lipases that preferentially release short-chain fatty acids (C4 to C8), with and without long-chain fatty acids, cause the most off-flavor in butter.

F. Milk Powder

The activity of heat-stable bacterial proteinases and lipases present in raw and pasteurized milk will not be denatured during the manufacture of spray-dried milk powder. The low A_w will prevent these enzymes from degrading the proteins and lipids in dry milk. Powdered whole milk, nonfat milk, and whey are used as ingredients in a wide variety of foods that can have high amounts of proteins and lipids and are expected to have a long shelf-life. Bakery products, ice cream, desserts, meat products, chocolate, cheese products, and condensed dairy products are some of the foods in which dry milks are used. These products can develop off-flavor and texture defects from the action of the heat-stable proteinases and lipases during storage.

G. Flesh Products

Limited studies have shown that bacterial extracellular proteinases are not directly associated with development of off-flavor in fish and meat. The initial off-flavors associated with degradation of nitrogenous compounds develop from the bacterial metabolism of nonprotein nitrogenous compounds present in these products. However, the bacterial extracellular proteinases once produced are able to act on tissue proteins and bring about texture defects (such as slime formation). This probably occurs when the nonprotein nitrogenous compounds have been used up and the population has reached about 10^8/g or cm^2 products. However, protein hydrolysis by the bacterial proteinases favors the development of a greater degree of putrefactive changes. Some proteinases of *Pseudomonas fragi* have been found to reduce oxymyoglobin and discolor meat. The enzyme(s) probably hydrolyzes the globin part or the peptide chain of the myoglobin that alters the reactivity of the heme group, leading to discoloration of meat. Muscle lipids can also be hydrolyzed by bacterial lipases, causing flavor defects in meat and fish. However, as the lipids are susceptible to rancidity caused by lipases of the flesh and by autoxidation, the contribution of bacterial lipases is probably very little.

The heat stability of the bacterial proteinases and lipases associated with flavor and texture defects in raw meat and fish is not known. However, these enzymes are produced by the same psychrotrophic bacteria, namely *Pseudomonas, Aeromonas,* and similar species, that are known to produce extracellular heat-stable enzymes. It can be assumed that some of these enzymes are heat stable. It will be important to determine if these enzymes are potentially able of causing spoilage of low heat-processed (pasteurized) meat products with an expected shelf life over 100 days.

REFERENCES

1. Law, B. A., Reviews on the progress of dairy science: enzymes of psychrotrophic bacteria and their effects on milk and milk products, *J. Dairy Res.*, 46, 573, 1979.
2. Cousin, M. A., Presence and activity of psychrotrophic microorganisms in milk and dairy products: a review, *J. Food Prot.*, 45, 172, 1982.
3. Kroll, S., Thermal stability, in *Enzymes of Psychrotrophic Bacteria in Raw Foods,* McKellar, R. C., Ed., CRC Press, Boca Raton, FL, 1989, 121.
4. Cousin, M. A., Physical and biochemical effects on milk components, in *Enzymes of Psychrotrophic Bacteria in Raw Foods,* McKellar, R. C., Ed., CRC Press, Boca Raton, FL, 1989, 205.
5. Mottar, J. F., Effect on the quality of dairy products, in *Enzymes of Psychrotrophic Bacteria in Raw Foods,* McKellar, R. C., Ed., CRC Press, Boca Raton, FL, 1989, 227.
6. Greer, G., Red meats, poultry and fish, in *Enzymes of Psychrotrophic Bacteria in Raw Foods,* McKellar, R. C., Ed., CRC Press, Boca Raton, FL, 1989, 267.
7. Venugopal, V., Extracellular proteases of contaminating bacteria in fish spoilage: a review, *J. Food Prot.*, 53, 341, 1990.

CHAPTER 19 QUESTIONS

1. Discuss the role of intracellular enzymes of spoilage bacteria in food spoilage.

2. Describe the role of extracellular enzymes of spoilage bacteria in food spoilage.

3. Briefly discuss the production and characteristics of heat-stable enzymes by psychrotrophic Gram-negative bacteria. List three bacterial genera that produce heat-stable proteinases and lipases.

4. Briefly describe the problems associated with heat-stable microbial enzymes in these products: UHT milk, cottage cheese, cheddar cheese, and cream.

5. Explain possible spoilage problems using the nonfat dry milk and cream made from a batch of raw milk in which a *Pseudomonas flourescens* grew to a 5×10^6 cells/ml; nonfat dry milk is used to make yogurt and cream to make sour cream.

6. What are the possibilities for the heat-stable enzymes of spoilage bacteria to cause spoilage of a low heat-processed, vacuum-packaged refrigerated meat product with a shelf life of 15 weeks?

CHAPTER 20

New Food Spoilage Bacteria in Refrigerated Foods

CONTENTS

I. MICROORGANISMS THAT GROW IN REFRIGERATED FOODS (PSYCHROTROPHS)

In food microbiology, the terms *psychrophiles* and *psychrotrophs* have been used to identify those microorganisms that can grow in food stored at low temperatures, namely "chilling" and "refrigeration," and could range between –1°C and +7°C. A domestic refrigerator is expected to maintain a temperature of about 40°F (4.4°C), while commercial refrigeration can be lower or higher than this, depending upon a particular food being stored and the expected shelf life. Highly perishable foods are either refrigerated or chilled on ice. Thus psychrotrophs can multiply in these foods provided other conditions for growth are not restricting.[1,2]

There is some confusion about the definitions of the above two terms, particularly in relation to their importance in food microbiology. The term *psychrophile* or *psychrophilic* is quite specific and includes those microorganisms that grow optimally at about 12 to 15°C and have a growth temperature range between ≤ –5 and 22°C. These microorganisms are thus capable of growing in refrigerated and chilled foods. The definition of psychrotrophs is, however, not clear-cut. Originally, in 1960, this term was introduced to include those microorganisms that grow at 0 to 5°C irrespective of their optimum temperature or the range of growth temperatures. Rather they seem to grow best at 25 to 30°C and a few might not grow above 35°C. Thus they appear to be a subgroup of mesophiles (growth temperature of mesophiles: optimum 30 to 40°C and range 5 to 45°C), but not a subgroup of psychrophiles. These groups include Gram-positive and Gram-negative, aerobic, anaerobic and facultative anaerobic, motile and nonmotile, sporeformers and nonsporeformers, coccus and rod-shaped bacteria, as well as yeasts and molds. In 1976, the International Dairy Foundation defined psychrotrophs as those microorganisms that grow at 7°C, irrespective of their optimal and range of growth temperatures. However, this does not indicate if it includes only the subgroup of mesophiles with the ability to grow at or below 7°C or both psychrophiles and the subgroups of mesophiles. More recently, psychrotrophs have been used for mesophilic subgroup capable of growing at 40°F or below (≤4.4°C, refrigerated temperature).[1,2]

Several studies have revealed that there are some mesophilic pathogens (e.g., *Yersinia enterocolitica* and *Listeria monocytogenes*) and spoilage bacteria (e.g., *Leuconostoc* spp., *Lactobacillus* spp., and *Serratia* spp.) that can grow in vacuum- and modified air (MA) packaged foods at 0 to 1°C. Incidence of spoilage of vacuum-packaged meats by psychrophilic *Clostridium* spp., with growth range between –2 and 20°C was also reported. It is important to have understandable communication among the people associated with food production, regulations, sanitation, academicians, and others without creating confusion by using "psychrotrophic spoilage" for the mesophilic subgroup and "psychrophilic spoilage" for the second group. For easier understanding, it is probably not unscientific to use "psychrotrophs" for organisms that can grow (in food) at refrigerated or chilled storage temperatures (≤40°F or 4.4°C). This will include both mesophile subgroups and psychrophiles. As both groups are important in spoilage and foodborne diseases and the methods used to detect and control them will not differentiate the two groups, there is no reason a single term cannot be used for both. The same system has been used in food microbiology for the term *thermoduric*, the microorganisms that survive pasteurization (or low heat treatment) and can include thermophiles, mesophiles, and psychrophiles (*Clostridium laramie*). In this book, psychrotroph is used to include microorganisms capable of growing in food stored at refrigeration temperature irrespective of their being psychrophilic or mesophilic.[1,2]

II. POPULARITY OF REFRIGERATED FOODS

The demand for refrigerated foods has increased dramatically in countries where refrigeration systems are economically available for the food processors, retailers, and consumers.[3] Three major factors can be attributed to this increase. One is the need and desire of the consumer for convenient foods. The changes in socioeconomic patterns, such as the increase in two-income families, single-parent households, singles, and the elderly, have created a demand for convenient foods. Second, there is an increase in awareness and belief that "harshly processed" and "harshly preserved" foods, as well as foods with high fat, cholesterol, and sodium, are not beneficial for a long and healthy life. Health-conscious consumers are interested in foods that are "natural" and fresh. These two factors have created a market for healthy foods that can be of high quality (restaurant quality) and convenient (i.e., will take the least time to prepare). Finally, the technologies necessary for the economical production of such foods and subsequent handling until they are consumed, are available, especially in developed countries. As a result, different types of refrigerated foods, including "new generation refrigerated and chilled foods," "sous vide foods," many of which are "ready-to-eat," "heat-n-eat," and "microwave-n-eat" types, are being commercially produced. Many of the "new generation" foods are given low heat treatment and contain minimal amounts of or no preservative and many are vacuum-packaged or packaged with modified atmosphere (MA; with 100% CO_2 or a mixture of CO_2 and nitrogen without or with some oxygen). They are expected to have a 20- to 60-d or more shelf life. The vacuum- or MA-packaged refrigerated unprocessed or raw foods are also expected to have a much longer shelf-life than the same foods stored in the presence of air. "Sous vide" foods are fresh foods (e.g., vegetables, meat, or fish) that are vacuum packaged, cooked at low heat, and stored in a refrigerator or over ice. They are warmed in the package prior to serving and eating, and are expected to have a 3-week shelf life.

Consumers prefer refrigerated foods because of their better taste, texture, quality, and convenience. They also perceive these foods to be relatively fresh, nutritious, and "close to natural," as compared to frozen, canned, or dried foods and "fast foods." The popularity of these refrigerated or chilled foods has convinced the food experts that this is not a fad; rather, it is here to stay and the demand will continue to increase in the future.[3]

III. MICROBIOLOGICAL PROBLEMS

The vacuum- or MA-packaged refrigerated ($\leq 5°C$) foods are expected to inhibit the growth of aerobic, most mesophilic, and thermophilic microorganisms. However, if the product contains dissolved or trapped oxygen and the modified air contains oxygen, aerobic and microaerophilic microorganisms capable of growing at refrigerated temperatures can also grow as long as the oxygen is available. Similar possibilities also exist if the packaging material is relatively permeable to atmospheric oxygen.[4]

Even under ideal conditions (no oxygen and refrigeration at ≤5°C), the unheated foods can normally harbor both anaerobic and facultative spoilage and pathogenic bacteria that can multiply during refrigerated storage. The heat-processed products are usually given a low heat treatment that, depending upon a product, varies from 60 to 74°C (140 to 170°F). At this temperature, pathogenic and spoilage bacterial spores and cells of some thermoduric bacteria can survive. Some of these survivors can be anaerobic and facultative anaerobic and capable of multiplying at refrigerated temperatures. Many of these heat-processed products are, however, handled extensively following heating and prior to final repackaging under vacuum or MA. Pathogenic and spoilage bacteria capable of growing during storage can also get into the products during this time as post-heat contaminants. Thus these products have the potential of harboring pathogens and spoilage bacteria capable of multiplying under vacuum or MA at refrigerated temperatures. As these products are expected to have a long shelf life, some up to 100 d, even a very low population ($\leq 10^1$/g) of psychrotrophic anaerobic and facultative anaerobic bacteria can multiply and reach a level that causes spoilage of the food or makes it unsafe. These products are often temperature abused during transportation, at the retail store display cases, as well as at the homes of the consumers. Depending upon the time and temperature of abuse, the shelf life and safety of the products can be drastically reduced due to accelerated growth rates of the psychrotrophs as well as by some mesophiles (anaerobic and facultative anaerobic) that do not grow at ≤5°C, but can do so at the abusive temperature. It has been suspected that even a few hours (4 to 6 h) at 12 to 15°C can reduce the shelf life of such products by 8 to 10 d.

There is concern among the regulatory agencies for the safety of these products as the incidence of foodborne diseases following consumption of pasteurized refrigerated foods has been recorded. Food processors are also encountering spoilage of these products in higher frequencies. To overcome the microbiological problems in these foods, the National Food Processors Association has recommended several guidelines. These include selection of good quality raw materials, installation of good sanitary procedures, incorporation of Hazard Analysis Critical Control Point (HACCP) at all phases between production and consumption of these foods, and where possible (see Chapter 38), application of highest permissible heat treatment to a product at the final step, and elimination or minimization of post-heat treatment contamination. Other recommendations are to incorporate multiple barriers or hurdles, along with refrigeration and vacuum or MA packaging. Some of these can be incorporated into the product formulation to reduce the pH and A_w. In addition, acceptable and suitable preservatives (e.g., biopreservatives; see Chapter 15) can be included to combat microbial survival and growth.

Many of the food processors are following one or more of these recommendations, especially good sanitation. This aids in bringing the initial microbial population of the products to a considerably low level. Yet the incidence of spoilage of vacuum- or MA-packaged refrigerated products are not infrequent and, at times with some processors, it is occurring in "epidemic proportions" both in raw and heat-processed products. Many of the bacteria frequently isolated from such spoiled products are being recognized either as new species or species that were not of major concern before. Some of these are *Clostridium laramie, Clostridium estertheticum, Clostridium algidicarnis, Carnobacterium* spp., *Leuconostoc carnosum, Leuconostoc gelidum, Lactobacillus sake, Lactobacillus curvatus,* atypical or unidentifiable lactobacilli and leuconostocs, *Brochothrix thermosphacta, Enterococcus* spp., *Serratia liquifaciens, Hafnia* spp., *Proteus* spp., and some other *Enterobacteriaceae.* Except for spores of *Clostridium* spp. and some thermodurics (some lactobacilli and enterococci), all are sensitive to heat treatment given to the processed products. Thus they are getting into heated products as post-heat contaminants. There is growing speculation and belief among scientists that although these bacterial species have become major causes of spoilage of vacuum- or MA-packaged refrigerated foods, they are neither new nor variants of the existing species. They are present in the environment but probably as minor flora. Three reasons may have helped them become major spoilage organisms. One could be the changes in the environment of food, such as efficient vacuum (or compositions used in MA) packaging and oxygen barrier systems, higher pH in some meat products (low fat, high phosphate), and long storage time at low temperature; these may have controlled the growth of "traditional" bacteria that were associated with food spoilage before. Another reason could be that in order to produce a safer food with a long shelf life, the food processors are using "super sanitation" that has efficiently eliminated the "traditional" microorganisms associated with spoilage before, but at the same time enabled the minor flora to establish in the environment to contaminate the products. Finally, the processors may be introducing equipment that is highly efficient for the production of a large volume of products, but may also be the "microbiologist's nightmare" for efficient cleaning and sanitation. As a result, the equipment harbors the microorganisms in "dead spots" and serves as a source of inoculation to the products. Some of the examples used here may justify these assumptions.[4]

IV. INCIDENCE OF SPOILAGE IN VACUUM-PACKAGED REFRIGERATED FOOD

A. Spoilage of Unprocessed (Fresh) Beef by *Clostridium* spp.

Large-scale spoilage of vacuum-packaged, refrigerated, unprocessed beef was recognized recently.[4-7] The spoilage is characterized by the accumulation of large quantities of hydrogen sulfide-smelling gas and purge (liquid) in the bag within 2 weeks at 4°C, and loss of texture (soft) and color (red) of meat, which in 10 to 12 weeks changed to greenish (Figure 20.1a). Enumeration by plating revealed about 10^8/ml leuconostocs cells in the purge. However, examination of a little purge under a phase contrast microscope revealed a great number of medium to large thick motile rods, some cells with large terminal spores (drumstick-shaped), as well as some leuconostocs-like cells (lenticular, small chains). The rods and the lenticular cells were Gram positive. The rod-shaped bacterium was suspected to be a *Clostridium* species, but could not be cultured in the agar and broth media recommended for *Clostridium* spp. Later, it was purified by heating the purge (from a 12-week-old

Figure 20.1 Bacterial spoilage of vacuum-packaged refrigerated foods. (a) Spoilage of beef by *Clo. laramie*: a spoiled sample showing gas accumulation and an unspoiled sample. (b) Spoiled tofu by a *Clostridium* sp. showing accumulation of liquid and gas. (c) Spoiled frankfurters and (d) spoiled luncheon meat predominantly by *Leuconostoc* spp. showing gas accumulation. (e) Spoiled sliced turkey roll by heterofermentative lactobacilli, leuconostocs, and *Ser. liquifaciens* showing gas accumulation (and pink discoloration) and an unspoiled sample. (f) Spoilage of ground beef chub by heterofermentative lactobacilli showing gas accumulation.

spoiled meat containing spores) at 80°C for 10 min and culturing the material in a broth under anaerobic conditions. Biochemical studies revealed it to be a new species and was designated as *Clo. laramie* NK1 (ATCC 51254). The cells are extremely sensitive to oxygen and grow optimally at 12 to 15°C, with growth temperature range of –2 to 22°C. The species can sporulate and germinate at 2°C. A limited in-plant sampling was done from the slaughtering area to the conveyor systems carrying the primal cuts for vacuum-packaging. Three of the 25 samples were positive for *Clostridium laramie,* and all three were from the conveyor system at the vacuum-packaging area. The conveyor system used by this plant consisted of small links. In general, the microbiological quality of the meat was very good ($\simeq 10^3$/g aerobic plate counts). It was suspected that the sanitation specifically designed to keep the microbial load low was doing a good job, but was unable to kill the spores, especially in the inaccessible joints (dead spot) and similar places in the links of the conveyor system. Thus the fabricated meats were being constantly inoculated with the spores, which in the absence of competition, due to low associated bacteria, germinated and grew to cause large-scale spoilage along with psychrotrophic *Leuconostoc* spp. Recent studies with meat samples from several sources have revealed that there is more than one strain of *Clo. laramie* involved in spoilage.

A similar spoilage of vacuum-packaged, refrigerated beef was also reported to be caused by *Clostridium estertheticum*. This species also could not be enumerated by agar plating methods and was first detected microscopically.[4-7]

B. Spoilage of Roasted Beef by *Clostridium* spp.

Clostridium spp. were isolated from at least two separate incidents of large-scale spoilage of roasted beef. The meats (10 to 12 lb each) were roasted at ≤160°F and then vacuum-packaged and refrigerated.[6,7] When the samples were received in the laboratory, they had large quantities of gas and purge, but due to spices used in the meat, only a strong spicy smell was evident. Phase contrast microscopy of the purge revealed large numbers of motile *Clostridium* cells, some with typical terminal spores. Isolation by agar and broth media failed to obtain colonies of this bacterium, but large numbers of leuconostoc colonies ($\simeq 10^8$/ml of purge) were obtained. The spore-containing purge was heated and cultured to purify the *Clostridium* spp. Inoculation of the pure culture to rare roasted beef, followed by vacuum packaging and refrigeration, showed the formation of H_2S gas, change of meat color from light brown to pink, and the accumulation of pink to red purge. Examination of the purge under a microscope revealed the presence of *Clostridium* cells and spores. Some of these isolates seem to have some differences in physiological pattern when compared with *Clo. laramie*.[6,7]

C. Spoilage of Pork Chops by *Clostridium Algidicarnis*

This new species was associated with the spoilage of cooked vacuum-packaged pork meat products during storage at 4°C.[8] It produces an offensive spoilage odor.

D. Spoilage of Tofu by *Clostridium* spp.

A vacuum-packaged soybean milk curd product (Tofu) during refrigerated storage revealed gas and liquid accumulation in the bag and off-flavor (Figure 20.1b).[7] Examination of the liquid showed the presence of large numbers of motile rods, many with terminal oval spores. The strain was isolated in pure form and found to have some differences in characteristics with *Clo. laramie* NK 1. However, it is psychrophilic and does not grow at 25°C or above.

E. Spoilage of Unripened Soft Cheese by *Leuconostoc* spp.

Spanish-style soft cheese, made by coagulating pasteurized milk with lactic acid and rennet and collecting the curd and vacuum packaging, showed gas formation and liquid accumulation in the package within 30 d at refrigerated temperatures (the product is expected to have 50+-d shelf-life).[7] The curd texture was soft. Examination of the liquid under a phase contrast microscope revealed short chains of leuconostocs-like cells. Enumeration in pH 5.0 plating media revealed 10^{8-9} cells per gram product. Purified cells from the colonies were Gram positive, heterofermentative, and biochemically identified as *Leuconostoc* spp. They possibly get in the product during extensive handling of the product following pasteurization. Even when they are initially present at a low level, during long storage they can multiply, form gas, and produce the defect. Any temperature abuse will accelerate the spoilage.

F. Spoilage of Low-Heat-Processed Meat Products by *Leuconostoc* spp.

Many types of commercially processed vacuum-packaged refrigerated meat products, following spoilage, were collected from the local supermarkets. The products were: restructured ham steak, chopped and formed ham, chopped sliced ham, chunked and formed turkey ham, cooked luncheon meat (sliced), several brands of frankfurters, wieners, cooked and smoked Polish sausage, summer sausage, and hot-link sausage. The common symptoms were accumulation of large quantities of gas and cloudy purge (Figures 20.1c,d).[4,7] The products did not have any disagreeable odor. The pH ranged from 5.0 to 6.0. Some were low-fat products. Phase contrast microscopy of purge revealed the presence of leuconostocs-like cells. Colony enumeration of the purge revealed about 10^8 to 10^9 per milliliter lactic acid bacteria, but lower aerobic plate counts. *Brochothrix thermosphacta*, when present, was present in $<10^3$ per milliliter. Biochemical tests revealed that the predominant lactic acid bacteria were *Leuconostoc carnosum* and *Leu. mesenteroides*, and many were capable of producing bacteriocins. It was suspected that bacteriocin-producing leuconostocs, by inhibiting the growth of other Gram-positive bacteriocin-sensitive contaminants, grew preferentially and produced CO_2 and acid and were responsible for gross distension (blowing) of the packs and release of liquid (purge). Most likely, the leuconostocs contaminated the products as post-heat contaminants from the equipment and other sources during handling prior to vacuum-packaging.[8a]

G. Ammonia Odor in Turkey Roll

Low-fat (2%), high pH (6.5) vacuum-packaged turkey breast rolls showed gas and liquid accumulation with strong ammonia odor (diaper smell) when opened.[7] Examination of the liquid showed predominantly leuconostoc-like cells, as well as motile single or small-chain rods and some lactobacilli-like cells. Gram staining revealed the leuconostocs-like and lactobacilli-like cells to be Gram positive and the motile rods to be Gram negative. Aerobic plate counts and lactic acid bacteria counts of liquid were above 10^8 per milliliter. Biochemical analysis revealed the Gram-negative species to be *Serratia liquifaciens,* and Gram-positive bacteria were *Leu. mesenteroides* and *Lab. sake*. It was suspected that these predominant bacterial types contaminated the products after heat treatment and before vacuum-packaging. *Leuconostoc* sp. produced gas (CO_2) and acid, and *Lab. sake* also produced acid but not gas as it is homofermentative. However, the pH did not drop, owing to the high phosphate content used in these low-fat products. Subsequently, *Serratia* sp. grew and caused deamination of amino acids, thus releasing ammonia. The products also had a slight pink color, the cause of which was not determined.

H. Yellow Discoloration of Luncheon Meat

Vacuum-packaged cooked luncheon meat, prepared from chopped ham, developed yellow-colored spots within 3 to 4 weeks of storage at 40°F (4.4°C).[9] Microbiological analysis indicated the presence of high numbers of *Enterococcus faecium* subsp. *casseliflavus* that were able to survive 71.1°C for 20 min. Thus the species probably survived cooking. The yellow discoloration was suspected to be due to the production of a carotenoid substance by the bacterial species.

I. Gray Discoloration of Turkey Luncheon Meat

Turkey luncheon meat slices prepared mainly from dark meat developed gray spots or patches within 2 to 3 d during aerobic storage at refrigeration temperature.[7] The causative bacteria was isolated and identified to be an H_2O_2-producing strain of *Lactobacillus* spp. The product formulation contained about 1% lactate. It is suspected that the bacterial strain, under aerobic growth conditions utilized lactate to produce H_2O_2 (lactate + $O_2 \xrightarrow{\text{L-lactate oxidase}}$ pyruvate + H_2O_2). H_2O_2 then oxidized the myoglobin to produce the white-gray color. The H_2O_2-producing strain could come as post-heat contaminants or survive low heat treatment, as some lactobacilli are thermoduric. Under vacuum storage, the strain will not produce H_2O_2 to cause discoloration.

J. Pink Discoloration of Sliced, Chopped, and Formed Roast Beef

Vacuum-packaged chopped and formed low-fat roast beef slices during refrigeration storage for 1 to 2 weeks developed pink to red patches, along with the normal brown color.[7] Initially, the pH of the product was 5.8 to 6.0, and had a slight off-odor. Within 4 to 5 weeks at refrigerated temperature, the slices developed a strong fishy to putrid odor; no H_2S was detected and the pH remained around 6.0. The meat color changed to deep pink to red with dark red-colored purge, but there were very small amounts of gas accumulation. Microbiological analysis revealed about 9×10^9 per milliliter purge of APC and 1×10^9 per milliliter purge of lactic acid bacteria (leuconostocs or lactobacilli). Examination of the purge under a phase contrast microscope revealed leuconostocs and lactobacilli-like cells, as well as motile rods. Streaking on xylose-lysine iron agar helped to isolate three types of Gram-negative rods that were biochemically identified as *Hafnia alvei, Proteus vulgaris,* and *Serratia liquifaciens*. Inoculation of the isolates in roast beef, followed by vacuum packaging and storage at refrigerated temperature, revealed that all three Gram-negative isolates were capable of changing the brown color of roasts to pink or red (like rare roast beef) with red-colored purge, but *Pro. vulgaris* changed the meat color to cherry red. The packages inoculated with *Ser. liquifaciens* had slight gas also. All three Gram-negative isolates produced small amounts of H_2S. Analysis of the absorption spectra of the extracts revealed that the color change occurs due to the partial reduction of metmyoglobin.

The lactic acid bacteria, as well as the Gram-negative species, isolated from these samples were all heat-sensitive psychrotrophic and facultative anaerobic. They definitely entered the product as post-heat contaminants, indicating poor sanitation in the production facilities. The absence of large amounts of gas in this product, as opposed to some products described before, could be due to the presence of very little metabolizable carbohydrate. The reason for the pH remaining unchanged (~pH 6.0) during deamination of amino acids could be the use of high amounts of phosphate in the formulation to bind water.

K. Gas Distension and Pink Discoloration of Sliced Turkey Rolls

Sliced, low-fat turkey rolls (initial pH 6.5) were packaged by modified atmosphere (drawing vacuum and then flushing 30% CO_2 + 70% N_2 mixture in small amounts) so that the packages remained slightly loose. The product is expected to have an 8-week shelf-life at refrigerated temperature. By 5 weeks, many packages developed gross distension with gas, little to moderate amounts of purge, and pinkish-red discoloration from the normal gray-white color of breast meat (Figure 20.1e).[7] The intensity of the color was more inside the slices than at the rim. The gas produced off-odor (like wet socks), but no H_2S was detected. Soon after opening a bag, the color of the meat slices changed to normal. Vacuum packaging of the meat slices again changed the color in 8 to 12 h to pinkish-red. Addition of ascorbic acid to the normal-color meat slices gave the similar pinkish-red color quickly (1 to 2 h). Examination of the liquid under a phase contrast microscope revealed leuconostocs-like and lactobacilli-like cells, as well as motile rods. Colony enumeration revealed the presence of lactic acid bacteria at about 10^{7-8} per gram and *Enterobacteriaceae* at about 10^6 per gram. While lactic acid bacteria were predominantly *Leuconostoc carnosum* (heterofermentative) and *Lactobacillus curvatus* (homofermentative), *Enterobacteriaceae* were *Serratia liquifaciens*. Inoculation studies revealed that *Ser. liquifaciens* was important for the off-flavor. The pink color was due to formation of reduced metmyoglobin; however, it is not clear how it formed. It seems that psychrotrophic gram-positive and -negative bacteria, entering the products as post-heat contaminants, grow during storage and reduce the O-R potential of the product. This changes the metmyoglobin from reversible oxidized to reduced state and color from grey-white to pink. In air (oxygen) metmyoglobin reverses from reduced to oxidized state, changing the color from pink to grey-white. Absorption studies with a spectrophotometer of extracted materials also supported the conclusion that the pink color was due to a reduced form of metmyoglobin. The large amounts of gas were produced by the *Leuconostoc carnosum* from the metabolism of glucose used in the formulation.

As the product had fairly large amounts of phosphate, pH did not change, even from the growth of lactic acid bacteria; thus there were only small amounts of purge present in the bags.

Examination of the production facility revealed that the cooked products were being contaminated by bacteria mainly from the slicer and the conveyors between the slicer and the packaging machine. However, the initial contamination level was very low (<10/g). The high number was reached during storage.

L. Gas Distension (Blowing) of Ground Beef Chubs

Vacuum-packaged, low-fat ground beef chubs were found to accumulate large quantities of gas within a few days of refrigerated storage (Figure 20.1f).[7] Initially (within 1 week of storage), there was a slight off-flavor in the gas or meat, the meat pH was 5.6, and microbiological analysis revealed large numbers (1 to 4 × 10^7/g) of psychrotrophic lactic acid bacteria, mainly two atypical *Lactobacillus* spp. and *Leuconostoc mesenteroides*. Aerobic plate counts (APC) was ≤1 × 10^5/g and Gram negatives ≤10^3/g. The most predominant *Lactobacillus* sp. is heterofermentative, has atypical morphology, and biochemically does not fit with the known species. Following storage of the meat for about 6 weeks at refrigerated temperature, psychrotrophic lactic acid bacterial counts increased to about 1.5 × 10^8/g (APC was 2.7 × 10^7/g), the pH increased to 6.2, the gas had slight H_2S, and the meat had a strong putrid odor. It was suspected that initial gas formation was due to metabolism of carbohydrates in meat by the heterofermentative psychrotrophic lactic acid bacteria. During extended storage, lactic acid bacteria as well as other bacteria (especially Gram-negative) assimilated amino acids, including sulfur-containing amino acids and produced the offensive odor.

The examples described here give some idea of how these products are being contaminated with different types of spoilage bacteria as postheat contaminants. Complicated machineries, handling of large volumes of products, and a desire for low initial microbial load by good sanitation have probably selected out those bacteria that can establish in the facilities and contaminate the products. In the absence of competition and being psychrotrophic and anaerobic or facultative anaerobic, they can multiply and reach high levels during long storage to cause spoilage. In the absence of effective preservative(s) as well as due to temperature abuse during storage transport and display, they can grow more rapidly and cause rapid spoilage of the products.

REFERENCES

1. Kraft, A. A., *Psychrotrophic Bacteria in Foods, Disease and Spoilage,* CRC Press, Boca Raton, FL, 1992, 3.
2. Olson, J. C., Jr. and Nottingham, P. M., Temperature, in *Microbial Ecology of Foods,* Vol. 1, Silliker, J. H., Ed., Academic Press, New York, 1980.
3. Ray, B., The need for food biopreservation, in *Food Biopreservatives of Microbial Origin,* Ray, B. and Daeschel, M. A., Eds., CRC Press, Boca Raton, FL, 1992, 1.

3a. Ray, B., Foods and microorganisms of concern, in *Food Biopreservatives of Microbial Origin,* Ray, B. and Daeschel, M. A., Eds., CRC Press, Boca Raton, FL, 1992, 25.
4. Ray, B., Kalchayanand, N., and Field, R. A., Meat spoilage bacteria: are we prepared to control them?, *The National Provisioner,* 206(2), 22, 1992.
5. Kalchayanand, N., Ray, B., and Field, R. A., Characteristics of psychrotrophic *Clostridium laramie* causing spoilage of vacuum-packaged refrigerated fresh and roasted beef, *J. Food Prot.,* 56, 13, 1993.
6. Collins, M.D., Rodrigues, U. M., Dainty, R. H., Edwards, R. A., and Roberts, T. A., Taxonomic studies on a psychrophilic *Clostridium* from vacuum-packaged beef: description of *Clostridium estertheticum* spp. no, *FEMS Microbiol. Lett.,* 96, 235, 1992.
7. Ray. B., Kalchayanand, N., Means, W., and Field, R. A., Spoilage of vacuum-packaged refrigerated fresh and roasted beef by *Clostridium laramie* is real. So are other spoilage bacteria in processed meat products, *Meat and Poultry,* 40(7), 12, 1995.
8. Lawson, P., Dainty, R. H., Kristiansen, N., Berg, J., and Collins, M. D., Characterization of a psychrotrophic *Clostridium* causing spoilage in vacuum-packed cooked pork: description of *Clostridium algidicarnis* spp. no, *Lett. Appl. Microbiol.,* 19, 153, 1994.
8A. Yang, R. and Ray, B., Prevalence and biological control of bacteriocin-producing psychrotropic leuconostoes associated with spoilage of vacuum-packaged meats, J. Food Prot., 57, 209, 199H.
9. Whiteley, A.M. and D'Souza, M. D., A yellow discoloration of cooked cured meat products: isolation and characterization of the causative organism, *J. Food Prot.,* 52, 392, 1989.

CHAPTER 20 QUESTIONS

1. Define psychrophilic and psychrotrophic microorganisms. Discuss why psychrotrophs are important in food spoilage.

2. Briefly discuss the reasons for the current popularity of refrigerated foods.

3. What are the possible causes of microbiological problems in refrigerated foods?

4. List six bacterial genera/species currently designated as new spoilage bacteria. Explain the possible reasons for their becoming important in the spoilage of refrigerated foods.

5. A processor recently started producing a new vacuum-packaged, low heat-processed, low-fat meat product with an expected shelf life of 60 d at 40°F. He found that about 10% of the product was getting spoiled within 30 to 40 d. The characteristics are: accumulation of gas and purge (cloudy), with acidic and cheesy odor. He asked you to help him find out the cause(s) and resolve the problem. Explain briefly, in stepwise fashion, how you plan to proceed in order to understand the problem (cause(s), source(s), sequence of events, etc.) and solve it.

CHAPTER 21

Indicators of Microbial Food Spoilage

CONTENTS

I. INTRODUCTION

Microorganisms are capable of causing food spoilage in two ways. The most important one is through the growth and active metabolism of food components by the live cells. The other one is produced, in the absence of live cells, by their extracellular and intracellular enzymes that react with the food components and change their functional properties, leading to spoilage. The loss of food due to microbial spoilage has economic consequences for the producers, processors, and consumers. With the increase in population in the world, loss of food due to microbial (and nonmicrobial) spoilage means less food is available for the hungry mouth. To fight against world hunger, efforts should be directed not only to increase food production, but also to minimize spoilage so that enough food is available for consumption. Many preservation methods have been devised to reduce microbial spoilage and these are discussed in Section VI. Under certain methods of preservation, both raw and partially processed (semipreserved, perishable, and nonsterile) foods are susceptible to microbial spoilage. This is more evident in foods that are expected to have a long shelf life. To reduce loss of raw and partially processed foods due to microbial spoilage, two things are important. One is to predict how long a food, following production, will stay acceptable under the condition(s) of storage normally used for that food; i.e., what is its expected shelf life under normal conditions of handling and storage? The other is to determine the current status, with respect to spoilage, of a food that has been stored for some time. This information needs to be available well before a food has developed obvious detectable spoilage and thus is unacceptable.

Many criteria have been evaluated to determine their efficiency as indicators to predict expected shelf life, as well as to estimate stages of microbial food spoilage. These criteria or indicators can be grouped as sensory, microbiological, and chemical (microbial metabolites). The sensory criteria (e.g., changes in color, odor, flavor, texture, and general appearance) have several drawbacks as indicators, especially if used alone. Changes in texture and flavor generally appear at the advanced stages of spoilage. Odor changes can be masked by the spices used in many products. Odor changes from volatile metabolites may not be detected in a product that is exposed to air, as compared to the same product in a package. Color changes, such as in meat exposed to air, may not be associated with microbial growth. Finally, individuals differ greatly in their perception for organoleptic criteria. However, sensory criteria can be used advantageously along with microbiological and/or chemical criteria.

Studies by a large number of researchers have clearly revealed that a single microbiological or chemical test is not effective in predicting either the shelf life of a product or its spoilage status. The contributing factors in microbial spoilage of a food include the type of product, its composition, methods used during processing, contamination during processing, nature of packaging, temperature and time of storage, and possible temperature abuse. As these factors differ with products, it may be rational to select indicator(s) on the basis of a product or a group of similar products. Some of the factors to be considered in selecting a microbial or chemical indicator(s) for a product (or several similar types of products) are:

1. In a good fresh product, it (or they) can be present in low numbers (microbial) or absent (chemical).
2. Under normal conditions of storage (temperature, time, packaging), it (or they) should increase (microbial or chemical) in quantity to reach a very high level.
3. When spoilage occurs under normal storage conditions, it (or they) should be the predominant causative agent(s) (microbial or chemical).
4. It (or they) can be detected rapidly (microbial or chemical).
5. It (or they) can be used reliably to predict shelf life and spoilage status (microbial or chemical).
6. It (or they) should have a good relationship with the sensory criteria of spoilage of the particular product (microbial or chemical).

Different microbial groups and their metabolites (chemicals) were evaluated for their suitability as indicators of food spoilage. As bacteria are the most predominant microbial group in food spoilage, the effectiveness of some bacteria and metabolites as indicators is briefly discussed. In addition, the effectiveness of testing microbial heat-stable enzymes in predicting the shelf life of products susceptible to spoilage by them is also discussed.

II. MICROBIOLOGICAL CRITERIA

Previous discussions indicated that spoilage microorganisms differ with the products or, more accurately, with the intrinsic and extrinsic environments of the products. It is rational to select the microorganism(s) predominantly involved in spoilage of a food (or a food group) as the indicator(s) of spoilage for that food. As an example, refrigerated ground meat during aerobic storage will normally be spoiled by the Gram-negative psychrotrophic aerobic rods, most importantly by the *Pseudomonas* spp. Thus the population level of the psychrotrophic Gram-negative rods should be the most appropriate indicator of spoilage for this product (or for raw meats stored under the same conditions), both for predicting the shelf life of this product and to estimate the status of spoilage during storage. Aerobic plate count (APC), which measures the mesophilic population, may not be a good indicator for this product since many mesophiles do not multiply at psychrotrophic temperature and, conversely, there are some psychrotrophic bacteria that do not multiply at 35°C in the 2 d used to enumerate APC in meats. However, APC (also standard plate, count, SPC, for dairy products) has special importance in food microbiology. In fresh products, it indicates the effectiveness of sanitary procedures used during processing and handling of the product. A high APC or SPC in a food product such as hot dogs and pasteurized milk is viewed with suspicion, both for stability and safety. Thus it is a good idea to include APC or SPC along with the method suitable to detect the load of an appropriate indicator group for a food.

Some of the specific microbial groups that can be used as indicators in different foods (or food types) are listed here. Details of the procedure can be found in books dealing with microbiological examination of foods (also see Appendix D).

1. Refrigerated raw (fresh) meats stored aerobically: Enumeration of colony-forming units (cfu/g or cm^2) of psychrotrophic aerobes, especially Gram-negative aerobes. The data may be available in 2 to 7 d depending upon the incubation temperature (10 to 25°C) and plating methods (pour or surface) used.
2. Refrigerated raw (fresh) meats stored anaerobically (vacuum-packaged): Enumeration of cfu/g or cm^2 of psychrotrophic lactic acid bacteria (by plating in a suitable agar medium adjusted to pH 5.0) as well as psychrotrophic *Enterobacteriaceae* (in violet-red bile glucose agar medium). Depending upon incubation temperature, the data can be available in 2 to 7 d. The plates may be incubated in a CO_2 environment. The products can also be tested for psychrotrophic *Clostridium* spp., such as *Clo. laramie*, using specific methods.
3. Refrigerated low heat-processed, vacuum-packaged meat products: Enumeration of cfu/g or cm^2 of psychrotrophic lactic acid bacteria (by plating in an agar medium adjusted to pH 5.0) as well as psychrotrophic *Enterobacteriaceae* (in violet-red bile glucose agar medium). Depending upon incubation temperature, the results can be available in 2 to 7 d. The plates may be incubated in a CO_2 environment.

The products can be tested for psychrotrophic *Clostridium* spp., such as *Clo. laramie*, using specific methods.
4. Raw milk: SPC, psychrotrophic Gram-negative rods, thermoduric bacteria.
5. Pasteurized milk: SPC, psychrotrophic bacteria (Gram negative and Gram positive).
6. Butter: Lipolytic microorganisms.
7. Cottage cheese: Psychrotrophic, especially Gram-negative rods.
8. Fishery products (raw): Psychrotrophic Gram-negative rods.
9. Beverages: Aciduric bacteria, yeasts, and molds.
10. Salad dressing and mayonnaise: *Lactobacillus* spp. (especially *Lab. fructivorans*) and yeasts.

It is quite evident that several days are necessary before the population levels of the indicator microorganisms from enumeration of colony-forming units become available. This is the major disadvantage of the microbiological methods. To overcome this problem, several indirect methods that indicate the probable population of microorganisms in foods have been devised. One such method is the determination of lipopolysaccharides (LPS) present in a food. LPS is specifically found in the Gram-negative bacteria. Thus, by measuring LPS concentration, an estimate of the level of Gram-negative bacteria in a food can be obtained. However, this method is not applicable for spoilage by Gram-positive bacteria. Several other indirect methods are measurement of ATP (ATP concentrations increase with high numbers of viable cells), impedance/conductivity (electric conductivity decreases with increase in cell numbers), and dye reduction time (the higher the population, the faster the reduction). However, each method has specific advantages and disadvantages.

III. CHEMICAL CRITERIA

As microorganisms (particularly bacteria) grow in foods, they produce many types of metabolic by-products associated with the spoilage characteristics. If a method is developed that is sensitive enough to measure a specific metabolite in very low concentrations and long before the spoilage becomes obvious, then the results can be used to determine the spoilage status of a food. Methods studied thus far to measure microbial metabolites include H_2S production, NH_3 production by colorimetric or titration methods, production of volatile reducing substances, CO_2 production, diacetyl and acetoin production, indole production, and others. However, different metabolites are produced by different species-strains of bacteria and the results are not consistent; thus they cannot be used for different types of products.

Change in food pH, especially in meat and meat products, due to microbial growth has also been used to determine the spoilage status of a food. In normal meats, with a pH of about 5.5, metabolism of amino acids by some spoilage bacteria generates NH_3, amines, and other basic compounds. They will shift the pH to the basic side (as high as pH 8.0). In contrast, the metabolism of carbohydrates (present or added) by some bacteria produces acids and reduces the pH further to the acidic side. Thus measurement of pH of a stored meat product can also give some indication of its spoilage status. As the pH increases, the proteins become more hydrated, i.e., its water-holding capacity (WHC) increases and, when pressed, this meat will have less extract-release volume (ERV). In contrast, when the pH shifts toward the acid side, the WHC will be lower and ERV will be higher. However, many low-fat products are formulated with high phosphate and generally have a pH close to 7.0 (to increase WHC). The buffering action of phosphate may not allow the pH to shift to the basic or acid side from the microbial metabolism of amino acids and carbohydrates, respectively. In these products, pH measurement (or the measurement of WHC or ERV) may not be good indicators of their spoilage status.

None of the microbiological and chemical criteria studied fulfill all the factors necessary for a good indicator that will indicate expected shelf life of a fresh product as well as its spoilage status during storage. More emphasis needs to be given to develop suitable indicators that reduce the loss of food due to microbial spoilage.

IV. ASSAY OF HEAT-STABLE ENZYMES

A. Heat-Stable Proteinases in Milk

Proteinases of some psychrotrophic bacteria, such as *Pseudomonas fluorescens* strain B52, when present as low as 1 ng/ml raw milk, can reduce the acceptance quality of UHT-treated milk during normal storage.[5] Because of this, it is very important that sensitive assay method(s) be used in their estimation to predict the shelf life of dairy products. Some of the earlier methods, such as UV absorbance, Folin-ciocalteu reagent reaction, and gel diffusion assay, are probably not sensitive enough for this purpose. Several new methods, such as the use of TNBS (trinitrobenzene sulfonic acid) and fluorescamine reagents are quite sensitive and being tested to assay proteinases in milk. In the TNBS method, the reagent reacts with free amino groups and, under the experimental conditions, develops color that can be colorimetrically measured to determine the extent of free amino acids present due to proteolysis. Fluorescamine reacts with amino acids to form fluorescent compounds at pH 9.0 and thus can be fluorimetrically measured to determine protein hydrolysis. Other methods, such as enzyme-linked immunosorbent assay (ELISA) and luciferase inactivation assay are extremely sensitive methods and need further development before they can be used reliably.

B. Heat-Stable Lipases in Milk

Due to the presence of natural lipases in milk, the measurement of lipases produced specifically by the psychrotrophic bacteria creates some difficulties. However, it can be overcome by heating the milk, which destroys milk lipases but not the bacterial heat-stable lipases. Assay methods that measure release of free fatty acids (FFA) due to hydrolysis of milk fat by the lipases can be titrated to determine the potential of lipolysis of the lipases. As milk contains FFA naturally, this method may not be accurate. Methods in which esterases of chromogenic and fluorogenic compounds react with lipases to produce color or fluorescent products have also been developed, but they have limitations. Recently, a rapid and sensitive sandwich

ELISA method was tested to determine lipases of *Pseudomonas* spp. An antibody produced against a *Pse. fluorescens* strain and linked to horseradish peroxidases reacted with lipases from many strains of *Pseudomonas* spp. The reliability and sensitivity of this technique are now being studied.

REFERENCES

1. Kraft, A. A., *Psychrotrophic Bacteria in Foods: Diseases and Spoilage,* CRC Press, Boca Raton, FL, 1992, 121.
2. Gill, C. O., Meat spoilage and evaluation of the potential storage life of fresh meat, *J. Food Prot.,* 46, 444, 1983.
3. Suhren, G., Producer microorganisms, in *Enzymes of Psychrotrophs in Raw Foods,* McKellar, R. C., Ed., CRC Press, Boca Raton, FL, 1989, 3.
4. Tompkin, R. B., Indicator organisms in meat and poultry products, *Food Technol.,* 37(6), 107, 1983.
5. Fairbairn, D., Assay methods for proteinases, in *Enzymes of Psychrotrophs in Raw Foods,* McKellar, R. C., Ed., CRC Press, Boca Raton, FL, 1989, 189.
6. Stead, D., Assay methods for lipases and phospholipases, in *Enzymes of Psychrotrophs in Raw Foods,* McKellar, R. C., Ed., CRC Press, Boca Raton, FL, 1989, 173.

CHAPTER 21 QUESTIONS

1. List the factors to be considered in selecting indicators of food spoilage.

2. List two spoilage aspects that a suitable food spoilage indicator should indicate to prevent loss of a food due to spoilage. Explain why different indicator(s) need to be selected for different groups of foods.

3. Suggest and justify suitable microbial spoilage indicators for: (a) tray-packed raw sausage; (b) vacuum-packaged, low-fat, refrigerated ground turkey; (c) fish fillet kept over ice in air; (d) salsa in a bottle stored at room temperature; and (e) ready-to-eat baby carrots packaged in a bag and stored at about 15°C (consult Chapter 18).

4. List four chemical methods that can be used for spoilage detection of fresh meat stored at refrigerated temperature under aerobic conditions.

5. Explain how pH changes in fresh meats can be used as an indicator of spoilage. How is it related to WHC and ERV?

SECTION V

MICROBIAL FOODBORNE DISEASES

Human illness from the consumption of foods contaminated with factors other than poisons or chemical toxic agents was recognized long before the understanding of the role of pathogens in foodborne diseases. Dietary guidelines by ancient civilizations, such as eating foods after cooking well and warm foods, and not eating spoiled foods, were probably invoked to protect people at least partly from the danger of microbial foodborne diseases. In the Middle Ages, several mass food poisoning incidences in Europe from the consumption of grains infested with toxin-producing fungi were recorded. Concern about the possible role of foods of animal origin in human diseases led to the introduction of proper hygienic methods in the handling of fresh meat. However, not until Pasteur's discovery of the role of microorganisms in foods, was their involvement in foodborne diseases understood. This helped in searching for and isolating pathogens by suitable techniques from food incriminated in foodborne diseases. *Salmonella* spp. and *Staphylococcus aureus*, due to the high incidence of salmonellosis and staphylococcal poisoning, and *Clostridium botulinum*, due to the high fatality rate from botulism, were isolated from foods incriminated with foodborne disease before the 20th century (Chapter 1). Subsequently, many other pathogenic bacteria, toxin-producing molds, and pathogenic viruses were recognized as causative agents in human foodborne illnesses. The involvement of a few new pathogenic bacteria and viruses has been recognized only recently. Even now, the exact role of many bacteria and viruses in foodborne diseases is not well understood, and several of them are designated as opportunistic pathogens. The changes in food production, processing, marketing, and consumption, together with our knowledge about the characteristics of these microorganisms and the development of efficient techniques for their detection, have enabled us to identify the role of some "new pathogens" in foodborne diseases. History suggests there will probably always be "new pathogens" and thus, as we develop methods to control the existing pathogens, we have to remain alert for the "new pathogens." In this section, an overview of the microorganisms associated with foodborne diseases is presented. The diseases caused by the foodborne pathogens have been divided into three groups, although the basis of division is not always clear-cut. Also included are some parasites and algae that can cause foodborne diseases.

The following topics are discussed in this section.

Chapter 22: Important Facts in Foodborne Diseases
Chapter 23: Foodborne Intoxications
Chapter 24: Foodborne Infections
Chapter 25: Foodborne Toxicoinfections
Chapter 26: Opportunistic Pathogens, Parasites, and Algal Toxins
Chapter 27: New and Emerging Foodborne Pathogens
Chapter 28: Indicators of Bacterial Pathogens

CHAPTER 22

Important Facts in Foodborne Diseases

CONTENTS

I. INTRODUCTION

The objective of this topic is to recognize the causes of foodborne diseases, the role of microorganisms and several other agents in foodborne diseases, and the importance of predisposing factors in the occurrence of a foodborne disease. This information will help not only to understand the causes of foodborne diseases, but also to develop means of controlling them.

II. HUMAN GASTROINTESTINAL DISORDERS

The causes of foodborne gastrointestinal disorders can be broadly divided into three groups[1]:

1. From the consumption of food and water containing viable pathogenic microorganisms or their preformed toxins
2. From the ingestion of pathogenic algae, parasites, and their preformed toxins through food
3. For reasons other than viable pathogens or their toxins

Some of the factors included in group "3" are:

1. Ingestion of toxins naturally present in food. This includes certain mushrooms, some fruits and vegetables, and some seafoods.
2. Toxins formed in some foods. Examples are some biological amines, (e.g., histamine) that form in some fish, cheeses, and fermented meat products.
3. The presence of toxic chemicals in contaminated food and water, such as heavy metals and some pesticides.
4. Allergy to some normal components of a food. There are individuals who are allergic to gluten in cereals and develop digestive disorders following consumption of food containing gluten.
5. Genetic inability to metabolize normal food components. The inability of some individuals to hydrolyze lactose in the small intestine, due to the lack of production of enzyme lactase, results in digestive disorders (lactose intolerance).
6. Nutritional disorder such as rickets due to calcium deficiency.
7. Indigestion from overeating or other reasons.

Among the various causes, the incidence of foodborne diseases of microbial origin is higher than all others combined. In the United States between 1972 to 1978, among the total number of reported cases, 94.4% were caused by pathogenic microorganisms, and only 1.1% and 4.3% were caused by parasites and chemicals, respectively. In recent years, foodborne diseases of microbial origin have become the number one food safety concern among consumers and regulatory agencies. This trend is probably true in most other developed and developing countries.[1]

III. INVESTIGATION OF A FOODBORNE DISEASE

In general, various regulatory agencies at the local, state, and national levels are empowered with the responsibility of investigating the cause of a reported foodborne disease (see Appendix E). In the United States, the regulatory agencies involved in the investigation of a foodborne disease include local health department, state health, food, and agriculture departments, the federal Food and Drug Administration (FDA), U.S. Department of Agriculture (USDA), Centers for Disease Control (CDC), and several others. Initially, a medical doctor, suspecting a patient or patients to have a foodborne disease, informs the local or state health officials about the incident. These agencies, following a preliminary investigation and recognizing the cause to be of food origin, report the incidence to the appropriate federal agencies, who then conduct an epidemiological investigation. The investigation at the state and federal levels involves examination of the suspected food(s), environmental samples and materials obtained from the patient(s) for pathogens, microbial and nonmicrobial toxins, and chemicals. Results of these tests provide direct evidence of the association of an agent (e.g., pathogen, toxin, parasite, chemicals) with the disease. In addition to testing suspected samples, both the sick and other people who consumed the same food from the same source are interviewed to establish an indirect association of the most likely food(s) with the disease. This information is collected, recorded, and reported by the CDC based in Atlanta, GA.

IV. FOODBORNE DISEASE OUTBREAK

In the United States, the federal regulatory agencies define a foodborne disease as an outbreak when two or more people become sick with a similar illness (symptoms) from the consumption of the same food(s) from the same source, and the epidemiological investigations implicate, either directly or indirectly, the same food(s) from the same source as the cause of the illness. However, in the case of botulism, due to a high fatality rate, even when only one person has the illness, it is considered an outbreak. For chemical poisoning, a single case is also considered an outbreak.

V. INCIDENCE OF FOODBORNE DISEASE OUTBREAK

The incidence of foodborne illnesses in most of the developed countries is lower than in many developing countries.[1-3] The major reasons for the low incidence are the implementation of necessary regulations in production and handling of foods, good sanitary practices, and the availability of necessary facilities to reduce abuse.

In the United States from 1983 to 1987, an average of 479 foodborne disease outbreaks were reported each year, involving 18,336 individuals (Table 22.1). However, on average, only 38% of the outbreaks involving about 10,908 individuals were confirmed.[2] This shows that even for the reported incidents, only some could be confirmed from direct evidence. It is suspected that even in the developed countries, only a small fraction of the actual incidents are reported. In many instances, an individual will not go to a doctor; and even if a person sees a doctor, the incidence may not be reported to the regulatory agencies. Based on the mechanisms involved in the surveillance system, it is estimated that in the United States, about 5 million individuals are affected by food- and waterborne diseases per year. Others consider this to be very conservative. According to them, if on an average, a person is affected once in 10 years, 10% of the population can become sick annually by foodborne illnesses. In the United States, this is equivalent to about 24 to 25 million people annually. There are other groups who think the numbers could go as high as 80 million per year.

Table 22.1 Foodborne Disease Outbreaks in the United States during a 5-year Period

	1983	1984	1985	1986	1987	Average/year
Reported outbreaks	505	543	495	467	387	479 (100%)
Total cases	14,898	16,420	31,079	12,781	16,500	18,336 (100%)
Confirmed outbreaks	187	185	220	181	136	182 (38%)
Confirmed cases	7,904	8,193	22,987	5,804	9,652	10,908 (59.5%)

VI. COST OF FOODBORNE DISEASES

Foodborne illnesses can be fatal as well as cause suffering, discomfort, and debilitation among the survivors. The economic losses from various factors could be very high.[1,4] The factors include medical treatment, lawsuits, lost wages and productivity, loss of business, recall and destruction of products, and investigation of the outbreaks. In the United States, the annual cost of foodborne diseases is estimated to be $1 billion to $9 billion. This is based on an estimate that the cost per case ranges from $200 to $2000, and that about 5 million people suffer from foodborne diseases yearly.

VII. PREDOMINANT ETIOLOGICAL AGENTS

As indicated before, gastrointestinal disorders can be caused by the consumption of food and water containing pathogenic microorganisms and their toxins, pathogenic algae and parasites, and their toxins, toxic chemicals, either naturally or as a contaminant, and by other factors. Among these, the largest number of outbreaks, the total number of cases, and the number of deaths are caused by pathogenic bacteria and their toxins. Data presented in Table 22.2 show that in the United States from 1983 to 1987, bacteria caused about 66% of the outbreaks affecting 92.2% of the cases and 96.4% of the fatalities.[1-3] Outbreaks caused by viruses, parasites, and chemicals were 4.5, 4.0, and 25.5%, respectively. The number of cases and deaths from these agents was also low. No outbreaks from molds were reported during this period. Results of a similar U.S. study between 1972 and 1978 showed that percentages of foodborne disease outbreaks and cases, respectively, were as follows: bacterial, 66.3% and 90.9%; viral, 2.7% and 3.5%; parasitic, 7.8% and 1.1%, and chemical 23.5% and 4.3%. This report did not include the number of fatalities. Both reports indicated that pathogenic bacteria are the major cause of foodborne diseases in the United States. This is probably also true for other countries. Several factors may be involved in the high incidence caused by pathogenic bacteria: many pathogenic bacteria are found in the raw food materials of animal and plant origin, many are present in the food environments, many grow very effectively in different foods, and many are not killed by the conditions used for processing different foods.

Table 22.2 Confirmed Foodborne Disease Outbreaks, Cases, and Deaths by Etiological Agents during 1983 to 1987 in the United States

Etiological agents[a]	Outbreaks		Cases		Deaths	
	No.	%	No.	%	No.	%
Bacterial	600	66.0	50,304	92.2	132	96.4
Viral	41	4.5	2,789	5.1	1	0.7
Parasitic[b]	36	4.0	203	0.4	1	0.7
Chemicals[c]	232	25.5	1,244	2.3	3	2.2
Total	909	100	54,540	100	137	100

[a] No incidence from mycotoxins was reported.
[b] Includes *Trichinella spiralis* and Giardia.
[c] Includes ciguatoxin, scombrotoxin, mushrooms, heavy metals, and other chemicals.

VIII. TYPES OF MICROBIAL FOODBORNE DISEASES

Foodborne diseases in humans result from the consumption of either food and water contaminated with viable pathogenic bacterial cells (or spores in case of infant botulism) or food containing toxins produced by toxigenic bacteria and molds. On the basis of mode of illnesses, these can be arbitrarily divided into three groups: intoxication or poisoning, infection, and toxicoinfection (Table 22.3).

A. Intoxication

Illness occurs as a consequence of ingestion of a preformed bacterial or mold toxin due to its growth in a food. A toxin has to be present in the contaminated food. Once the microorganisms have grown and produced toxin in a food, there is no need

Table 22.3 Microbial Foodborne Diseases and Causative Pathogens

Type of disease	Causative microorganism	Microbial group	Major symptom(s) type
Intoxication			
Staph poisoning	*Staphylococcus aureus* strains	Bacteria, Gm+[a]	Gastric
Botulism	*Clostridium botulinum* strains	Bacteria, Gm+	Nongastric
Mycotoxin poisoning	Mycotoxins producing mold strains, e.g., *Aspergillus flavus*	Molds	Nongastric
Infection			
Salmonellosis	Over 2000 *Salmonella* species (except *Sal. typhi* and *Sal. paratyphi*)	Bacteria, Gm–[a]	Gastric
Campylobacter enteritis	*Campylobacter jejuni* and *Cam. coli* strains	Bacteria, Gm–	Gastric
Yersiniosis	Pathogenic strains of *Yersinia enterocolitica*	Bacteria, Gm–	Gastric
Enterohemorrhagic *Esc. coli* colitis	*Esc. coli* 0157:H7	Bacteria, Gm–	Gastric and nongastric
Nonhemorrhagic *Esc. coli* colitis	Shiga-like toxin (verotoxin) producing *Esc. coli* strains like *Esc. coli* 026:H11	Bacteria, Gm–	Gastric
Listeriosis	*Listeria monocytogenes* (pathogenic strains)	Bacteria, Gm+	Gastric and nongastric
Shigellosis	Four *Shigella* species, e.g., *Shi. dysenteriae*	Bacteria, Gm–	Gastric
Vibrio parahaemolyticus gastroenteritis	Pathogenic strains of *Vib. parahaemolyticus*	Bacteria, Gm–	Gastric
Vibrio vulnificus infection	*Vibrio vulnificus* strains	Bacteria, Gm–	Gastric and nongastric
Brucellosis	*Brucella abortus*	Bacteria, Gm–	Gastric and nongastric
Viral infections	Pathogenic enteric viruses, e.g., Hepatitis A virus	Viruses	Gastric and nongastric
Toxicoinfection			
Clostridium perfringens gastroenteritis	*Clostridium perfringens* strains	Bacteria, Gm+	Gastric
Bacillus cereus gastroenteritis	*Bacillus cereus* strains	Bacteria, Gm+	Gastric
Esc. coli gastroenteritis	Enteropathogenic and enterotoxigenic *Esc. coli* strains	Bacteria, Gm–	Gastric
Cholera	Pathogenic strains of	Bacteria, Gm–	Gastric

Table 22.3 Microbial Foodborne Diseases and Causative Pathogens

Type of disease	Causative microorganism	Microbial group	Major symptom(s) type
	Vibrio cholerae		
Gastroenteritis by opportunist pathogens			
Aeromonas hydrophila gastroenteritis	*Aeromonas hydrophila* strains	Bacteria, Gm–	Gastric
Plesiomonas shigelloides gastroenteritis	*Plesiomonas shigelloides* strains	Bacteria, Gm–	Gastric

[a] Gm+, Gm–: Gram-positive and Gram-negative, respectively.

of viable cells during the consumption of the food for illness to occur. Example: Staph food poisoning.

B. Infection

Illness occurs as a result of the consumption of food and water contaminated with enteropathogenic bacteria. It is necessary for the cells of enteropathogenic bacteria to remain alive in the food or water during consumption. The viable cells, even if present in small numbers, have the potential to establish and multiply in the digestive tract to cause the illness. Example: Salmonellosis.

C. Toxicoinfection

Illness occurs from the ingestion of a large number of viable cells of some pathogenic bacteria through contaminated food and water. Generally, the bacterial cells either sporulate or die and release toxin(s) to produce the symptoms. Example: *Clostridium perfringens* gastroenteritis.

In addition to the pathogenic microorganisms associated with foodborne illnesses, there are some bacterial species and strains, normally considered nonpathogenic, that are capable of causing gastroenteritis, especially in susceptible individuals. They are designated as opportunistic pathogens. They are normally required to be alive and present in large numbers when consumed through a contaminated food.

IX. PREDOMINANT BACTERIAL AND VIRAL PATHOGENS ASSOCIATED WITH FOODBORNE DISEASES

Although many pathogenic bacterial species and viruses have been implicated in foodborne (and waterborne) disease outbreaks, there are some that have occurred at higher frequency than others. This can be seen from the data presented in Table

22.4. Among the two most common pathogens associated with foodborne intoxication from 1983 to 1987, the number of outbreaks as well as deaths was higher for *Clo. botulinum*, but the total number of cases was much higher for *Sta. aureus*, and the number of deaths was very high for *Clo. botulinum*. Among the enteric pathogens, the largest number of outbreaks, cases, and fatalities resulted from the foodborne infections caused by *Salmonella* spp. Toxicoinfection outbreaks and number of cases were higher for *Clo. perfringens* than for *Bac. cereus*. Among the two most common viral diseases from contaminated food and water, the number of outbreaks was higher for hepatitis A but Norwalk virus affected more people. From 1983 to 1987, salmonellosis was associated with the highest number of outbreaks of all foodborne diseases, affecting the largest number of individuals and causing the most deaths. In contrast, botulinum affected the least number of people but caused the highest number of deaths among the affected people (17 out of 140). No death was reported from staphylococcal intoxication, *Bacillus cereus* gastroenteritis, or Norwalk viral infection from 1983 to 1987. Some of the reasons for higher incidence of some pathogens, such as *Salmonella* spp., over others could be due to their occurrence in higher frequency in the food environments, their greater ability to produce the disease, and their ability to grow more rapidly under abusive conditions. With some pathogens, a single outbreak can involve many cases. This is typical for salmonellosis, shigellosis, and staphylococcal intoxication. For other pathogens, such as *Clo. botulinum*, only a few individuals are affected in a single outbreak. This is because of the predisposing factors associated with a particular outbreak. This aspect is explained later with respect to the specific disease. *Lis. monocytogenes*, not included in this data, caused 70 deaths among 259 cases in three outbreaks.[2]

Table 22.4 Predominant Bacterial and Viral Pathogens Associated with the Confirmed Foodborne Diseases between 1983 to 1987 in the United States

Bacteria and viruses[a]	Outbreaks		Cases		Deaths	
	No.	%	No.	%	No.	%
Staphylococcus aureus	47	7.6	3,181	6.2	0	0
Clostridium botulinum	74	11.9	140	0.3	10	17.0
Salmonella spp.	342	55.1	31,245	61.1	39	66.0
Shigella spp.	44	7.1	9,971	19.5	2	3.4
Escherichia coli	7	1.1	640	1.3	4	6.8
Campylobacter spp.	28	4.5	727	1.4	1	1.7
Clostridium perfringens	24	3.9	2,743	5.4	2	3.4
Bacillus cereus	16	2.6	261	0.5	0	0
Hepatitis A virus	29	4.7	1,067	2.0	1	1.7
Norwalk virus	10	1.5	1,164	2.3	0	0
	621	100	51,139	100	59	100

[a] Not included in the table are: *Brucella* spp., *Streptococcus* spp., *Vibrio* spp., and several others that combiningly were associated with 18 (2.8%) outbreaks. *Lis. monocytogenes* caused 3 outbreaks affecting 259 people with 70 deaths.

Table 22.5 Predominant Food Types Associated with Confirmed Foodborne Disease Outbreaks of Bacterial and Viral Origin from 1983 to 1988 in the United States

Food types	No. of outbreaks	%	Predominant pathogen(s) (% No. of outbreaks)
Meat products[a]	91	14.0	*Salmonella* spp. (53%), next *Sta. aureus*
Fish products[b]	20	3.0	*Clo. botulinum* (50%)
Egg products	11	2.0	*Salmonella* spp. (82%)
Dairy products	26	4.0	*Salmonella* spp. (27%)
Salads[c]	33	5.0	*Salmonella* spp., *Sta. aureus, Shigella* spp.
Baked foods	8	1.0	*Sta. aureus*
Fruits and vegetables	44	7.0	*Clo. botulinum*
Mushrooms	2	0.5	*Clo. botulinum*
Beverages	3	0.5	*Salmonella* spp.
Ethnic foods[d]	19	3.0	*Clo perfringens, Bac. cereus, Salmonella* spp.
Multiple foods[e]	123	19.0	*Salmonella* spp. (59%)
Unknown foods[f]	254	40.0	*Salmonella* spp. (68%), *Shigella* spp., viruses

[a] Includes: beef, ham, pork, sausages, chicken, turkey, stews.
[b] Includes: shellfish and fin fish.
[c] Includes: potato, chicken, fish, egg and other salads.
[d] Includes: fried rice, Chinese food, Mexican food.
[e] More than one food was involved in an outbreak.
[f] Although food(s) was implicated with an outbreak, confirmation about the involvement of a food or foods could not be made with certainty.

X. PREDOMINANT FOOD TYPES ASSOCIATED WITH FOODBORNE DISEASES OF BACTERIAL AND VIRAL ORIGIN

Certain food types or foods prepared under specific conditions and environments have been implicated more frequently with foodborne disease than others. Some of the factors could be the presence of a pathogen in the raw materials in higher frequency with a greater chance of contamination of the finished products, the ability of a pathogen to grow advantageously in a particular type of food, a greater chance of failure in quality control in a specific environment, and a higher possibility of contaminating the finished products by food handlers. Some of these aspects can be explained by the foodborne disease outbreak data presented in Table 22.5. Foods of animal origin (meat, fish, eggs, and dairy products) were implicated in 23% of the outbreaks, and *Salmonella* spp. were involved in large proportions (except for fish); the latter may arise because many food animals harbor *Salmonella* spp. as carriers in the digestive tract and thus can contaminate meat, eggs, and dairy products. In fish products, the incidence of botulism outbreak was high. This is because fishery products can be contaminated with *Clo. botulinum* (type: nonproteolytic E) present in some marine environments. Similarly, fruits and vegetables can be contaminated with *Clo. botulinum* from the soil. In contrast, salads, which are handled extensively, can be contaminated with several pathogens of human origin. The high incidence of *Clo. perfringens* in Mexican foods and *Bac. cereus* in Chinese fried rice is likely due to the methods used in the preparation of certain ethnic foods. The food types in 40% of the outbreaks could not be determined, because the food samples were not available for testing in many cases. However, one can speculate from the high incidence of *Salmonella* spp. in both multiple food and unknown food categories that many were probably foods of animal origin.[2]

XI. PREDOMINANT PLACES OF FOOD CONSUMPTION ASSOCIATED WITH CONFIRMED FOODBORNE DISEASE OUTBREAKS OF BACTERIAL AND VIRAL ORIGIN

An analysis of the relationship between the places of food consumption and the number of foodborne disease outbreaks revealed that, at least in the United States, the highest number of incidences occurred with foods served at food establishments (Table 22.6). These include fast food services, restaurants, cafeterias, and school lunches. Several factors could be associated with this.[2] The total number of meals served is very high. Also, the high proportion of the foods is served within a very short period of time. In addition, the number of people involved in handling the foods is very high, many of which may not have training in safe food handling. This can result in a greater chance of contamination and failure to observe proper sanitation. Improper cooling of foods, cross-contamination, and contamination of food from pets could be the reasons for the foodborne diseases with foods served in homes. Mishandling of foods, temperature abuse, and improper sanitation could be associated with the outbreaks in picnics. The exact sources could not be identified in 30% of the outbreaks; however, from the available information it is most likely that many of them were probably with foods served at food establishments. Although not listed in the table, the relative frequency of foodborne disease outbreaks with foods from large commercial processors is low. This is generally because of their use of good quality control, sanitation, and testing of the products for the pathogens. However, when contamination of food by a pathogen occurs in a big processing establishment, a large number of people over a wide area are affected.[2]

XII. PREDOMINANT CONTRIBUTING FACTORS ASSOCIATED WITH CONFIRMED FOODBORNE DISEASE OUTBREAKS FROM PATHOGENIC BACTERIA AND VIRUSES

Before consumption, foods are exposed to many different environments and conditions. These dictate if a pathogen present initially will survive or be killed, if recontamination can occur, or if a pathogen can multiply to reach a high population to cause disease. The foodborne disease outbreak data between 1983 and 1987 in the United States revealed that the predominant cause of outbreak was improper holding temperature (refrigerated temperature) of the food (Table 22.7); the temperature abuse resulted in the growth of pathogens (bacteria) to reach a level that causes illness following consumption. The other factors, in order of importance, were

Table 22.6 Predominant Establishments of Food Consumption Associated with Confirmed Foodborne Diseases Due to Pathogenic Bacteria and Viruses between 1983 and 1987 in the United States

Establishment	No. Outbreaks	%
Homes	149	24.8
Food services[a]	225	37.5
Picnic[b]	45	7.5
Processing plants[c]	NA	NA
Unknown[d]	181	30.2
	600	100

[a] Includes: Fast food places, restaurants, cafeterias, school lunch programs.
[b] Includes: Family picnics, church picnics, camp.
[c] Data not available.
[d] Exact place(s) could not be identified.

Table 22.7 Predominant Contributing Factors Associated with Confirmed Foodborne Outbreaks from Pathogenic Bacteria and Viruses from 1983 to 1987 in the United States

Contributing factor	1983	1984	1985	1986	1987	Total (%)
Improper holding temperature	52	68	57	40	27	244 (34.6%)
Poor personal hygiene	21	42	38	25	19	145 (20.5%)
Inadequate cooking	24	19	36	20	19	118 (16.7%)
Contaminated equipment	17	33	25	21	12	108 (15.3%)
Food from unsafe sources	15	3	10	5	6	39 (5.5%)
Other	9	12	12	11	8	52 (7.4%)

contamination of foods due to poor personal hygiene, survival of pathogens due to cooking at lower temperatures than specified, cross-contamination of foods from equipment (previously contaminated with pathogens), and food from unsafe sources, such as raw foods. These five factors, in combination, were responsible for 92.6% of the outbreaks caused by pathogenic bacteria and viruses.

XIII. INFLUENCE OF MONTH (OF THE YEAR) TO NUMBER OF FOODBORNE DISEASE OUTBREAKS CAUSED BY PATHOGENIC BACTERIA AND VIRUSES

In general, foodborne outbreaks of pathogenic bacterial and viral origin are more prevalent during the summer months. The compiled data between 1983 and 1987, as presented in Table 22.8, show that in the United States about 66% of the outbreaks occurred during May to September, with the highest incidence in August.[2] Between November and April, the incidence was reduced to about 34%. During commercial processing, the food products are exposed to indoor temperature, which may not vary greatly in the summer or the winter. However, both raw products and processed products could be exposed to outside temperature during transportation, displays in stores, at food establishments, and at home. Also, during the summer, frequency of picnics and outdoor eating is higher. A high temperature during the summer can stimulate rapid growth of the pathogens to reach high levels, even from a low initial level, within a relatively shorter period. During the winter months, the growth rate can be greatly reduced.

Table 22.8 Influence of the Month of the Year to Confirmed Bacterial and Viral Foodborne Disease Outbreaks from 1983 to 1987 in the United States

Months	No. of outbreaks			Comments
	Total (%)	Average	Range	
May to October	413 (65.7%)	69	57 to 86	Highest in August
November to April	216 (34.3%)	36	26 to 45	Lowest in February

Adapted from Bean, N., et al., *J. Food Prot.*, 53, 711, 1990. With permission.

XIV. INFLUENCE OF LOCATION ON FOODBORNE DISEASES OF PATHOGENIC BACTERIAL AND VIRAL ORIGIN

It has been mentioned before that in general the incidence of foodborne disease outbreaks is lower in the developed countries than in the developing countries. Several socioeconomic reasons and climatic conditions have major roles in this difference. However, the frequency of outbreaks can vary greatly among regions, both in the developed and developing countries. In the United States, foodborne disease outbreaks have been recorded more frequently in some states than in others. States with high populations, large numbers of industry, and warm climates have more incidences. Some of the states in this group are New York, Washington, California, Hawaii, and Pennsylvania (Table 22.9). Some other factors, such as a large ethnic population, greater frequency of migration and traveling of people, as well as better surveillance systems, could also be the reasons for the high incidence rate. In comparison, states with lower populations had very few incidents.

Table 22.9 Influence of Location on Confirmed Foodborne Disease Outbreaks from Pathogenic Bacteria and Viruses in the United States from 1983 to 1987

High incidence			Low incidence		
State	Total	Range	State	Total	Range
New York[a]	744	129–174	Wyoming	2	0.1
Washington	243	38–55	Montana	3	0.2
California	166	30–38	S. Dakota	4	0.2
Hawaii	138	20–36	Nevada	1	0.1
Pennsylvania	96	15–24	Utah	6	0–1

[a] New York City had 216 outbreaks with a range of 24 to 80 during the same period.

XV. HUMAN FACTORS IN FOODBORNE DISEASE SYMPTOMS

Consumption of a food contaminated with live cells of pathogens or their toxins by a group of people does not make everybody develop disease symptoms.[1] Also, among those who develop the symptoms, all may not show either the same symptoms or the severity of any one symptom. This is probably due to the difference in resistance among individuals. One of the factors involved in developing symptoms from the consumption of a contaminated food is the susceptibility of an individual to the contaminants. In general, infants and old, sick, and immunodeficient people are more susceptible than normal and healthy individuals. The chance of developing disease symptoms is directly related to the amount of contaminated food consumed. This is related to the number of viable cells of a pathogen or the amount of a toxin consumed by an individual. The virulence of a pathogen or a toxin consumed through a food also determines the onset of a disease and severity of symptoms. For some highly virulent pathogens, such as *Esc. coli* 0157:H7, consumption of as low as ten viable cells can cause disease in an infant. In contrast, for some pathogens, such as *Yersinia enterocolitica*, consumption of as high as 1 million or more viable cells is necessary for the symptoms to develop.[1]

XVI. ACCEPTANCE QUALITY OF FOOD DUE TO GROWTH OF PATHOGENS

Pathogenic viruses need viable host cells for growth; in prepared foods, they cannot grow and thus do not affect the food quality. Pathogenic bacteria can grow in many foods. When the environment is suitable, only a few viable cells present initially can reach a high level, maybe several millions per gram or milliliter. However, growth of some pathogens, even to a high level, may not alter the color, texture, and odor of a food. People consume this food without suspicion and develop symptoms of a foodborne illness.

XVII. SEQUENCE OF EVENTS IN A FOODBORNE DISEASE

For a foodborne disease to occur, several events must happen in sequence (Figure 22.1).[2] An understanding of these sequences is helpful in investigating the cause of a foodborne disease. It also helps in recognizing how the sequence can be broken in order to stop a foodborne disease. Initially, there has to be a source of pathogen. Next, the pathogen has to contaminate a food. Consumption of the food contaminated with a pathogenic virus may lead to viral infection. For bacterial pathogens (and toxicogenic molds) the contaminated food has to support growth and be exposed for a certain period of time at a suitable temperature to enable the pathogens to grow. For intoxication, the growth should reach a sufficient level to produce enough toxin so that when the food is consumed, the individual(s) develops the symptoms. For bacterial infection, viable cells of a pathogen need to be consumed in sufficient numbers, which vary greatly with pathogens, to survive stomach acidity, establish in the digestive tract, and cause illness. In case of toxicoinfection, viable cells should be consumed either in very high numbers (for those that cannot multiply in the digestive tract, such as *Bac. cereus*) or in reasonable numbers (for those that can multiply in the digestive tract, such as *Vibrio cholerae*) so that toxins released by them into the digestive tract can produce the symptoms.

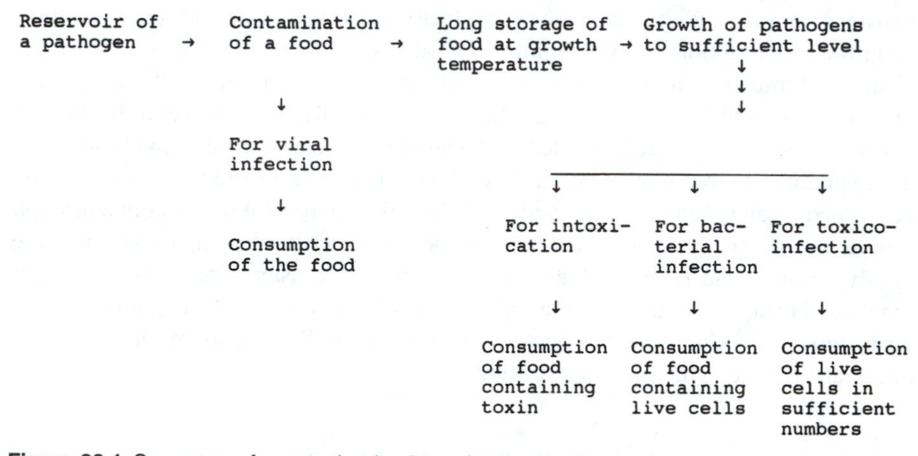

Figure 22.1. Sequence of events that lead to a foodborne disease by pathogenic bacteria and viruses.

XVIII. CONCLUSION

The material presented in this chapter shows that foodborne disease outbreaks can be caused by different microorganisms, some of which are more predominant than others. The different parameters associated with the predominant factors (pathogenic bacteria and viruses) were also described. Finally, the sequence of events necessary for a foodborne disease to occur were discussed. This information will be helpful in understanding the material presented in the next chapters.

REFERENCES

1. Garvani, R. B., Food science facts, *Dairy Food Sanitation*, 7, 20, 1987.
2. Bean, N., Griffin, P. M., Goulding, J. S., and Ivey, C. B., Foodborne disease outbreak 5-year summary, 1983–1987, *J. Food Prot.*, 53, 711, 1990.
3. Bean, N. and Griffin, P. M., Foodborne disease outbreaks in the United States, 1973–1987: pathogens, vehicles, and trends, *J. Food Prot.*, 53, 804, 1990.
4. Todd, E. C. D., Preliminary estimate of costs of foodborne disease in the U.S., *J. Food Prot.*, 52, 595, 1989.

CHAPTER 22 QUESTIONS

1. List the factors associated with a foodborne disease outbreak.
2. Define or explain an outbreak, a reported case, and a confirmed outbreak.
3. Survey 25 of your friends and relatives about the number of times each had a digestive disorder in the last 5 years (or 2 years). From this, calculate approximately how many people can be affected with foodborne disease in the United States per year. Is it close to 20,000, 5 million, 25 million, or 80 million?
4. Define foodborne intoxication, infection, and toxicoinfection, and give two examples for each.
5. Discuss why pathogenic bacteria are the predominant agents in foodborne diseases.
6. "Foods of animal origin are more frequently associated with foodborne disease outbreaks." Justify or explain the statement.
7. List the possible reasons of higher incidence of foodborne diseases with food served at food service establishments.
8. List, in order of importance, the five most important contributing factors in the occurrence of foodborne disease of pathogenic bacterial and viral origin.
9. List the sequence of events in the case of a salmonellosis outbreak.
10. Briefly discuss what you have learned from the material in this section.

CHAPTER 23

Foodborne Intoxications

CONTENTS

I. INTRODUCTION

Foodborne intoxication or food poisoning of microbial origin occurs from the ingestion of a food containing preformed toxin. Two of bacterial origin, staphylococcal intoxication and botulism, and mycotoxicosis of mold origin are briefly discussed in this chapter. The discussion includes the relative importance of a disease, characteristics of the microorganism(s) involved, predominant types of food, nature of the toxin(s), disease and the symptoms, preventative measures, and, for some, analysis of actual outbreak.

Some general characteristics of food poisoning include:

1. Toxin is produced by a pathogen while growing in a food.
2. A toxin can be heat labile or heat stable.
3. Ingestion of a food containing active toxin, not viable microbial cells, is necessary for poisoning (except for infant botulism, in which viable spores need to be ingested).
4. Symptoms generally occur quickly, as early as 30 min after ingestion.
5. Symptoms differ with type of toxin; enterotoxins produce gastric symptoms and neurotoxins produce neurological symptoms.
6. Febrile symptom is not present.

II. STAPHYLOCOCCAL INTOXICATION (STAPHYLOCOCCAL GASTROENTERITIS; STAPHYLOCOCCAL FOOD POISONING; STAPH FOOD POISONING)

A. Importance

Staphylococcal food poisoning, caused by toxins of *Staphylococcus aureus,* is considered to be one of the most frequently occurring foodborne diseases worldwide.[1,2] In the United States, at least before the 1980s, it was implicated in many outbreaks. However, in recent years, the number of staphylococcal food poisoning outbreaks has declined. CDC reports indicate that during 1972 to 1976, it was associated with 21.4% of the foodborne disease outbreaks affecting 29.7% of total cases; in contrast, between 1983 and 1987, there were 5.2% staphylococcal foodborne outbreaks affecting 5.8% of the cases, with no deaths. This decline is probably a reflection of the better use of refrigerated temperatures to store food and improved sanitary practices that can control contamination and growth of *Sta. aureus*. Even then, the number of outbreaks and number of cases of staphylococcal gastroenteritis is much higher than several other microbial foodborne disease outbreaks.

B. Characteristics

1. Organisms

Sta. aureus are Gram-positive cocci, occur generally in bunches, and are nonmotile, noncapsular, and nonsporulating.[3-5]

2. Growth

Most strains ferment mannitol and produce coagulase, thermonuclease, and hemolysin, but differ in their sensitivity to bacteriophages. The cells are killed at 66°C in 12 min, and at 72°C in 15 s. *Sta. aureus* are facultative anaerobes, but grow rapidly under aerobic conditions. They can ferment carbohydrates and also cause proteolysis by the extracellular proteolytic enzymes. They are mesophiles with a growth temperature range of 7 to 48°C, with fairly rapid growth between 20 and 37°C. Other important growth characteristics are their ability to grow at low A_w (0.86), low pH (4.8), and high salt and sugar concentrations of 15% and in the presence of NO_2. However, their growth can be reduced by combining two or more parameters. Because of their ability to grow under several adverse conditions, *Sta. aureus* can grow in many foods. Normally, they are poor competitors in the presence of many other microorganisms found in foods. But their ability to grow in adverse environments gives them an edge in growth in many foods where others do not grow favorably.[3-5]

3. Habitat

Enterotoxin-producing *Sta. aureus* strains have generally been associated with staphylococcal food intoxication. Although strains of several other *Staphylococcus* species are known to be enterotoxin producers, their involvement in food poisoning is not fully known. *Sta. aureus*, along with many other staphylococci, are naturally present in the nose, throat, skin, and hair (feathers) of healthy humans, animals, and birds. *Sta. aureus* can be present in infections, such as cuts in the skin and abscesses in humans, animals, and birds, and cuts in hands and facial erupted acne in humans. Food contamination generally occurs from these sources.[3-5]

C. Toxins and Toxin Production

Enterotoxigenic strains of *Sta. aureus* produce six different enterotoxins: A, B, C1, C2, D, and E (also designated as SEA, SEB, etc.).[3,4,6] They are serologically distinct, heat-stable proteins of molecular weight 26 to 30 kDa and differ in toxicity. The toxins vary in heat stability, SEB being more stable than SEA. Normal temperature and time used in processing or cooking foods will not destroy the potency of the toxins. Outbreaks from SEA are more frequent; one reason could be because of its high potency.

Rate of toxin production by a strain is directly related to its rate of growth and cell concentrations. Optimum growth occurs around 37 to 40°C. Under optimum conditions of growth, toxins can be detected when a population has reached over a few million per gram or milliliter of food and generally in about 4 h. Some of the lowest environmental parameters of toxin production are 10°C, pH 5.0, A_w 0.86. However, by combining two or more parameters, the lowest ranges can be adversely affected.

D. Disease and Symptoms

Staphylococcal toxins are enteric toxins and cause gastroenteritis.[2-6] A normal, healthy individual has to consume about 30 g or ml of a food containing toxins produced by 10^6 to 10^7 cells per gram (or milliliter); infants and old and sick individuals need less. The symptoms occur within 2 to 4 h, with a range of 30 min to 8 h, and are directly related to the potency and amount of toxin ingested and the resistance of an individual. The disease lasts for about 1 to 2 d and is rarely fatal. Only four deaths were recorded between 1973 and 1987.[2-6]

The primary symptoms are salivation, nausea and vomiting, abdominal cramps, and diarrhea. Some secondary symptoms are sweating, chills, headache, and dehydration.[2-6] However, the symptoms and their severity vary among individuals in an outbreak.

E. Food Association

Many foods have been implicated in staphylococcal foodborne outbreaks.[1,2,5,7] In general, the bacterium grows in the food and produces toxins without adversely affecting its acceptance quality. Many protein-rich foods, foods that are handled extensively, foods in which associated bacteria grow poorly, and foods that have been temperature abused are associated with staphylococcal gastroenteritis. Some of the foods that have been more frequently implicated include ham, corned beef, salami, bacon, barbecued meat, salads, baking products containing cream, sauces, and cheeses. The relative frequencies of different foods involved in staphylococcal food poisoning in the United States between 1973 and 1987 are presented in Table 23.1. Pork, baking products, beef, turkey, chicken, and eggs are associated with the high percentages of outbreaks.[1] Different types of salads, due to extensive handling and high chance of temperature abuse, have been implicated in relatively high numbers in staphylococcal food poisoning. From 1983 to 1987, salads were associated with 13% of total staphylococcal food poisoning outbreaks in the United States. They are included in the table in the "other food" category. Three major contributing factors in these outbreaks from 1983 to 1987 were improper holding temperature (51.6%), poor personal hygiene (23.4%), and contaminated equipment (17.2%). Major places (where foods were prepared or served) were food services (24.7%), homes (14.9%), and picnics (8.5%). A high percentage of outbreaks occurred between May and October (63.8%), with the highest frequency in August (21.3%). In the case of imported foods, a raw or processed food exported from a country can have *Sta. aureus* toxins, but can cause food poisoning in a different country. This is exemplified by at least six outbreaks of staphylococcal gastroenteritis in 1989 in the United States from the consumption of dishes prepared using enterotoxin containing canned mushrooms processed in a plant in the People's Republic of China. *Sta. aureus* probably grew prior to canning and the enterotoxin, being heat stable, remained potent after canning.

Table 23.1 Food Types Involved in Confirmed Staphylococcal Food Poisoning Outbreaks in the United States from 1973 to 1987

Food type	% Outbreaks[a]	Food type	% Outbreaks
Pork	16.2	Fish	1.3
Bakery products	7.1	Dairy products	1.7
Beef	6.0	Fruits and vegetables	1.1
Turkey	5.5	Ethnic foods	1.1
Chicken	3.8	Other	37.2[b]
Eggs	2.5	Unknown	6.5

[a] Out of a total of 367 outbreaks.

[b] The foods included in the "other" category are not known. However, salads were implicated in many outbreaks. Between 1983 and 1987, salads were involved in about 13% of the total staphylococcal foodborne disease outbreaks.

F. Prevention (Reduction) of the Disease

The normal occurrence of *Sta. aureus* in raw food materials, among food handlers, and many food environments makes it impossible to produce nonsterile foods that are free of this bacterium. Thus, a "0" tolerance is not economically possible to achieve. One needs to recognize that many foods can contain *Sta. aureus* and consumption of a food containing 100 or 500 cells per gram (or milliliter) is, in all probability, not going to make a person sick (unless the food has large amounts of preformed toxin). To reduce the incidence of staphylococcal food poisoning, the aim will be to reduce initial load of *Sta. aureus* in a food by proper selection of raw materials and ingredients, sanitation of the food environments, and proper personal hygiene among the food handlers.[3-5] People with respiratory disease, acute types of facial acne, skin rash, and cuts in hands should not handle the food. Where possible, the products should be heat treated to ensure killing of the live cells. Following heating, recontamination of the products should be avoided. The most important aim should be that the processed products should be chilled to ≤5°C quickly. Suitable preservatives can also be used to kill or arrest growth. Care should be taken so that the inside of the food, not just the surface, reaches the chilled temperature, preferably within 1 h. Finally, the food should not be subjected to temperature abuse and stored for a long period of time at growth temperature prior to eating. One should recognize that once the heat-stable toxins are formed, heating prior to eating will not ensure safety.

G. Identification Methods

[See Appendix D] To associate a food implicated in staphylococcal food poisoning, the food, foods, or vomit samples are analyzed for the presence of high levels of enterotoxigenic *Sta. aureus* cells and enterotoxin(s). Enumeration techniques in one or more selective differential agar media to determine the load of viable cells of *Sta. aureus*, followed by several biochemical tests, such as hemolysis, coagulase, thermonuclease reactions, or the ability of a pure culture to produce enterotoxin, are performed to link the potential cause of the food poisoning outbreaks.

The enterotoxin(s) from the food or vomit samples are extracted and tested, either by biological means or by serological means to associate it with the outbreak. In the biological means, animals (such as cats, monkeys, or dogs) are given the enterotoxin preparation orally or injected intraperitoneally or intravenously. Vomiting symptoms by the test animals is a positive indication of the presence of staphylococcal enterotoxin.

In the serological methods, the enterotoxins are purified and examined by one of the several recommended methods. Not only are these very sensitive tests, but they allow one to identify the type(s) of enterotoxin(s) involved in a food poisoning case.

H. Analysis of an Outbreak

A foodborne disease outbreak affecting 52 of 101 people who attended a dinner (foods prepared at home) was reported on December 6, 1986, in Riverton, Wyoming.[8] Of these, 49 people needed immediate medical attention. The following symptoms developed between less than 1 to 7 ½ h after the meal, were recorded among the affected cases: nausea (100%), vomiting (98%), diarrhea (90%), abdominal cramps (83%), prostration (62%), chills (52%), sweating (35%), and blood pressure–temperature depression (21%). When the regulatory people came to investigate after 36 h, samples of foods or vomit of the patients from the doctors were not available. However, one vomit sample frozen by a patient was available and, when analyzed, was found to have 12 to 19 × 10^6/g coagulase-positive *Staphylococcus aureus*. Meat obtained from a leftover turkey carcass had 1 × 10^6 coagulase-positive *Sta. aureus*. An investigation revealed that a person who deboned and handled the turkey had erupting facial rash (acneiform) during food preparation. The turkeys (three total) were improperly cooled following cooking and held on an improperly heated steam table for 4 to 4 ½ h prior to serving.

This is a classic case of staphylococcal food poisoning outbreak, in which over half the people who ate foods involving several preparations developed many of the classic symptoms, some within 30 min. Although the food samples served in the dinner or vomit samples from the attending physicians or medical facilities were not available, analysis of a vomit sample saved by a patient revealed the presence of very high numbers of coagulase-positive *Sta. aureus* and meat available from a leftover turkey carcass also had fairly large numbers of *Sta. aureus*. So, from the symptoms and indirect evidence, the outbreak was concluded to be an incidence of staphylococcal food poisoning.

The sequences of events were most likely as follows: the erupting facial rash of the food handler was most probably the source of the pathogen. The pathogen was transmitted by the individual through the hands during the deboning of the cooked turkey. *Sta. aureus* is capable of growing in meat under a suitable environment. The turkey, following cooking and prior to serving, was temperature abused for a long period of time, thus enabling the contaminants to grow and reach very high populations. However, the acceptance quality of the meat was not adversely affected. The stage was set and many consumed enough toxin with turkey (and possibly with other servings also) to develop the symptoms.[8]

III. BOTULISM

A. Importance

Botulism results following consumption of food containing the toxin botulin of *Clostridium botulinum*. It is a neurotoxin and produces neurological symptoms along with some gastric symptoms.[1,2] Unless prompt treatment is administered, it is quite fatal. Infant botulism occurs from the ingestion by the infant of *Clo. botulinum* spores that germinate, grow, and produce toxins in the GI tract and produce specific symptoms.

In the United States, the average number of outbreaks per year is about 15 to 16. From 1973 to 1983, there were a total of 23 outbreaks. This represents about 8% (231 out of 2841) of the total foodborne outbreaks involving less than 4% (4984 out of 124,994) of the total cases. However, out of a total of 247 deaths (2 per 1000 cases) from all foodborne disease, 47 died of botulism during this period (9.5 per 1000 cases), even with the medical treatment available in this country. A majority of the outbreaks occurs with foods prepared at home. Ethnic food preparations have been implicated in many cases.

B. Characteristics

1. Organism

Cells of *Clo. botulinum* strains are Gram-positive rods, occur as single cells or in small chains; many are motile, obligate anaerobes and form single terminal spores. Cells are sensitive to low pH (<4.6), lower A_w (0.93), and moderately high salt

(5.5%). Spores do not germinate in the presence of nitrite (250 ppm). Spores are highly heat resistant (killed at 115°C); but the cells are killed at moderate heat (pasteurization). Toxins form during growth. Strains can either be proteolytic or nonproteolytic.[8-10]

2. Growth

Clo. botulinum strains, on the basis of the type of toxin production, have been divided into six types: A, B, C, D, E, and F. Of these, A, B, E, and F have been associated with human foodborne intoxications. Type A strains are proteolytic, type E are nonproteolytic, but types B and F strains can be either proteolytic or nonproteolytic. The proteolytic strains can grow between 10°C and about 48°C, with the optimum at 35°C. The nonproteolytic strains grow optimally at 30°C, with a range between 3.3 and 45°C. Optimum growth facilitates optimum toxin production. Anaerobic conditions are necessary for growth. Spore germination and outgrowth prior to cell multiplication is favored in the environment of cell growth. Similarly, conditions that prevent cell growth also adversely affect spore germination. As indicated before, either pH 4.6, A_w 0.93, or NaCl (5.5%) can prevent cell growth; but by using two or more parameters along with lower temperature, the lower growth limits of any of the above parameters can be greatly reduced.[8-10]

3. Habitat

Spores of *Clo. botulinum* are widely distributed in soil, sewage, mud, sediments of marshes, lakes and coastal waters, plants, and intestinal contents of animals and fishes. Fruits and vegetables can be contaminated with spores from soil, fishes from water and sediments, and various other foods from many of the above sources. While type A and B spores are more prevalent in soil, sewage, and fecal matter of animals, type E spores are generally found in marine environments. While type A spores are predominant in the western United States, type B spores are found predominantly in the eastern United States and different parts of the world.[8-10]

C. Toxins and Toxin Production

The toxins of *Clo. botulinum* are neurotoxic proteins. In general, toxins associated with food intoxication in humans (types A, B, E, and F) are extremely potent, and only a small amount of toxin is required to produce the symptoms and cause death. Following ingestion, toxins are absorbed through the intestinal wall and reach the nerve cells, stopping signal transfers and causing paralysis of all involuntary muscles. The toxins move slowly through the body. Toxins produced by nonproteolytic strains are not fully activated; trypsin treatment is necessary to activate them. The toxins are heat labile and can be destroyed in a contaminated food by high and uniform heat, such as 90°C for 15 min or boiling for 5 min. Radiation at 5 to 7 mrad can also destroy them.[9,10]

Cell growth is necessary for toxin production. At optimum growth temperature, toxins are produced in large amounts. However, at both extremes of growth range enough toxin can be produced by a strain in a food to cause disease and death following ingestion.

D. Disease and Symptoms

Botulism is caused by the ingestion of the neurotoxin botulin present in a food. However, at the initial stage (generally 12 to 36 h, but can be 2 h), some gastrointestinal disorders (e.g., nausea, vomiting, diarrhea, and constipation) may be evident. Neurological symptoms develop within a short time, especially if the amount of botulin consumed is high. As they are highly potent toxins, only a very small amount (1 ng/kg body weight) is necessary for severe symptoms and even death. In general, neurological symptoms include blurred or double vision, difficulty in swallowing, breathing, and speaking, dryness of the mouth, and paralysis of different involuntary muscles that spreads to the lungs and heart. Death usually results from respiratory failure.

The toxins are antigenic; thus antitoxins are available. Soon after the onset of the symptoms and if the amount of a toxin consumed is fairly low, antitoxins can be successfully administered to treat the disease. But in some advanced cases, especially if the diagnosis is delayed, antitoxin administration may not be successful. In the United States, even with the available facilities, botulism accounts for about 19% of the total foodborne fatal cases.[9-11]

E. Food Association

The events in a foodborne botulism involve contamination of a food with *Clo. botulinum* spores, survival of the spores during processing, and the ability of the spores to germinate, outgrow, and multiply when the product is abused (temperature and time). Results presented in Table 23.2 show that the largest number of outbreaks are associated with fruits and vegetables.[1] These were mainly low-acid vegetables (e.g., green beans, corn, spinach, asparagus, pepper, and mushrooms) and fruits (e.g., figs and peaches). The next highest incidence was with fin fish and they include fermented, improperly cooked, and smoked fish and fish eggs. Type E was associated predominantly with fish, while type A and B were associated with vegetables. The major cause of outbreaks was improper home canning of the contaminated products. Between 1983 and 1987, out of 231 botulism outbreaks in the United States, 56 occurred at home. The occurrence of botulism from meat, poultry, and dairy products is low. This is probably because they are mostly heated and eaten quickly. Several outbreaks from "unlikely" foods (sauteed onions or baked potatos) have been recorded; but in many instances, the foods were subjected to temperature abuse (held for a long time at warm temperature). Some condiments, such as chili peppers, relish, and sauce, have also been associated with outbreaks.

Growth of proteolytic strains in meats, and low-acid, high-protein vegetables generally produce obnoxious odors and gas. In low-protein vegetables, these characteristics may not be obvious. The growth of nonproteolytic strains, even when growing in meat, fish, and other high-protein foods, do not produce spoilage characteristics. As the toxins are heat labile, high uniform heating (90°C for 15 min or

Table 23.2 Food Types Involved in Confirmed Botulism Outbreaks from 1973 to 1987 in the United State

Food types	No.	%	Food types	No.	%
Beef stew	2	0.9	Mushrooms	5	2.2
Chicken	1	0.4	Beverages (nondairy)	5	2.2
Dairy products	1	0.4	Pork	1	0.4
Fin fish	35	15.2	Turkey	1	0.4
Fruits and vegetables	99	42.9	Other[a]	40	17.3
Mexican food	3	1.3	Unknown	38	16.4

[a] Not identified, but probably include many types of food, many of which were of "unlikely" categories, and no single type was involved in a large number of outbreaks.

boiling for 5 min) of a suspected food will destroy the toxins and make the food theoretically safe. But as only a small amount of toxin is enough to cause disease, it is better not to consume a suspected food.[1,2,11]

F. Prevention of Botulism

The single most important control method is the use of proper temperature and time in home canning of low-acid products.[13] Commercial processors use the 13 D concept. Directions for pressure canning of foods at home are available and they should be strictly followed. Proper and uniform cooking of some foods (e.g., fish) at high temperatures should be followed. Foods cooked at temperatures in which spores survive should be stored at low temperatures (at 3°C or below); at refrigerated temperature (4 to 5°C), storage should not be prolonged unless some additional precautions are used, such as NO_2, low pH, low A_w, NaCl, and others. Suspected foods should be properly heated prior to consumption but it is better not to eat them. Even tasting a small amount of suspected food without giving high and uniform heat treatment can be dangerous.

G. Identification Methods

In a suspected food, the presence of *Clo. botulinum* can be determined by enumeration techniques using selective agar media and anaerobic incubation. The presence of toxins in the food is more often tested. This involves injection of a food extract intraperitoneally to mice. Development of characteristic neurological symptoms, followed by death in 92 h, suggest the presence of toxin. The people engaged in testing for the organism or the toxins need to immunize themselves prior to handling the materials.

H. Infant Botulism

Clo. botulinum spores, ingested by human infants through food and the environment, can germinate in the intestine and produce toxin to cause infant botulism.[12] The spores fail to produce the same disease, generally in individuals above 1 year of age. In these individuals, probably the well-established normal population of gastrointestinal flora discourage spore germination and cell multiplication by *Clo. botulinum*. Both type A and type B have been identified in infant botulism cases. The symptoms consist of general weakness, inability to control the head, loss of reflexes, and constipation. Foods such as honey and corn syrup and dirt have been linked as sources of *Clo. botulinum* spores in infant botulism cases.[11,12]

I. Analysis of a Foodborne Botulism Case

On May 4, 1992, in New Jersey, a man of Egyptian origin developed dizziness, facial drooping, dry mouth, weakness, and respiratory difficulties. Within 2 days, two more family members developed similar symptoms. The cases were diagnosed as botulism and treated with trivalent (A, B, E) botulinal antitoxins. The source of toxin was traced to an ethnic fish preparation made with uneviscerated, salt-cured fish. On May 3, the family had obtained the product from outside and consumed it without cooking.

The sequence of events was most probably as follows. The fish had *Clo. botulinum* spores in the gut (probably type E) from the water and water sediment. The uneviscerated fish were cured in salt, the concentration of which was not high enough to prevent germination, cell growth, and toxin production. The product was consumed, probably in different amounts, without cooking by the family members who developed symptoms at different times. The outbreak could have been avoided by removing the viscera, curing the fish in recommended salt concentrations, and heating prior to eating.[13]

IV. MYCOTOXICOSIS

A. Importance

Many strains of molds, while growing in a suitable environment (including in foods), produce metabolites that are toxic to humans, animals, and birds, and are grouped as mycotoxins.[14,15] Consumption of foods containing mycotoxins causes mycotoxicosis. They are secondary metabolites and not proteins or enteric toxins. Many are carcinogens and, when consumed, can cause cancer in different tissues in the body. Some cause toxicity of organs by unknown mechanisms. Incidence of mycotoxicosis in humans has not been recorded in recent years in many countries, at least in the developed countries. This is because their presence in many foods are critically regulated and evaluated. Incidences of mycotoxicosis have been recorded in some developing countries, in recent years. Some of the well-known mycotoxicosis incidences in humans in the past include ergotism from the consumption of bread made from rye infected with *Claviceps purpurea* in Europe between the 14th and 16th centuries, yellow rice disease from the consumption of rice infested with toxigenic strains of several *Penicillium* species in Japan during the 17th century, and alimentary toxic aleukia from the ingestion of grains infested with toxigenic strains of *Fusarium* species in Russia in the early 20th century. In recent years, several incidences were reported in animals and birds, which included the death of thousands of turkeys from liver necrosis in the 1960s in England following feeding peanut meal in which *Aspergillus flavus* grew and produced the toxin aflatoxin.

B. Characteristics

1. Organisms

Toxigenic species and strains of molds from many genera are known to produce mycotoxins. Some of the toxigenic species from several genera and the toxins they produce include *Aspergillus flavus, Asp. parasiticus* (both produce aflatoxins), *Asp. nidulans* and *Asp. versicolor* (sterigmatocystin), *Penicillium viridicatum* (ochratoxin), *Pen. patulum* (patulin), *Pen. roquefortii* (roquefortin), and *Claviceps purpurea* (ergotoxin). The toxigenic strains cannot be differentiated from nontoxigenic strains just from their morphological characteristics. It is necessary to grow a strain under suitable conditions and test the material for the presence or absence of a mycotoxin. This is particularly important for the mold strains from the different genera used in food production.[14]

2. Growth

In general, molds grow best in humid and warm environments. They are aerobic and thus need air for growth. They can grow, though slowly, at very low A_w (0.65), low temperature (refrigerated temperature), and low pH (3.5). These conditions are often used to extend the shelf life of many foods.[14] Unless other methods (such as vacuum packaging) are used, they can grow in these foods and, if toxigenic, can produce toxins in the foods.[14]

3. Habitat

The spores are present in soil, dust, and the environment. Many foods can have viable spores or mycelia, especially before a heat treatment.[14]

C. Toxins and Toxin Production

Mycotoxins include a large number of toxins produced by different toxigenic species and strains of molds.[14,15] Many have not yet been identified. Some of the toxins have been listed before. They are small molecular heterocyclic organic compounds and some have more than one chemical type. An example of this is aflatoxin, which has two major types, B1 and G1, and each has several subtypes. Aflatoxin B1 is considered the most potent.

Mycotoxins are produced by the toxigenic mold strains as secondary metabolites. Toxin production, in general, is directly related to the growth rate of a mold strain. In microbiological media suitable for growth of molds, the *Asp. flavus* strain is capable of producing optimum concentrations of aflatoxin at 33°C, pH 5.0, and A_w 0.99.[14,15]

D. Food Association

The growth of toxigenic mold strains and the presence of specific mycotoxins have been detected in many foods. These include corn, wheat, barley, rye, rice, beans, peas, peanuts, bread, cheeses, dry sausages, apple cider, grain meals, dough cassava, cotton seeds, and spaghetti.[14,15] Consumption of mycotoxin-contaminated food can cause mycotoxicosis in humans. Feeding moldy products to food animals (including moldy silage) and birds can also produce foods of animal origin (milk, eggs) that are contaminated with mycotoxins. Many of the mycotoxins are resistant to heat used in the normal preparation of foods. Thus their elimination by heating is not used as a means to remove them from foods.

E. Prevention of Mycotoxicosis

In preventing human mycotoxicosis, the contamination of food with toxigenic mold strains (or all molds) should be reduced.[14,15] This is relatively difficult to achieve, but proper packaging can be used to reduce the incidence. Heat treatment, where possible, can also reduce the load by killing the molds and their spores. Preventing growth in food (and feeds) should be a major consideration in reducing the incidence of human mycotoxicosis. This can be achieved by using anaerobic packaging, reducing A_w where possible to 0.6, freezing, and by using specific preservatives against mold growth. A product in which molds have grown should not be consumed. Although the trimming of foods showing mold growth is a common practice, one cannot be sure that the remaining portion is free of mycotoxins. Under commercial operations, removal of some mycotoxins (aflatoxins) from food or food ingredients has been studied. Generally, solvent extraction methods have been used. Finally, treatment of some foods or food ingredients with suitable chemicals to inactivate mycotoxins, such as aflatoxins, has also been studied. Some chemicals (e.g., ammonia, hydrogen peroxide, and sodium hypochlorite) were found to inactivate aflatoxins.[14,15]

F. Detection Methods

Several methods have been studied to detect mycotoxins; more particularly, aflatoxins in food. These include solvent extraction of a suspected food sample, thin layer chromatographs of the extract, and visualizing under a UV light or fluorescent light. Chemical tests and analysis by mass spectral methods are used for identification of the specific type of aflatoxin.[14,15]

REFERENCES

1. Bean, N. H. and Griffin, P. M., Foodborne disease outbreaks in the United States, 1973–1987, *J. Food Prot.*, 53, 804, 1990.
2. Bean, N. H., Griffin, P. M., Goulding, J. S., and Ivey, C. B., Foodborne disease outbreaks, 5 year summary, 1983–1987, *J. Food Prot.*, 53, 711, 1990.

3. Tatini, S. R., Influence of food environments on growth of *Staphylococcus aureus* and production of various enterotoxins, *J. Milk Food Technol.*, 36, 559, 1973.
4. Smith, J. L., Buchanan, R. L., and Palumbo, S. L., Effects of food environment on staphylococcal enterotoxin synthesis: a review, *J. Food Prot.*, 46, 545, 1983.
5. Garvani, R. B., Bacterial foodborne diseases, *Dairy Food Sanitation,* 2, 77, 1987.
6. Halpin-Dohnalek, M. and Marth, E. M., *Staphylococcus aureus*: production of extracellular compounds and behavior in foods: a review, *J. Food Prot.*, 52, 262, 1989.
7. Hard-English, P., York, G., Stier, R., and Cocotas, P., Staphylococcal food poisoning outbreaks caused by canned mushrooms from China, *Food Technol.*, 44(12), 74, 1990.
8. Anonymous, Food and environmental health, *Dairy Food Sanitation,* 7, 413, 1987.
9. Zottola, E. A., Botulism, *Agric. Expt. Sta. Service, Univ. Minnesota,* St. Paul, MN 55108, Ext. Bull. No. 372.
10. Pierson, M. D. and Reddy, N. R., *Clostridium botulinum, Food Technol.*, 42(4), 196, 1988.
11. Foster, E. M., *Clostridium botulinum, Food Technol.*, 40(8), 16, 1986.
12. Kautler, D. A., Lilly, T., Solomon, H. M., and Lynt, R. K., *Clostridium botulinum* species in infant foods: a survey, *J. Food Prot.*, 45, 1028, 1982.
13. Communicable Disease Center, Outbreak of type E botulism associated with an uneviscerated, salt-cured fish products — New Jersey, *Morbidity Mortality Weekly Rep.*, 41, 521, 1992.
14. Bullerman, L. B., Significance of mycotoxins to food safety and human health, *J. Food Prot.*, 42, 65, 1979.
15. Moorman, M., Mycotoxins and food supply, *Dairy Food Environ. Sanit.*, 10, 207, 1990.

CHAPTER 23 QUESTIONS

1. List five characteristics of foodborne intoxication.

2. Describe some growth characteristics of *Staphylococcus aureus* strains that give them an advantage for growth and toxin production in a food. Name two known food types that have the highest incidence of staphylococcal poisoning and state the reasons.

3. List the symptoms of staphylococcal food poisoning, and discuss how the disease can be differentiated from other food poisoning of microbial origin.

4. From a case of staphylococcal food poisoning outbreak (the instructor will provide), determine the sequence of events in the outbreak. What precautions could have been taken to avoid the incidence?

5. Describe how foodborne botulism differs from infant botulism.

6. List the types of *Clostridium botulinum* associated with food intoxication, and describe the changes that can occur from their growth in a food high in protein and a food low in protein.

7. Discuss how the combination of several environmental parameters can be used to control growth of *Clostridium botulinum* in a food.

8. List two food groups frequently associated with botulism in the United States. Suggest four methods that could reduce the incidence.

9. Explain the terms: Mycotoxins, mycotoxin-producing strain of a *Penicillium* species, aflatoxin, and growth characteristics of molds.

10. List five foods that directly or indirectly can be contaminated with mycotoxins. Discuss mycotoxin concerns in foods and food ingredients imported to the United States from developing countries.

CHAPTER 24

Foodborne Infections

CONTENTS

I. INTRODUCTION

Foodborne infection occurs from the consumption of food (and water) contaminated with pathogenic enteric bacteria and viruses. Many pathogens can be included in this group. However, several of them are involved more frequently than others and they are discussed in this chapter. The discussions include relative importance, characteristics, food association, toxins, disease symptoms, and prevention. For some, detection methods and case histories are also included.

Some characteristics of foodborne infections include

1. Live cells of the enteric pathogens (bacteria and viruses) must be consumed through food.
2. The pathogens penetrate through the membrane and establish in the epithelial cells of the intestines, multiply, and produce toxins (infection).
3. Dose levels to cause infection vary greatly. Theoretically, one live cell has the potential to produce the disease. The experts estimate that consumption of about ten cells (for an extremely virulent species and strain, such as *Esc. coli* 0157:H7) to about 10^5 cells or more (for a less virulent species and strain, such as *Yer. enterocolitica*) may be required for the disease.
4. Symptoms generally occur after 24 h, which, depending upon a pathogen, can be both enteric and nonenteric in nature.
5. Enteric symptoms are local and due to enteric infection and the effect of toxins. The symptoms include abdominal pain, diarrhea (sometimes accompanied with blood), nausea, vomiting, and fever. Examples of pathogens include *Salmonella*, *Shigella*, enteroinvasive *Esc. coli* (EIEC), *Vib. parahaemolyticus*, *Cam. jejuni*, and *Yer. enterocolitica*.
6. Nonenteric symptoms result when the pathogens or their toxins pass through the intestine and invade or affect other internal organs and tissues. Symptoms depend on the type of organ(s) and tissue affected but are accompanied by fever. Examples of pathogens include *Lis. monocytogenes*, enterohemorrhagic *Esc. coli* (EHEC), *Vib. vulnificus,* and hepatitis A virus.

II. SALMONELLOSIS BY *SALMONELLA* SPP.

A. Introduction

Prior to the 1940s, *Salmonella typhi* and *Sal. paratyphi* were considered the major causes of worldwide foodborne and waterborne diseases caused by *Salmonella*. However, with the pasteurization of milk and chlorination of water supplies, the spread of typhoid and paratyphoid fever through food and water was greatly reduced, at least in developed countries. As efficient techniques for the isolation and identification of other *Salmonella* species from foods and environmental samples were developed, it became apparent that the worldwide incidence of foodborne salmonellosis caused by other *Salmonella* species is quite high. Since the 1950s, foodborne salmonellosis has been the major cause of all foodborne diseases caused by bacteria and viruses, both in number of incidents (sporadic and outbreaks) and number of cases. Although scientific information about their habitats, mode of transmission in foods, growth characteristics, and survival parameters are available, and methods to control their contamination of foods have been developed, foodborne salmonellosis is still the leading cause of foodborne bacterial and viral diseases in the United States and other developed countries. This is quite puzzling. The control measures seem to be working for several other foodborne pathogens, at least in the United States, such as *Clostridium perfringens, Bacillus cereus, Yersinia enterocolitica, Vibrio parahaemolyticus,* and probably *Staphylococcus aureus*. Even incidences from *Staphylococcus aureus* and *Listeria monocytogenes* seemed to have declined since the 1990s. It is somewhat astonishing why it is not working against *Salmonella*. Between 1969 and 1976, the average number of foodborne salmonellosis outbreaks was about 37 per year in the United States. In contrast, between 1983 and 1987, the average number of outbreaks per year was over 68. Not only is the incidence not decreasing, it continues to increase at a high rate. It is difficult to point out the exact cause(s) of this increase. Maybe it is related to the large number of species, present in high frequency in food animals, birds, pets, insects, humans, and their ability to grow in foods; or maybe the way the food animals and birds are raised, processed, and marketed; or it may be due to better surveillance systems by the regulatory agencies; or maybe our lifestyles and food habits give these pathogens an edge.[1-3]

Tauxe[3] indicated that the present increase in salmonellosis (including foodborne salmonellosis) in the United States could be related to four recent factors: the increase in number of antimicrobial-resistant *Salmonella* isolates, the increase in individuals with immunodeficiency virus infection who are extremely susceptible to *Salmonella*, the increase in egg-associated *Salmonella enteritidis* contamination due to the increase in contaminated laying hens, and food production in centralized facilities that can lead to, if contamination occurs, extremely large and widespread outbreaks. It is necessary to understand the importance of these factors in the increase of salmonellosis and to develop corrective measures to control the incidence.

There are over 2000 serovars (based on somatic, flagellar, and capsular antigen types) of *Salmonella* (each of which has been given at present a species status) potentially capable of causing salmonellosis in humans. Along with fecal-oral direct transmission, contaminated food and water can cause salmonellosis. Recently, a better system to group these serovars in a few species was proposed, a system that may be adopted in the future.

B. Characteristics

All *Salmonella* spp. are Gram-negative, nonsporulating, facultative anaerobic motile rods. They form gas while growing in media containing glucose.[4,5] Generally, they ferment dulcitol, but not lactose, utilize citrate as carbon source, produce hydrogen sulfide, decarboxylate lysine and ornithine, do not produce indole, and are negative for urease. They are mesophilic, with optimum growth temperatures between 35 and 37°C, but generally have a growth range of 5 to 46°C. They are killed by pasteurization temperature and time, sensitive to low pH (4.5 or below), and do not multiply at $A_w \leq 0.94$, especially in combination with a pH at 5.5 and below. The cells survive under frozen and dried states for a long time. They are capable of multiplying in many foods without affecting the acceptance qualities.[4,5]

C. Habitat

Salmonellae are natural inhabitants of the gastrointestinal tracts of domesticated and wild animals, birds, and pets, such as turtles and frogs, and insects. In animals and birds, they can cause salmonellosis and then persist in a carrier state. Humans can also be carriers following an infection and shed the pathogens through feces for a long time. They have also been isolated from soil, water, and sewage contaminated with fecal matter.[4,5]

D. Toxins

Following ingestion of *Salmonella* cells, the pathogens invade mucosa of the small intestine, proliferate in the epithelial cells, and produce a toxin resulting in an inflammatory reaction and fluid accumulation in the intestine. The ability of the pathogens to invade and damage the cells is attributed to the production of a thermostable cytotoxic factor. Once in the cells, the pathogens multiply and produce a thermolabile enterotoxin that is directly related to the secretion of fluid and electrolytes. Production of the enterotoxin is directly related to the growth rate of the pathogens.[4,5]

E. Disease and Symptoms

Human salmonellosis is different from typhoid and paratyphoid fever caused by *Sal. typhi* and *Sal. paratyphi*. Although there are some *Salmonella* species (or serovar) specific against different animals and birds, all are considered to be potential human pathogens capable of causing salmonellosis. For foodborne salmonellosis, an individual generally has to consume about 10^5 to 10^6 cells; however, there are some virulent species and strains where ingestion of fewer cells can cause the disease.

Strains that are sensitive to gastric acidity generally need more cells to establish in the intestine and cause the disease. Following ingestion of the pathogens, symptoms appear within 8 to 42 h, generally in 24 to 36 h. The symptoms last for 2 to 3 days, but in certain individuals can linger for a longer time. An individual remains in a carrier state for several months following recovery.

Not all individuals ingesting the same contaminated foods will develop symptoms, nor will those who develop symptoms have all the symptoms in the same intensity. It varies with the state of health and natural resistance of an individual. The general symptoms are abdominal cramps, diarrhea, nausea, vomiting, chills, fever, and prostration. It can be fatal to the sick, infants, and the elderly.[4,5]

F. Food Association

Foods of animal origin have been associated with large numbers of outbreaks. These include beef, chicken, turkey, pork, eggs, milk, and products made from them. In addition, many different types of foods have been implicated in both sporadic cases and outbreaks (Table 24.1). These foods were contaminated directly or indirectly with fecal matter from carriers (animals, birds, and humans) and eaten either raw or improperly cooked, or contaminated following adequate heat treatment. Cross-contamination at home and at food services are the major sites of contamination of heated foods with *Salmonella*. Salmonellae have also been isolated from many foods of plant origin (due to use of sewage as fertilizer or washing products with polluted water), seafood, and fin fish (harvested from polluted water).[1,2]

Table 24.1 Foods Associated with Salmonellosis Outbreaks in the United States between 1973 and 1987

Food	Number of outbreaks[a]	Food	Number of outbreaks
Beef	77	Bakery products	12
Chicken	30	Fruits and vegetables	9
Turkey	36	Beverages	4
Pork	25	Chinese food	2
Eggs	16	Mexican food	10
Dairy products	50	Other foods	191
Shellfish and fin fish	8	Unknown	320

[a] Total number of outbreaks, 790; number of cases, 55,864; and number of deaths, 88.

Although there are over 2000 serotypes (species) of *Salmonella*, only a small number of them have been associated with foodborne illnesses. This could be due to the geographical distribution of the serotypes, as well as pathogenicity of a species or a strain of a species. *Sal. typhimurium* has been associated in the United States as the major causative agent of foodborne salmonellosis (over 20% of the total cases). However, since the 1980s, foodborne salmonellosis from *Sal. enteritidis* has increased mainly from contaminated Grade A shell eggs; in recent years, it has been involved in the same number of cases as *Sal. typhimurium*. The exact causes of the predominance of *Sal. enteritidis* is not yet clearly understood.[4,5]

G. Prevention and Control

Raw foods of animal origin that are heat treated before consumption can have *Salmonella*. However, in the United States (and other developed countries), as per regulatory requirements, heat-treated and ready-to-eat foods that contain *Salmonella* in portions tested are considered to be adulterated and should not be sold. Many food processing industries have in-house *Salmonella* surveillance programs to control the presence of *Salmonella* in their products. The regulatory agencies also have programs to educate consumers at home and food handlers in food service places to control *Salmonella* contamination in foods. These include proper cooking of foods (minimum to pasteurization temperature and time, such as 71.7°C for 15 s or equivalent) and prompt cooling (to 3 to 4°C or freezing, if not used in 2 h); prevention of cross-contamination of ready-to-eat food with a raw food through cutting boards, equipment, utensils, and hands; use of proper sanitation and personal hygiene; not handling a food while sick; and properly reheating a food refrigerated for a long time.[4,5]

H. Detection Methods

The methods involve preenrichment of a sample of food in a nutrient broth, followed by selective enrichment streaking on a selective-differential agar medium, and biochemical and serological confirmation (see Appendix D). Several rapid methods, based on specific immunological characteristics and nucleotide base sequence in the nucleic acids, have been developed.

I. A Case Study

In July of 1989, 21 of 24 people who attended a baby shower at a home in New York had gastroenteritis with severe diarrhea, vomiting, fever, and cramps within 6 to 57 h after the party.[6] Twenty people needed medical help, of whom 18 were hospitalized; one at 38 weeks pregnancy delivered while ill and the infant developed septicemia. *Salmonella enteritidis* was isolated from all 21 people (rectal swab) and the infant. All 21 ill attendees, but not the 3 who remained well, ate a homemade baked ziti pasta dish consisting of one raw egg, ricotta cheese, cooked tomato, and meat sauce, mixed together in a pan and refrigerated overnight. Before serving, the preparation was baked for 30 min at 350°F (176.7°C). Several attendees commented that the center of the ziti was cold when served. *Sal. enteritidis* was isolated from the left-over baked ziti and from the unused eggs from the carton. *Sal. enteritidis* was also isolated in laying chickens at the poultry farm that supplied the eggs.

The poultry farm should have tested the birds and culled those carrying the pathogen as a regulatory requirement. However, at home, proper methods should have been taken in the preparation of the dish to avoid the incidence. It is likely, as only one egg was used for a serving of 24+ people, that *Sal. enteritidis* was present initially in the food in small numbers, and multiplied before or during (slow) refrigeration. Also, if the dish was taken out of the refrigerator a long time prior to baking, *Salmonella* cells could have multiplied. During 30 min baking, the content was not thoroughly and evenly heated (the center was cold), so some *Sal. enteritidis* cells survived, and when ingested caused salmonellosis. The dish could have been baked immediately after mixing for a longer time for the food to attain a high and uniform temperature and then refrigerated quickly. It could have been reheated to a high and uniform temperature prior to serving to prevent this salmonellosis outbreak.

III. LISTERIOSIS BY *LISTERIA MONOCYTOGENES*

A. Importance

Human listeriosis has been recognized for a long time. However, the presence of *Listeria monocytogenes* in many foods of animal and plant origin and illnesses resulting from the consumption of contaminated foods were recognized rather recently.[7-11] Human listeriosis is considered by some to be an opportunistic disease. Individuals with normal health may not develop the symptoms or show a very mild enteric form of the disease. However, it is highly fatal (30 to 40%) to fetuses, newborns, infants, the elderly, pregnant women, and immunocompromised people, such as those with cancer, renal disease, heart disease, and AIDS. In addition, its ability to grow in many foods at refrigerated temperature helps the organism reach from a low initial level to an infective dose level during storage of refrigerated foods that include those that originally harbored the pathogen and those that were post-heat contaminated. The increase in consumption of many types of ready-to-eat foods that are stored for fairly long periods of time, and the fact that many of these foods are consumed without properly reheating or by microwave heating, has given an edge for this pathogen to cause the disease. Many technological developments used for the production of these ready-to-eat foods may have steps that can contaminate food with the pathogens in low levels, which then can reach a higher level during subsequent refrigerated storage prior to consumption. Any temperature abuse, even for a short time, can accelerate the growth rate.

It is quite clear that many of the above conditions have given an advantage to *Lis. monocytogenes* to become a newly emerging foodborne pathogen in many countries. However, an understanding of the type of foods that are mostly involved in listeriosis, the food processing steps that can contaminate ready-to-eat foods with this pathogen, and the special groups of people that are most susceptible to the disease, helped the regulatory agencies develop procedures to reduce foodborne listeriosis quite effectively. These have been achieved through the changes in the processing steps and testing of the foods so that contaminated ready-to-eat foods do not reach the consumers and in educating the susceptible consumer groups about food choices, eating habits, and sanitary practices in food preparation. As a result, in developed countries, the number of listeriosis cases has dropped dramatically; in the United States, the number of human listeriosis cases has fallen from 2000 cases per year in 1986 to about 1000 cases per year in 1991.

The genus *Listeria* contains seven species, of which *Lis. monocytogenes* is considered to be pathogenic. The species has several serogroups, namely, 1/2a, 1/2b,

1/2c, 3a, 3b, 3c, and 4b. While 1/2a and 1/2b were the predominant species isolated in foodborne human listeriosis in Europe, 4b was predominant in Canada and the United States.

B. Characteristics

Lis. monocytogenes is a Gram-positive, psychrotrophic, facultative, nonsporulating, motile small rod. In fresh culture, the cells may form short chains. It is hemolytic and ferments rhamnose but not xylose.[7,8,12]

Lis. monocytogenes is a psychrotroph and grows between 1 and 44°C, with optimum growth at 35 to 37°C. At 7 to 10°C, it multiplies relatively rapidly. It ferments glucose without producing gas. It can grow in many foods and environments. The cells are also relatively resistant to freezing, drying, high salt, and pH 5.0 and above. It is sensitive to pasteurization temperature (71.7°C for 15 s or 62.8°C for 30 min); but when inside the white blood cells, a temperature of 76.4 to 77.8°C for 15 s is required to kill the cells.[7,8,12]

C. Habitat

Lis. monocytogenes is isolated from many environmental samples, such as soil, sewage, water, and dead vegetation.[8,12] It is isolated from the intestinal contents of domesticated animals and birds. Humans can also carry the organisms in the intestine without any symptoms. A large proportion of uncooked meat, milk, egg, seafoods, and fish, as well as leafy vegetables and tubers (potatos and radishes, in particular), are found to contain *Lis. monocytogenes*. Many heat-processed foods, such as pasteurized milk and dairy products, and ready-to-eat meat preparations also have been found to contain the organism. *Lis. monocytogenes* is isolated in high frequency in different places of food processing and storage areas.[8,12]

D. Toxin

The virulence factor of *Lis. monocytogenes* is a specific type of hemolysin, listeriolysin O. It is produced during the exponential growth of the cells. The pathogens invade different body tissues and multiply inside the body cells, releasing toxin. The toxin causes death of the body cells.[7,8,12]

E. Disease and Symptoms

People with normal health, following ingestion of a *Lis. monocytogenes* contaminated food, may or may not produce symptoms. Most often, the symptoms appear within 1 to 7 d following ingestion and include mild flu-like symptoms with slight fever, abdominal cramps, and diarrhea. The symptoms subside in a few days, but the individual sheds *Lis. monocytogenes* in the feces for some time.[7,8,12]

The symptoms among sensitive groups are quite different. These groups include pregnant women, unborn fetuses, infants, elderly people with reduced immunity due to diseases, and people taking special medications, such as steroids and chemotherapeutic agents. Initial symptoms are enteric, with nausea, vomiting, abdominal cramps, and diarrhea with fever and headache. The pathogens then invade tissues in different vital organs, including the central nervous system, through the blood. In pregnant women, the pathogen can invade organ tissues of the fetus. Symptoms include bacteremia (septicemia), meningitis, encephalitis, endocarditis, and others. The fatality rate among fetuses, infected newborn infants, and immunocompromised individuals is very high. The infective dose is considered to be about 100 to 1000 cells, particularly for the sensitive groups.

F. Food Association

Foodborne listeriosis in humans is usually sporadic; however, outbreaks were reported from the consumption of contaminated cole slaw, pasteurized milk, raw milk and dairy products, soft cheeses (Mexican style, Brie, and Liederkranz), meat paté, turkey franks, coldcut meats, improperly cooked chicken, and smoked mussels. Sporadic listeriosis was also documented from the consumption of these foods either at home or at delicatessen counters. Heat-treated foods were either not properly heated or were contaminated following heating. As many raw foods of both animal and plant origin harbor *Lis. monocytogenes*, consumption of raw foods or recontaminated heat-processed foods have been the cause of listeriosis. Growth during long refrigerated storage and temperature abuse prior to eating have been implicated in many cases. Major groups that suffered from listeriosis and fatality from the disease were the sensitive groups.[7,8,12]

G. Prevention and Control

Due to the ubiquitous presence of *Lis. monocytogenes*, it is impossible to have foods free of this pathogen. However, a strong *Listeria* control program at commercial production facilities has been imposed by the regulatory agencies and industries in many countries. This has helped to greatly reduce the number of listeriosis cases since 1991. In the United States, this includes the absence of *Lis. monocytogenes* in 25-g portions of cooked, ready-to-eat meat and poultry products from each lot; a positive lot is considered adulterated and discarded. A plant producing contaminated products is then thoroughly tested and subjected to proper sanitation (hazard analysis critical control point; see Appendix C) until products free of *Lis. monocytogenes* are produced. Due to the effectiveness of the current regulatory procedures, there is a possibility that in the future the current stringent methods may be slightly relaxed.[8-12]

In addition to the control measures in the processing facilities, the regulatory agencies have implicated consumer education to reduce foodborne listeriosis. This includes thoroughly cooking raw foods of animal origin; thoroughly washing raw vegetables before eating; keeping uncooked meats separate from vegetables, cooked foods, and ready-to-eat foods; not consuming raw milk or foods made with raw milk; washing hands, knives, and cutting boards after handling uncooked foods. In addition, special recommendations have been provided to high-risk individuals: avoid soft cheeses (Mexican style, Feta, Brie, Camembert, blue-veined; not hard, cream, or cottage cheeses); reheat (until steaming) all leftover and ready-to-eat foods before eating; pregnant women, elderly people, and immunocompromised people should avoid foods from delicatessen counters.[8-12]

H. Detection Methods

The most commonly used method involves preenrichment and enrichment steps in recommended broths and streaking on specific selective-differential agar media plates. The suspected colonies are then tested for biochemical and serological profiles. Several rapid methods have also been developed based on immunological characteristics and nucleic acid base sequences (see Appendix D).

I. A Case Study

In December of 1988 in Oklahoma, a cancer patient developed septic listeriosis after consumption of turkey franks heated in a microwave oven.[12,13] *Lis. monocytogenes* serotype 1/2a was isolated from the patient, from the remaining franks in the opened package (in high numbers) in the refrigerator, and several other foods in the same refrigerator. Isolation of the same serotype from unopened packages (in low numbers) of the same brand of turkey franks facilitated the regulatory agencies to associate the contaminated turkey franks as the cause of this foodborne listeriosis. Subsequent investigations, after 4 months, by the regulatory agencies revealed that the plant processing this brand of franks had the same serotype of *Lis. monocytogenes* (as well as other serotypes) in the processing environment, and the heated franks were recontaminated from the conveyor belt following peeling of casings and before repackaging (see Chapter 27).

This incident reveals several important aspects of foodborne human listeriosis: (1) the person affected with listeriosis was immunocompromised; (2) ready-to-eat franks were the source of the pathogenic serotype; (3) microwave heating, the way used by the individual, did not kill the pathogen; (4) the pathogen grew during refrigerated storage in the franks; (5) several other foods in the same refrigerator were also contaminated; (6) the pathogenic strain was present in the processing environment (along with other serotypes) even after 4 months; and (7) recontamination of franks following heating occurred from contaminated equipment prior to packaging.

To control this incidence of listeriosis, proper sanitation of the processing plant would be the primary requirement. At the consumer level, the incident indicates the importance of proper heating of the product before eating (and prevention of cross-contamination of refrigerated ready-to-eat food, especially by the high-risk individual).

IV. PATHOGENIC *ESCHERICHIA COLI*

A. Importance

Since its discovery in 1885, *Escherichia coli* was considered a harmless, Gram-negative, motile, nonsporulating, rod-shaped, facultative anaerobic bacterium, a normal inhabitant of the intestinal tract of humans and warm-blooded animals and birds. Because it is normally present at very high levels (in millions per gram of the contents of the large intestine), for a long time it has been used as an index organism of possible fecal contamination and the presence of enteric pathogens in food and water. Since the mid-1940s, evidence has accumulated that certain *Esc. coli* strains cause diarrhea, particularly in infants, and they were designated as enteropathogenic *Esc. coli*. However, current evidence indicates that pathogenic strains of *Esc. coli* could be of more than one type. Currently, they are subdivided into four groups:[14,15]

1. Enteropathogenic Esc. Coli (EPEC)

These strains are important in infant diarrhea worldwide, especially in places with poor sanitation. They are transmitted directly or indirectly through human carriers. Several serotypes are implicated in waterborne and foodborne disease outbreaks in different countries. The mechanism of pathogenesis is not clearly known. One needs to ingest high numbers of cells (10^6 to 10^9) to develop the symptoms. The predominant symptom is gastroenteritis.

2. Enterotoxigenic Esc. Coli (ETEC)

These strains are the major cause of diarrhea among travelers, as well as in infants in many developing countries with poor sanitation. The presence of disease is due to the ability of the pathogens to produce an invasive factor and either a heat-labile (HL), a heat-stable (HS), or both HL and HS, enterotoxin(s). The symptom is gastroenteritis, like a mild form of cholera. The pathogen is spread directly or indirectly by human carriers. Both food and water have been implicated in outbreaks and sporadic cases in humans. In 1983, imported Brie cheese contaminated with 027:H7 serotype caused outbreaks in several countries, including the United States. Ingestion of large numbers of cells (10^8 to 10^9) are necessary for an individual to develop the symptoms.

3. Enteroinvasive Esc. Coli (EIEC)

These strains are known to cause dysentery, like shigellosis. The ability of the strains to produce an invasive factor is thought to be the cause of the disease. Human carriers, directly or indirectly, spread the disease. Ingestion of as many as 10^6 cells may be necessary for an individual to develop the symptoms. An outbreak in the United States as early as 1971 was recognized from the consumption of imported Camembert cheese contaminated with serotype 0124:B17.

4. *Enterohemorrhagic Esc. Coli (EHEC)*

The strains in this group (a principal serogroup is 0157:H7) have been recognized relatively recently as the cause of severe bloody diarrhea (hemorrhagic colitis) and hemorrhagic uremic syndrome (HUS) in humans. Animals, particularly dairy cattle, are thought to be the carriers. Ingestion of as few as 10 to 100 cells can produce the disease. The ability of this serotype to produce three enterotoxins (verotoxins) has been recognized to be the causative agent of the disease symptoms.

Although there is no clear differentiation for these four subgroups, both EIEC and EHEC strains seem to fit with the enteric pathogens associated with foodborne infection. Thus these two groups are discussed in this chapter. Both EPEC and ETEC groups are included in the chapter discussing toxicoinfection (Chapter 24).

B. Gastroenteritis Due to EIEC

1. *Toxins*

The pathogens produce several polypeptides, the genes of which are encoded in a plasmid. These are considered to be the invasive factors that enable the pathogen to invade epithelial cells and set up infection in the colon. Separate toxin(s) have not yet been identified.[14,15]

2. *Disease and Symptoms*

The disease and symptoms are like shigellosis. Following ingestion of the pathogen (about 10^6 cells) and incubation period, symptoms appeared as abdominal cramps, profuse diarrhea, headache, chills, and fever. A large number of pathogens are excreted in the feces. The symptoms can last for 7 to 12 d, but a person can remain a carrier and shed the pathogens in feces for a long time [14,15]

3. *Food Association*

Only humans are known to be the host of the pathogen, and a food can get contaminated directly or indirectly through fecal contamination. Outbreaks from the ingestion of foods contaminated with this pathogen have been recorded. The 1971 outbreak in the United States from the ingestion of an imported cheese was traced to contamination of the processing plant equipment from a malfunctioning water filtration system. In 1983, another outbreak on a cruise ship was related to potato salad contaminated by a carrier food handler.[14,15]

4. *Prevention*

The pathogen is sensitive to pasteurization temperature. Thus proper heat treatment, elimination of post-heat contamination for a ready-to-eat food, and refrigeration of a food soon after preparation are necessary to control the disease. In addition, proper sanitation at all stages of food processing and handling will be an important factor. Finally, individuals suspected of being carriers should not handle food, especially ready-to-eat food.[14,15]

C. Gastroenteritis Due to EHEC

1. Characteristics

The principal serotype associated with enterohemorrhagic colitis is *Esc. coli* 0157:H7. As opposed to other *Esc. coli,* it does not ferment sorbitol or have glucuronidase activity. Like other *Esc. coli,* it grows rapidly at 30 to 42°C, grows poorly at 44 to 45°C, and does not grow at 10°C or below. The organism is destroyed by pasteurization temperature and time and killed at 64.3°C in 9.6 s. The cells survive well in food at –20°C.[15-18]

2. Toxins

Esc. coli 0157:H7 produces a verotoxin (VTI). There can be more than one toxin involved in the disease and the symptoms related with the disease. It is not known if the pathogen also produces invasive factors or not. The organism probably colonizes in the intestine and produces toxins, which then act on the colon.[15-18]

3. Disease and Symptoms

Esc. coli 0157:H7 causes hemorrhagic colitis, hemolytic uremic syndrome (HUS), and thrombotic thrombocytopenic purpura (TTP). Symptoms occur within 3 to 9 d after ingestion and generally last for about 4 d. The colitis symptoms include a sudden onset of abdominal cramps, watery diarrhea (which in 35 to 75% of cases turns to bloody diarrhea), and vomiting. Fever may or may not be an associated symptom. Damage to the lining of the large intestine is the cause of bleeding. Toxin(s) also causes the breakdown of red blood cells, clotting in small blood vessels of the kidney, causing kidney damage, and occasional kidney failure, producing the symptom called HUS. It can be fatal, particularly in children. TTP results from a blood clot in the brain, with seizures, coma, and often death.[15-18]

4. Food Association

The pathogen is thought to be present in the intestine of animals, particularly in dairy cattle, without producing symptoms. Food of animal origin, particularly ground beef, has been implicated in many outbreaks in the United States, the U.K., and Canada. The affected people were found to consume improperly cooked, contaminated hamburgers. In a 1993 outbreak, affecting over 500 people with four deaths, consumption of hamburgers served by a fast-food chain in Washington, Nevada, Oregon, and California was implicated. The hamburgers, contaminated with *Esc. coli* 0157:H7, were cooked at a temperature that failed to kill the pathogen. In addition to ground beef, other foods, such as apple cider and uncooked sausages, have been implicated. Investigations revealed the presence of *Esc. coli* 0157:H7 in many different types of foods of animal origin, such as ground beef, pork, poultry, lamb, and raw milk, in low percentages. The organism was isolated in low frequencies from dairy cows as well as calves and chickens.[15-18]

5. Prevention

Proper sanitation, cooking or heating at appropriate temperatures, proper refrigeration, and prevention of cross-contamination should be practiced in order to control the presence of *Esc. coli* 0157:H7 in ready-to-eat food. The Food Safety Inspection Service (FSIS) in the United States has provided the following guidelines to control foodborne illness from this pathogen: use only pasteurized milk; quickly refrigerate or freeze perishable foods; never thaw a food at room temperature or keep a refrigerated food at room temperature over 2 h; wash hands, utensils, and work areas with hot soapy water after contact with raw meat and meat patties; cook meat or patties until the center is gray or brown; and prevent fecal-oral contamination through proper personal hygiene.[15-18]

V. SHIGELLOSIS (BACILLARY DYSENTERY)

A. Importance

The genus *Shigella* contains four species: *Shi. dysenteriae, Shi. flexneri, Shi. boydii,* and *Shi. sonnei* and each species has several serovars. Only humans and some primates are their hosts. The organisms are either transmitted directly through fecal-oral routes or indirectly through fecal-contaminated food and water. While in most developed countries transmission through drinking water has been reduced, in developing countries contaminated drinking water is a major cause of shigellosis. The disease is prevalent in some geographic locations, particularly in Asia, Mexico, and South America. It occurs more frequently in places with poor sanitation. In the United States, shigellosis occurs more among migrant workers, on the reservations, in poor urban institutions, and in day-care centers. Children below 5 years of age are more affected. In developing countries, there is a high fatality rate among children suffering from shigellosis. In the United States, the predominant species prior to the 1950s was *Shi. dysenteriae*. At present, the predominant species is *Shi. sonnei*. In the United States between 1983 and 1987, there were 44 outbreaks affecting 9971 people, with 2 deaths. In one outbreak in 1987, several thousand people were affected by eating commercial meals prepared under poor hygienic conditions that resulted in the contamination of food with *Shi. sonnei*. In general, food service establishments have been implicated in more outbreaks, and poor personal hygiene has been the major cause. The disease is more prevalent during late spring to early fall.[12,19]

B. Characteristics

The cells of the species are Gram-negative, nonmotile, facultative anaerobic rods. They are generally catalase positive and oxidase and lactose negative. They ferment sugars, usually without forming gas. On the basis of DNA homology, both *Shigella* and *Escherichia* can be included in one genus, and, due to many biochemical similarities, the separation between two genera is not clear.

The strains grow between 7 and 46°C, with an optimum at 37°C. The cells are not as fragile as once was thought. They survive for days under different physical and chemical stresses. These include refrigeration, freezing, 5% NaCl, and pH 4.5. They are killed by pasteurization. The strains are able to multiply in many types of food when stored at the growth temperature range.[19,20]

C. Habitat

The intestines of humans and some primates is the only habitat known. Humans can carry the organism in the intestine and shed it in the feces without showing any symptoms. Following recovery from shigellosis, an individual can remain a carrier for months.[19,20]

D. Toxins

The strains are believed to carry plasmid-encoded invasive traits that enable the shigellae cells to invade epithelial mucosa of the small and large intestine. Once engulfed by the epithelial cells, they are capable of producing an exotoxin that has the enterotoxigenic property. The toxin is designated as Shiga toxin. The invasive trait is expressed at 37°C but not at 30°C. Shigellae cells growing at 30°C need a few hours of conditioning at 37°C before they can invade the intestinal epithelial cells. The engulfed shigellae cells kill the epithelial cells and then attack fresh cells, causing ulcers and lesions.[19,20]

E. Disease and Symptoms

The infective dose is very low, about 10^1 to 10^3 cells per person. Following ingestion of a contaminated food, the symptoms occur within 12 h to 7 d, but generally within 1 to 3 d. In cases of mild infection, the symptoms last about 5 or 6 d; but in severe cases, the symptoms can linger for 2 to 3 weeks. Certain individuals may not develop symptoms. An infected person sheds the pathogens long after the symptoms have stopped. The symptoms are the consequence of both the invasiveness of epithelial mucosa and the enterotoxin and include abdominal pain, diarrhea often mixed with blood, mucus and pus, fever, chills, and headache. Generally, children are more susceptible to the disease than adults.[19-21]

F. Food Association

Shigella cells are present in a food only through fecal contamination, directly or indirectly, from a person either suffering from the disease, or a carrier or a person who has not developed the symptoms yet but is shedding the pathogens in feces. Direct contamination occurs from poor personal hygiene. Indirect contamination occurs from the use of fecal-contaminated water to wash foods that will not be subsequently heat processed. Also, cross-contamination of ready-to-eat foods can be involved in an outbreak. Foods often implicated in shigellosis are those that are handled too much and ready-to-eat.

In many developed countries, the most frequently involved foods are different types of salads (potato, tuna, shrimp, and chicken), with potato salads ranked at the top. Foods that are chopped, diced, or cut prior to eating, such as vegetables used in salads, have also been involved in outbreaks. Shellfish harvested from sewage-polluted water and eaten raw have been associated with shigellosis. Many foods support their growth. As the infective dose is very low, growth in food is probably not an important factor for the disease.[19-21]

G. Prevention

Foodborne shigellosis, at least in developed countries, is caused by contamination of foods by food handlers shedding the pathogen in the feces and having poor personal hygiene. To prevent contamination of ready-to-eat food by these individuals, it is necessary to forbid them to handle such foods. Quite often, this is impossible, especially if the individual is an asymptotic carrier. Proper education of the food handlers about the importance of good personal hygiene and the need to not handle food if one suspects having a digestive disorder will be important. Use of rigid sanitary standards to prevent cross-contamination of ready-to-eat food, use of properly chlorinated water to wash vegetables to be used for salads, and refrigeration of the foods are necessary to reduce foodborne shigellosis.[19-21]

VI. CAMPYLOBACTERIOSIS (*CAMPYLOBACTER* ENTERITIDIS AND *CAMPYLOBACTER JEJUNI* ENTERITIDIS)

A. Importance

Several *Campylobacter* species have the capability of causing human gastroenteritis; however, *Cam. jejuni* and *Cam. coli* are considered the most common causes of human diarrheal disease in many countries around the world. It is thought that in many countries the number of cases of campylobacteriosis far exceeds the combined number of salmonellosis and shigellosis. Epidemiological data have confirmed it to be true in Canada, the United Kingdom, and Scotland. Isolation of *Campylobacter* spp. from a suspected sample requires specific methods. After the development of this method and its incorporation in the isolation of suspected foodborne pathogens, *Cam. jejuni* has been confirmed as a causative agent in many foodborne illnesses. Since it was first recognized as the cause of an outbreak in 1979, *Cam. jejuni* has been implicated in 53 foodborne disease outbreaks in the United States between 1979 and 1987 affecting 1547 individuals, with 2 deaths. The foods implicated most often in campylobacteriosis were raw milk and improperly cooked chicken. Although several *Campylobacter* spp. have been associated with foodborne campylobacteriosis, *Cam. jejuni* has been isolated in most incidences; the discussion here is for *Cam. jejuni*.[1,2,22,23]

B. Characteristics

Cam. jejuni is a Gram-negative, motile, nonsporulating, rod-shaped bacterium. The cells are small, fragile, and spirally curved. The strains are microaerophilic, and catalase and oxidase positive. The strains require a microaerophilic environment of about 5% oxygen, 8% CO_2, and 87% N_2 for growth. Growth temperature ranges between 32 and 45°C, with optimum around 42°C. They grow better in amino acids than in carbohydrates. They generally grow slowly and are not a good competitor while growing with other bacteria. They do not generally grow well in many foods. They are sensitive to many environmental parameters, including oxygen (in air), NaCl (above 2.5%), low pH (below pH 5.0), temperature (below 30°C), heat (pasteurization), and drying. However, they survive well under refrigeration and for a while in the frozen state.[24-26]

C. Habitat

Cam. jejuni is an enteric organism. It has been isolated from feces of animals and birds with high frequency. Human carriers were also found to shed the organisms in feces. Fecal materials from poultry were found to contain 10^6 cells or more per gram in some instances. Water, sewage, vegetables, and foods of animal origin are easily contaminated with *Cam. jejuni* excreted through feces.[25,26]

D. Toxins

Cam. jejuni has a thermolabile enterotoxin that is responsible for the enteric disease symptoms. The toxin cross-reacts with cholera toxin, and the toxin production trait is plasmid linked. In addition, the strains produce an invasive factor that enables the cells to invade and establish in epithelial cells in both the small and large intestines in humans.[25,26]

E. Disease and Symptoms

The infective dose for campylobacteriosis can be considerably low, maybe only about 500 cells. Following ingestion, the symptoms of the disease occur within 2 to 5 d. The symptoms generally last for about 2 to 3 d, but can linger for 2 weeks or more. Persons with no visible symptoms can shed the cells in feces for a long time. The main symptoms are enteric and include abdominal cramps, profuse diarrhea, nausea, and vomiting. Other symptoms include fever, headache, and chills. In some cases, bloody diarrhea has been reported. An individual can have a relapse of symptoms after a short interval.[25,26]

F. Food Association

Due to the presence of the organism in high frequency among animals, birds, and in environments, many foods, from both plant and animal sources, can be contaminated with *Cam. jejuni*. The foods can be contaminated directly with fecal material from animals and infected humans, or indirectly from sewage and contaminated water. *Cam. jejuni* has been isolated from raw meats (beef, lamb, pork, chicken, and turkey), milk, eggs, vegetables, mushrooms, and clams, with very high frequency. In heat-processed food, their presence has been related to cross-contamination following heat treatment or to improper heating. The use of animal feces as fertilizer was found to contaminate vegetables. Outbreaks of campylobacteriosis result from the consumption of raw milk, improperly cooked chicken, dairy products, bakery products, turkey products, Chinese food, eggs, and others. Consumption of raw milk and chicken were implicated in many outbreaks. Although the organism is a poor competitor against other microorganisms present in a food and generally does not grow well in foods, enough cells can survive in a contaminated food to provide the dose required for the disease.[25,26]

G. Prevention

It is rather difficult to control their access in raw foods, particularly foods of animal origin. However, proper sanitation can be used to reduce their load in raw foods during production, processing, and future handling. Prevention of consumption of raw foods of animal origin, heat treatment of a food, when possible, and prevention of postheat contamination are important to control campylobacteriosis in foods of animal origin. Contamination of vegetables can be controlled by not using animal feces as fertilizer and not using contaminated water to wash vegetables (especially ready-to-eat types). Contamination from humans can be reduced by establishing good personal hygiene and not allowing sick individuals to handle foods, especially ready-to-eat foods.[25,26]

VII. YERSINIOSIS (*YERSINIA ENTEROCOLITICA* GASTROENTERITIS)

A. Importance

Foodborne yersiniosis was first confirmed in the United States in 1976 following an outbreak among a large number of school children from the consumption of chocolate milk contaminated with *Yersinia enterocolitica*. In many other countries, foodborne yersiniosis was recorded earlier. In Denmark, yersiniosis is one of the most common forms of gastroenteritis. Following the first outbreak in the United States, *Yer. enterocolitica* was designated as an emerging foodborne pathogen. Since the first one in the 1970s, several outbreaks were recorded until the early 1980s. In 1982, three outbreaks were recorded, one from the consumption of contaminated tofu packed with contaminated spring water, and the other two from the consumption of contaminated pasteurized milk. But yersiniosis is not a frequent cause of foodborne infection in the United States. No incidents were reported between 1982 and 1987, but one was reported in 1988. The two major aspects about the organism and the disease are that *Yer. enterocolitica* is a pychrotroph and can grow at 0°C, and the symptoms of the disease include a sharp abdominal pain with fever and mimics appendicitis. *Yer. pseudotuberculosis* is quite common in animals and has been isolated in foods. However, its involvement in foodborne illnesses is not confirmed. As such, only *Yer. enterocolitica* and its association in foodborne illnesses is discussed here.[1,2,27]

B. Characteristics

Yer. enterocolitica cells are Gram-negative short rods, nonsporeforming, motile below 37°C, and facultative anaerobic. The strains grow between 0 and 44°C with an optimum growth at 25 to 29°C. Growth occurs in milk and raw meat at 1°C, but at a slower rate. Cells can grow in 5% NaCl and at a pH above 4.6. The cells are sensitive to pasteurization.[27,28]

C. Habitat

Yer. enterocolitica is a normal inhabitant of intestines of food animals and birds, pets, wild animals, and humans. Human carriers do not show any disease symptoms. Different types of food can be contaminated from these sources.[27,28]

D. Toxin

Not all strains are capable of producing yersiniosis. Most strains isolated from the environment are nonpathogenic. The pathogenic strains are more predominant in pigs. Both the pathogenic and nonpathogenic strains produce a heat-stable toxin; thus toxin production is not directly related to the ability of a strain to cause yersiniosis. The pathogenic strains also carry an invasive factor that enables the cells to colonize intestinal epithelial cells and lymph nodes. Only after colonization is the heat-stable toxin capable of causing the disease. The pathogenic strains vary in serological characteristics; in the United States the most common serovar implicated in yersiniosis is 08.[27,28]

E. Disease and Symptoms

Young children are more susceptible to foodborne yersiniosis. Generally, a high dose (about 10^7 cells) is required for the disease. The symptoms are severe abdominal pain in the lower quadrant of the abdomen, diarrhea, nausea, vomiting, and fever. The symptoms generally appear within 24 to 30 h following consumption of a contaminated food and last about 2 to 3 d. The disease can be fatal in rare cases.[27,28]

F. Food Association

Since *Yer. enterocolitica* strains are found in the environment, many foods can have this organism. It has been isolated from raw milk, processed dairy products, raw and improperly cooked meats, fresh vegetables, and improperly chlorinated water. Foods implicated in yersiniosis include raw and pasteurized milk, ice cream, and improperly cooked meats. As the cells are heat sensitive, a properly pasteurized or a heated food can have this pathogen only due to contamination following heat treatment. A food can also be contaminated from a human carrier or a pet. As the cells can grow at refrigerated temperature, even a low initial load can reach a high level during extended storage of refrigerated foods.[27,28]

G. Prevention

Since the strains are psychrotrophs, refrigeration cannot be used to control their growth. Good sanitation at all phases of handling and processing and proper heat treatment are important in controlling the occurrence of yersiniosis. Consumption of raw milk or meat cooked at low temperatures should be avoided.[27,28]

VIII. GASTROENTERITIS BY *VIBRIO* SPECIES

In the genus *Vibrio*, four species have been implicated in foodborne illnesses. These include *Vibrio cholerae* (01 and non 01 serogroups), *Vibrio mimicus, Vibrio parahaemolyticus,* and *Vibrio vulnificus*. Of these, *Vib. parahaemolyticus* and *Vib. vulnificus* are discussed in this section, and *Vib. cholerae* and *Vib. mimicus* are included in the toxicoinfection section. Although there is no clear-cut basis for this differentiation, the absence of fever among the individuals affected by both *Vib. cholerae* and *Vib. mimicus* is used to separate these two species from *Vib. parahaemolyticus* and *Vib. vulnificus* infection.[29,30]

A. *Vibrio Parahaemolyticus* Gastroenteritis

1. Importance

Gastroenteritis caused by *Vib. parahaemolyticus* is quite common in Japan and accounts for 40 to 70% of the total bacterial foodborne diseases. The high incidence is directly related to the consumption of raw seafood. In the United States, its involvement in foodborne infection was first recognized in 1971 from a large outbreak associated with the consumption of steamed crabs contaminated with the pathogen following heat treatment. Several other large-scale outbreaks were confirmed in the 1970s. However, in the 1980s the number of outbreaks and number of cases per outbreak reduced greatly. Between 1980 and 1987, there were only 12 outbreaks involving a total of 75 people and no fatalities.[29-31]

2. Characteristics

The cells are Gram-negative, nonsporulating, motile, curved rods. They are generally catalase and oxidase positive. The strains grow in medium containing glucose without production of gas, but are unable to ferment lactose and sucrose. They can grow over a temperature range between 5 and 42°C with the optimum around 30 to 37°C. The cells multiply rapidly in the presence of 3 to 5% NaCl but are sensitive to 10% salt. Under optimum growth conditions, the cells can multiply in about 15 min. Growth is restricted at pH 5.0 or below. The cells are extremely sensitive to drying, heating (pasteurization), and refrigeration and freezing.[29-31]

3. Habitat

Vib. parahaemolyticus strains are halophilic bacteria distributed in coastal waters worldwide. They are found in estaurine environments and show a seasonal variation, being present in the highest numbers during the summer months. During the winter months, they remain in the estaurine bottom on chitinous materials of plankton.[29-31]

4. Toxin and Toxin Production

Not all strains of *Vib. parahaemolyticus* are pathogenic. The foodborne pathogenic strains are capable of causing hemolysis due to the presence of a heat-stable hemolysin and are designated Kanagawa positive. At present, the heat-stable hemolysin is considered to be the toxin. Most strains isolated from natural sources (estaurine water, plankton, shellfish, and fin fish) are Kanagawa negative. However, some Kanagawa-negative strains have been associated with foodborne outbreaks also. All pathogenic strains were found to adhere to human fetal intestinal cells in cell cultures. The toxin production rate and its level are directly related to cell growth, cell concentrations, and the pH of the environment. If the toxin forms (in a food), heating does not destroy it.[29-31]

5. Disease and Symptoms

The cells are sensitive to low stomach pH. Generally, an individual has to consume about 10^5 to 10^7 cells of a Kanagawa-positive strain for symptoms to develop. However, an increase in pH due to consumption of bicarbonates and foods can reduce the infective dose. The symptoms appear in 10 to 24 h following ingestion of live cells and last for 2 to 3 d. The symptoms include nausea, vomiting, abdominal cramps, diarrhea, headache, fever, and chills. The disease is not normally fatal.[29-31]

6. Food Association

Vib. parahaemolyticus strains have been isolated in high numbers from various types of seafoods harvested from the estaurine environments, especially during the summer months. The outbreaks, as well as sporadic cases of gastroenteritis, were linked to the consumption of raw, improperly cooked, or postheat-contaminated seafoods, including fish, oysters, crabs, shrimp, and lobster. In unrefrigerated raw and cooked seafoods, *Vib. parahaemolyticus* can grow rapidly, especially at 20 to 30°C. In temperature-abused seafoods, the cells can reach an infective dose level very rapidly, even from a low initial population. Many outbreaks in the United States were identified as due to inadequate cooking and cross-contamination of cooked seafoods followed by improper holding temperatures.[29-31]

7. *Prevention*

Several factors need to be considered in controlling gastroenteritis from *Vib. parahaemolyticus*. The seafoods harvested from an estuary should be assumed to contain *Vib. parahaemolyticus*, some strains of which can be pathogenic. In unrefrigerated raw, improperly heated, or postheat-contaminated seafood, the cells can multiply rapidly. Once the pathogenic strains grow to infective numbers, even heat treatment will not destroy the toxin. With this understanding, the control methods should include: no consumption of raw seafoods, proper heat treatment of seafoods, proper sanitation to avoid cross-contamination of heated foods, proper refrigeration of the raw and heated products, and consumption of the food within a reasonable period of time. Temperature abuse, even for a short duration, of seafood should be avoided.[29-31]

B. *Vibrio Vulnificus* Septicemia

Vib. vulnificus is a lactose-positive, salicin-positive *Vibrio* species found in the estuarine environment in the coastal waters. It is considered a highly lethal pathogen. *Vib. vulnificus* has been associated with fulminating septicemia from the consumption of contaminated seafood, as well as a progressive cellulitis resulting from wound infection. Following consumption of contaminated seafood, in many cases raw oysters, the cells penetrate the intestinal wall and produce primary septicemia within 20 to 40 h. The symptoms are chills, fever, and prostration, with occasional vomiting and diarrhea. The fulminating septicemia sets up rapidly, which can be fatal in many cases. The infection and fatality rate is very high (40 to 60%) among people with liver and gastric diseases and immunodeficiencies. The control measures should include methods discussed for *Vib. parahaemolyticus*. Susceptible individuals should avoid consumption of raw seafoods, such as oysters.[29,30,32]

IX. ENTERIC VIRUSES

A. Importance

Enteric viruses have the potential of becoming a major cause of foodborne illness in the United States and many other countries. However, unlike bacterial pathogens, they are difficult to detect and recover from a contaminated food; for some viruses, suitable methods to isolate from a food have not been developed and for some other viruses very little research has been conducted to develop methods to detect them in foods. Also, unlike bacteria, human enteric viruses do not multiply in food systems; some may die off rapidly under various conditions of food storage and preservation. Thus very seldom is a food routinely examined for contamination by enteric viruses. Even with all the difficulties in confirming a virus as an etiological agent in a foodborne disease outbreak, in the United States between 1973 and 1987, viral foodborne infection was the fifth leading cause in both number of outbreaks

(5%) and number of cases (9%) of all the reported foodborne incidents. Between 1986 and 1987, viral infection ranked third, behind salmonellosis and shigellosis, in relation to total number of cases affected by the foodborne disease outbreaks. During this period, confirmed numbers of outbreaks and cases were 29 and 1067 for hepatitis A, and 10 and 1174 for Norwalk-like viruses, respectively. With the improvement in detection techniques, it is anticipated that the incidence of viral foodborne infection will increase. In developing programs to reduce foodborne diseases, this possibility should be considered.[1,2]

B. Characteristics

Foodborne viral infections can occur only from enteric pathogenic human viruses. Prior to the 1940s, polio viruses transmitted through raw milk were considered to be the only virus involved in foodborne viral infections. In recent years, hepatitis A, Norwalk-like viruses, Roto viruses, small round viruses, and non-A–non-B hepatitis viruses have been associated with foodborne infections in different parts of the world. However, among them, hepatitis A and Norwalk-like viruses were more predominant. Both hepatitis A and Norwalk-like viruses are small (about 27 to 28 nm) RNA viruses. Hepatitis A virus can be grown in cell cultures and can be used to produce vaccine. Norwalk-like viruses could not be cultivated in cell cultures; their presence in patients' stools is detected by enzyme-linked immunosorbant assay.[33-35]

C. Habitat

They are of enteric origin and are excreted in very high numbers in human feces. They do not multiply outside the human body, including in contaminated foods. Pasteurization can effectively kill the enteric viruses.[33-35]

D. Disease and Symptoms

Following ingestion of hepatitis A viruses through contaminated food (or other sources), an individual may or may not develop symptoms. In affected individuals, the symptoms occur after about 4 weeks, with a range of about 2 to 7 weeks. The general symptoms are fever, malaise, nausea, vomiting, abdominal discomfort, and inflammation of the liver, which may follow with jaundice. The symptoms may last for 1 to 2 weeks or longer. The viruses are shed via feces, generally during the last half of the incubation period. Norwalk-like viruses cause gastroenteritis characterized by vomiting and diarrhea. The symptoms appear within 12 to 24 h after ingestion and last for 1 to 2 d. The viruses are excreted in the feces of infected persons.[33,35]

E. Food Association

Foods contaminated with fecal matters of infected people directly (from food handlers) or indirectly (via sewage and polluted water) are the main sources of both hepatitis A and Norwalk-like virus outbreaks. Infected food handlers, even without symptoms, can contaminate ready-to-eat food with fecal matter. Vegetables (salads) can be contaminated with polluted water. Shellfish (oysters, clams, mussels, and cockles) harvested from water polluted with sewage and eaten raw or improperly heated prior to eating have been implicated in many outbreaks of both types of viruses. The virus can survive in shellfish for a long time. Depuration (in tanks filled with disinfected saline water) or relaying (kept in unpolluted water in the sea) for self-cleaning of the viruses from the digestive tracts of the shellfish may not be very effective for hepatitis A and Norwalk-like viruses.[33,35] In the United States, the major cause of foodborne viral disease outbreaks is the contamination of ready-to-eat foods due to poor personal hygiene and contaminated equipment used with foods served at delicatessens, cafeterias, and restaurants.[1,2]

F. Prevention

The two major preventative methods of foodborne virus infections should be to kill the viruses in contaminated foods and adopt good sanitation and personal hygiene habits to control contamination. Proper heat treatment, such as pasteurization, will be enough to kill the viruses. Steaming lightly to open the shellfish may not be an effective heat-treatment procedure. As indicated before, depuration and relaying cannot be considered effective. Sanitation, using oxidative agents such as hypochlorite, can kill viruses in contaminated equipment or in water used in food processing. Good personal hygiene and keeping suspected individuals away from handling ready-to-eat food will also be important in controlling the viral foodborne infections.[33]

X. OTHER INFECTIVE FOODBORNE PATHOGENS

Several other pathogens are known and confirmed to cause foodborne infections in humans. However, their incidence, especially in outbreaks, is quite low. A brief discussion of some of these pathogens is included.

A. Brucellosis

Human brucellosis is caused by *Brucella* spp., namely *Bru. abortus, Bru. suis,* and *Bru. melitensis*.[36] They are Gram-negative, nonmotile, nonsporeforming, aerobic small rods and pathogenic to animals and humans. In infected animals, the organisms are located in the uterus of pregnant animals, and in the mammary glands of lactating females. Thus the pathogens can be excreted in milk. People working with animals can become infected with *Brucella* spp. Also, people working with meat can be infected. Consumption of raw milk and some products made from raw milk (such as some imported cheese) have been implicated in foodborne brucellosis. The cells survive for a long time in milk and milk products. However, pasteurization of milk and milk products kills *Brucella* cells. The symptoms of brucellosis in humans include undulant fever with irregular rise and fall of temperature, profuse sweats, body aches, aching joints, chills, and weakness. The symptoms appear in 3 to 21 d following consumption of a contaminated food. Between 1983 and 1987, there were two outbreaks in the United States from the consumption of imported cheese affecting 38 people, with 1 death. The control measures for foodborne brucellosis include pasteurization of milk, the manufacturing of dairy products from pasteurized milk, and proper sanitation to prevent recontamination of pasteurized products.

B. Streptococcal Infection

Streptococcus pyogenes, in group A, is a pathogen and has been isolated from lactating animals with mastitis. It is a Gram-positive coccus and has been associated with human pharyngitis with symptoms of sore throat, fever, chills, and weakness.[36] In some cases nausea, vomiting, and diarrhea can be present. Some strains can cause Scarlet fever. Foodborne infections have been recorded from the consumption of contaminated raw milk and products made with raw milk and different types of salads (contaminated by infected food handlers). In the United States between 1983 and 1987, there were seven outbreaks affecting 1019 individuals. In one outbreak in 1983, 553 people were affected from eating contaminated potato salad. In addition to group A *Streptococcus*, some other streptococci have been implicated in foodborne infections. Control measures include pasteurization of dairy products and avoiding the consumption of raw milk and products made with raw milk. People, either suffering from streptococcal infections or a carrier, should not handle ready-to-eat foods, such as salads. Proper sanitation in all phases of processing and proper refrigeration will help to reduce the incidence.

C. Q Fever

Q fever in humans is caused by a rickettsia, *Coxiella burnetii*.[36] Animals carry this organism without any symptoms. People handling animals and raw milk and meat can be infected by the rickettsia and develop symptoms of Q fever. The symptoms include fever, malaise, anorexia, muscular pain, and headache. The symptoms appear 2 to 4 weeks after infection and last for about 2 weeks. Foodborne infection occurs from the consumption of raw or improperly pasteurized milk and milk products. *Cox. burnetii* is more resistant to heat than many pathogenic bacteria. Because of this, the current temperature and time of pasteurization has been set to either 145°F (62.8°C) for 30 min or 160°F (71.1°C) for 15 s so that *Cox. burnetii* is killed. In the United States, no incidences of foodborne outbreaks from *Cox. burnetii* have been recorded since 1983.

REFERENCES

1. Bean, N. H. and Griffin, P. M., Foodborne disease outbreaks in the United States, 1973–1987, *J. Food Prot.*, 53, 804, 1990.
2. Bean, N. H., Griffin, P. M., Goulding, J. S., and Ivey, C. B., Foodborne disease outbreaks, 5 year summary, 1983–1987, *J. Food Prot.*, 53, 711, 1990.
3. Tauxe, R. V., *Salmonella*: a postmodern pathogen, *J. Food Prot.*, 54, 563, 1991.

4. Flowers, R. S., *Salmonella, Food Technol.*, 42(4), 182, 1988.
5. D'Aoust, J.-Y., *Salmonella,* in *Foodborne Bacterial Pathogens,* Doyle, M. P., Ed., Marcel Dekker, New York, 1989, 327.
6. Anonymous, *Salmonella enteritidis* infection and grade A shell eggs — United States, 1989, *Dairy Food Environ. Sanitation,* 10, 507, 1990.
7. Marth, E. H., Disease characteristics of *Listeria monocytogenes, Food Technol.,* 42(4), 165, 1988.
8. Rocourt, J., *Listeria monocytogenes*: the state of the science, *Dairy Food Environ. Sanitation,* 14, 70, 1994.
9. Marsden, J. L., Industry perspectives on *Listeria monocytogenes* in foods: raw meat and poultry, *Dairy Food Environ. Sanitation,* 14, 83, 1994.
10. Madden, J. M., Concerns regarding the occurrence of *Listeria monocytogenes, Campylobacter jejuni* and *Escherichia coli* 0157:H7 in foods regulated by the U.S. Food and Drug Administration, *Dairy Food Environ. Sanitation,* 14, 262, 1994.
11. Teufel, P., European perspective on *Listeria monocytogenes, Dairy Food Environ. Sanitation,* 14, 212, 1994.
12. Lovett, J., *Listeria monocytogenes*, in *Foodborne Bacterial Pathogens*, Doyle, M. P., Ed., Marcel Dekker, New York, 1989, 283.
13. Anonymous, Listeriosis associated with consumption of turkey franks, in *Morbidity Report,* April 21, 1989 (referred to in *Dairy Food Environ. Sanitation,* 10, 718, 1989).
14. Kornacki, J. L. and Marth, E. L., Foodborne illness caused by *Escherichia coli*: a review, *J. Food Prot.,* 45, 1051, 1982.
15. Doyle, M. P. and Padhye, V. V., *Escherichia coli,* in *Foodborne Bacterial Pathogens*, Doyle, M. P., Eds., Marcel Dekker, New York, 1989, 235.
16. Doyle, M. P. and Padhye, V. V., *Escherichia coli* 0157:H7: epidemiology, pathogenesis, and methods of detection in food, *J. Food Prot.,* 55, 555, 1992.
17. Doyle, M. P., *Escherichia coli* 0157:H7 and its significance in foods, *Int. J. Food Microbiol.,* 12, 289, 1991.
18. Anonymous, FSIS background of *Escherichia coli* update: *E. coli* 0157:H7, *Media Relation Office, Food Service and Inspection Service,* USDA, January, 1993.
19. Smith, J. L., *Shigella* as a foodborne pathogen, *J. Food Prot.,* 50, 788, 1987.
20. Wachsmuth, K. and Morris, G. K., *Shigella,* in *Foodborne Bacterial Pathogens*, Doyle, M. P., Ed., Marcel Dekker, New York, 1989, 448.
21. Flowers, R. S., Bacteria associated with foodborne diseases: *Shigella, Food Technol.*, 42(4), 185, 1988.
22. Lior, H., Campylobacters: epidemiological markers, *Dairy Food Environ. Sanitation*, 14, 317, 1994.
23. Stringer, M. F., Campylobacter — a European perspective, *Dairy Food Environ. Sanitation,* 14, 325, 1994.
24. DeMol, P., Human campylobacteriosis: clinical and epidemiological aspects, *Dairy Food Environ. Sanitation,* 14, 314, 1994.
25. Stern, N. J. and Kazami, S. U., *Campylobacter jejuni,* in *Foodborne Bacterial Pathogens,* Doyle, M. P., Ed., Marcel Dekker, New York, 1989, 71.
26. Doyle, M. P., Bacteria associated with foodborne diseases: *Campylobacter jejuni*, *Food Technol.,* 42(4), 187, 1988.
27. Doyle, M. P., Foodborne pathogens of recent concern, *Annu. Rev. Nutrition*, 5, 25, 1986.
28. Schiemann, D. A., *Yersinia enterocolitica* and *Yersinia pseudotuberculosis,* in *Foodborne Bacterial Pathogens,* Doyle, M. P., Ed., Marcel Dekker, New York, 1989, 601.
29. Anonymous, New bacteria in the news: *Vibrio, Food Technol.*, 40(8), 22, 1986.

30. Hackney, C. R. and Dicharry, A., Seafood-borne bacterial pathogens of marine origin, *Food Technol.,* 42(3), 104, 1988.
31. Twedt, R. M., *Vibrio parahaemolyticus,* in *Foodborne Bacterial Pathogens,* Doyle, M. P., Ed., Marcel Dekker, New York, 1989, 543.
32. Oliver, J. D., *Vibrio vulnificus,* in *Foodborne Pathogens,* Doyle, M. P., Ed., Marcel Dekker, New York, 1989, 569.
33. Cliver, D. O., Viral foodborne disease agents and concern, *J. Food Prot.,* 57, 176, 1994.
34. Cliver, D. O., Epidemiology of viral foodborne disease, *J. Food Prot.,* 57, 263, 1994.
35. Gerba, C. P., Viral disease transmission by seafoods, *Food Technol.,* 42(3), 99, 1988.
36. Stiles, M. E., Less recognized or presumptive foodborne pathogenic bacteria, *Foodborne Bacterial Pathogens*, Doyle, M. P., Ed., Marcel Dekker, New York, 1989, 673.

CHAPTER 24 QUESTIONS

1. List some characteristics of foodborne microbial infections. Define enteric and nonenteric infections and give examples of three pathogens in each group.

2a. Explain the possible reasons of the current increase in foodborne salmonellosis in a developed country.

b. Discuss the regulatory requirements and suggestions directed to control foodborne salmonellosis in the United States.

c. From the epidemiology of a recent foodborne salmonellosis outbreak (instructor will provide), analyze the causes of the incidence and suggest methods that could be implemented to prevent this.

3a. How does foodborne listeriosis differ from foodborne salmonellosis?

b. How can ready-to-eat foods be contaminated with *Lis. monocytogenes*? What are the implications of this contamination in a refrigerated ready-to-eat food?

c. "Since 1991 foodborne listeriosis cases have declined." What could have helped in this reduction in the United States?

4a. Define nonpathogenic and pathogenic *Esc. coli*. How are the pathogenic strains grouped?

b. Describe the pathogenicity of foodborne EHEC infection in humans.

c. In a 1993 multistate foodborne outbreak, EHEC was isolated from the victims, improperly cooked hamburger patties, and raw hamburger. A large amount of ground beef (several thousand pounds) was suspected of being contaminated with this pathogen. Some food microbiologists believe that when a large volume of a product is contaminated with a pathogen, there is a greater likelihood that a step in the processing operation is directly contaminating the product (which in turn could have been previously contaminated from a small amount of contaminated raw material or environment). With this idea in mind, develop a scenario that might have been involved in the EHEC contamination in large amounts of hamburger patties in the above incident.

5a. List two similarities and two differences between foodborne salmonellosis and shigellosis. Define "Shiga toxin."

b. Describe food association for a shigellosis outbreak and preventative measures to control the disease.

6a. "Campylobacteriosis is emerging as a major foodborne illness in many countries." Explain with proper reasons your justification of this statement.

b. Discuss the food association of and preventative measures against the causative pathogen associated with campylobacteriosis.

7a. Describe two important characteristics of the causative pathogen associated with foodborne yersiniosis.

b. List the symptoms of yersiniosis and indicate how it is often misdiagnosed.

c. Discuss the control measures to reduce the incidence of foodborne yersiniosis.

8a. List the *Vibrio* species considered to be human pathogens.

b. Discuss how foods can be the vehicle of gastroenteritis caused by *Vib. parahaemolyticus*.

c. Describe the pathogenicity of *Vib. vulnificus* in humans.

9a. List the enteric viruses that are implicated in foodborne disease outbreaks.

b. How are foods contaminated with hepatitis A virus, and what control measures should be adopted to reduce the incidence?

10. Briefly describe foodborne infection caused by *Brucella abortus*, pathogenic *Streptococcus*, and *Coxiella burnetii*.

CHAPTER 25

Foodborne Toxicoinfections

CONTENTS

I. INTRODUCTION

The pathogenesis and disease symptoms of several pathogens associated with foodborne and waterborne gastroenteritis are somewhat different from classical food poisoning or foodborne infection caused by the pathogens described in Chapters 23 and 24. Although the differences are not always very clear, in this chapter the gastroenteritis caused by *Clostridium perfringens, Bacillus cereus, Vibrio cholerae,* and enteropathogenic *Escherichia coli* are described as toxicoinfection. While the first two are Gram-positive sporeformers, the last two are Gram-negative small rods. For each pathogen, the importance, characteristics, nature of toxin(s), food association, and control measures are discussed. For some, detection methods and analysis of an outbreak are also included.

A. Some Characteristics of Foodborne Intoxication

1. For sporeformers, ingestion of large numbers of live vegetative cells are usually necessary.
2. Vegetative cells of sporeformers do not multiply in the digestive tract, but sporulate and release toxins.
3. For Gram-negative bacteria, the live cells can be ingested in moderate numbers.
4. The Gram-negative cells rapidly multiply in the digestive tract.
5. Many cells also die, releasing toxins.
6. Toxins of both groups produce the gastroenteritis symptoms.

II. *CLOSTRIDIUM PERFRINGENS* GASTROENTERITIS

A. Importance

Gastroenteritis caused by *Clostridium perfringens* has several specific characteristics. In most outbreaks, large numbers of cases are involved. In the United States in the 1960s and 1970s, it was involved in over 7% of the total foodborne outbreaks and over 10% of the total number of cases. In the 1980s, the incidence dropped to about 3% of total outbreaks, affecting about 5% of the total cases. The outbreaks generally occur with some foods that were prepared in advance by heating and then kept warm for several hours before serving. In the majority of the instances, these situations are associated with feeding many people within a short period of time in cafeterias, restaurants, schools, and banquets (banquet disease). Between 1983 and 1987 out of a total of 24 outbreaks, 12 were associated with such institutions, while only one outbreak occurred with food prepared at home. As the disease produces mild symptoms, many incidents are probably not reported. Thus the reported cases could be just a fraction of the actual numbers.[1,2]

B. Characteristics

The cells are Gram-positive rods, motile, and sporeformers. Cells vary in size and can form small chains. *Clo. perfringens* is anaerobic but can tolerate some air. The vegetative cells are sensitive to low heat treatment (pasteurization), but the spores are extremely heat resistant and some can survive boiling for several hours. The cells are resistant to D-cycloserine. In the presence of suitable substrates, H_2S is formed during growth. The cells need several amino acids for growth. Thus they can grow very effectively in many protein foods. The temperature of growth of vegetative cells and germination of spores and outgrowth range between 10 and 52°C. The optimum growth occurs around 45°C. At optimum conditions, cell multiplication can be very rapid, in about 9 min. The cells fail to grow well at below pH 5.0, in NaCl concentrations above 5%, at A_w below 0.93, and in 500 ppm nitrite.[3-5]

C. Habitat

Spores and vegetative cells are found in soil, dust, the intestinal contents of animals, birds, and humans, and sewage. Many foods, particularly raw foods, are contaminated from these sources.[3-5]

D. Toxins and Toxin Production

Among the five types of *Clo. perfringens*, type A strains are involved in foodborne toxicoinfection. The enterotoxin associated with the foodborne disease is a heat-labile protein. It is an intracellular protein, produced by the cells during sporulation in the intestine and released. Unlike toxins of food-poisoning microorganisms, the enterotoxin is produced in the digestive tract. The environmental parameters for the production of enterotoxin are directly related to the sporulation environment. There are some reports that, in addition to the intestine, sporulation and enterotoxin production to certain levels can also occur in some foods.[3-5]

E. Disease and Symptoms

The enterotoxin causes only gastroenteritis. The symptoms appear 8 to 24 h following ingestion of a large number of viable cells through a food (about 5×10^5 or more per gram). The main symptoms are diarrhea and abdominal pain. Nausea, vomiting, and fever can also occur, but are less common. Fatality, although rare, can occur among the very young, elderly, and sick. The symptoms generally disappear within 24 h. It is considered a mild disease and is very seldom reported.[3-5]

F. Food Association

Raw meats from animals and birds are most commonly contaminated with the spores and cells from the digestive tract content, while vegetables and spices commonly get them from soil and dust. As the bacterium has some amino acid requirements for growth, meat and meat-containing foods provide good environments for cell growth. The foods commonly incriminated with the outbreaks include meat stews (beef and poultry), roasts, meat pies, casseroles, gravies, sauces, bean dishes, and some Mexican foods (tacos and enchiladas). The food types involved in 190 *Clo. perfringens* gastroenteritis outbreaks in the United States from 1973 to 1987 are listed in Table 25.1. Among the foods listed, protein-rich foods were involved in high percentages.[1] The three most important contributing factors in these outbreaks, in order of importance, were improper holding temperature, inadequate cooking, and contaminated equipment. The most important predisposing cause was cooking food in large volumes in advance, allowing it to cool slowly, then holding for a long period and often serving without reheating. Cooking kills the vegetative cells, but the spores survive. Large volumes of the food provide an anaerobic environment and slow cooling provides the temperature for the spores to germinate and outgrow, and the cells to multiply rapidly to reach high population levels prior to eating. The food may not show any loss of acceptance quality.[3]

Table 25.1 Food Types Involved in *Clostridium perfringens* Outbreaks from 1973 to 1987 in the United States

Food types	No. of outbreaks[a]	%	Food types	No. of outbreaks[a]	%
Beef	51	26.3	Fin fish	3	1.6
Mexican foods	23	11.9	Shellfish	2	1.0
Turkey	19	9.8	Vegetables	1	0.5
Chicken	9	4.6	Other	46	23.8
Pork	8	4.1	Unknown	28	14.4

[a] Total cases were 12,234.

G. Prevention

The presence of *Clo. perfringens* spores and cells in foods is not uncommon. Unless viable cells are present in high numbers prior to eating, they cannot cause gastroenteritis; thus the aim will be to keep the cell numbers low. This can be achieved by using proper sanitation in all phases of food preparation and handling. Food should be cooked to the highest temperature recommended to kill the cells and as many spores as possible. The food should be cooled quickly and uniformly (preferably within 1 h) to refrigerated temperature. A shallow container, in contrast to a deep container, will facilitate rapid cooling (below 10°C). If a food is stored for a long time, it is important to reheat it quickly and uniformly (to kill vegetative cells) and keep it hot (above 60°C) while being served.[3,5]

H. Detection Methods

The detection method involves enumeration of the incriminated food and fecal samples for *Clo. perfringens* in selective agar medium and incubation of plates under anaerobic conditions.[3,5]

I. Analysis of an Outbreak

In March 1989, about 300 of 420 employees became ill following attendance at a corporate luncheon the previous day. Among the 113 people interviewed, diarrhea, abdominal cramps, and nausea were reported. The symptoms started between 2 and 18 h (mean average, 9.5 h) following the luncheon and the disease lasted about 12 h. Stool specimens contained high numbers of *Clo. perfringens*. The menu consisted of chicken, roast beef and gravy, lasagna, and mixed vegetables. Interviews revealed that 98% of those who ate the roast beef became ill; thus it was implicated as the likely source. Examination of the catering facilities revealed poor sanitation and refrigeration facilities. The roasts, each about 19 lb, were cooked 3 d prior to the luncheon date and kept in the refrigerator. Examination of the remaining roast beef showed the presence of *Clo. perfringens* at over 2×10^6 per gram.

The roasted meat, prepared in advance under inadequate sanitary conditions, stored at improper temperature, and served without further heating, was associated with this *Clo. perfringens* gastroenteritis outbreak. The most likely sequence of events was as follows. The raw beef (or the facilities in the kitchen) served as the source of the bacterium (most probably spores). Roasting of the large-size meat did not kill the spores. Subsequently, during storage for 3 d under improper refrigeration, it took a long time for the temperature in the large roasts to drop below 10°C. Thus the spores germinated and the cells multiplied rapidly in the protein food and reached a high population before the temperature reached below 10°C. When the roast was served, it had a high population of *Clo. perfringens* cells. Following consumption of the roast, the cells sporulated, releasing enterotoxins and producing the gastroenteritis symptoms within 2 to 18 h that lasted for about 12 h. This outbreak could have been avoided by proper sanitation, proper cooking, proper and quick refrigeration (10°C or below), and warming the roast above 60°C prior to serving.

III. *BACILLUS CEREUS* GASTROENTERITIS

A. Importance

The incidence of foodborne gastroenteritis of *Bacillus cereus* origin is relatively high in some European countries.[1,2] In contrast, the incidence is relatively low in the United States. In the United States, it was recognized as a causative gastroenteritis agent in 1969. Between 1973 and 1987, 58 outbreaks involving 1123 cases were reported. Both the incidence of outbreaks and number of cases per outbreak were low and no fatalities have been reported. However, as the symptoms are not severe and last for about 12 h, many cases may not be reported.

B. Characteristics

The cells are Gram-positive, motile rods that form endospore in the middle of the cells. Cells are sensitive to pasteurization. Spores can survive the high heat treatment used in many cooking procedures. *Bac. cereus* is aerobic, but can also grow under some degree of anaerobic environment. The cells can multiply in the temperature range between 4 and 50°C with the optimum being around 35 to 40°C. Other parameters of growth are pH between 4.9 and 9.3, A_w 0.95 and above, and NaCl concentrations below 10%.

C. Habitat

Spores and cells of *Bac. cereus* are common in soil and dust and can be readily isolated in small numbers from many foods that include both raw and finished products. The intestinal tract of 10% of healthy adult humans have *Bac. cereus* under normal conditions.

D. Toxins and Toxin Production

The strains produce at least two types of enterotoxins. Each one is probably associated with specific types of symptoms.[3,6] The toxins are produced during growth of cells at the growth temperature range and retained in the cells. Only when the cells are lysed are the toxins released. This occurs in the intestinal tract but can also occur in foods.

E. Disease and Symptoms

In general, a large number of cells (10^6 to 10^7 per gram) need to be ingested to produce gastroenteritis. The two types of enterotoxin produce two types of symptoms. The enterotoxin associated with the diarrheal form is a heat-labile protein, and that associated with the emetic form is a heat-stable protein. In the diarrheal form, the symptoms occur within 6 to 12 h following consumption of a food containing the viable cells. The symptoms include abdominal pain, profuse watery diarrhea, and maybe nausea, but no vomiting or fever. Recovery is usually within

24 h. These symptoms are similar in many respects to those produced by *Clo. perfringens*.

In the emetic form, the symptoms occur within 1 to 5 h following ingestion of a food containing the viable cells. As the toxin is heat stable, once the toxin forms in cells, heating of food containing a large number of cells prior to eating can produce the symptoms. The symptoms are characterized by nausea and vomiting; abdominal pain and diarrhea may also be present. The symptoms last for about 24 h. In some respects, these symptoms are similar to staphylococcal gastroenteritis.

F. Food Association

Many types of food can contain small numbers of cells and spores of *Bac. cereus*.[6,7] Consumption of these foods will not cause the disease. However, when these foods are abused to facilitate spore germination and cell multiplication, the population of live cells can reach the high levels necessary for the disease.[6,7] In diarrheal outbreaks, a variety of food, including vegetables, salads, meats, pudding, casseroles, sauces, and soups, have been implicated, mostly due to improper cooling. However, in the emetic form, the outbreaks mostly involve rice and sometimes other starchy foods. A list of foods associated with *Bac. cereus* gastroenteritis outbreaks in the United States from 1973 to 1987 is presented in Table 25.2. Although many foods are listed, the most predominant was fried rice in a Chinese restaurant (Chinese rice syndrome). Investigations revealed that following boiling for preparation, the rice is kept at room temperature for a long time. The surviving (or contaminated) spores germinate, and the cells multiply to high levels before the rice is served.

Table 25.2 Food Types Associated with *Bacillus cereus* Gastroenteritis between 1973 and 1987 in the United States

Food type	No. of outbreaks	%	Food type	No. of outbreaks	%
Chinese food	24	41.5	Fin fish	1	1.7
Mexican food	5	8.6	Shellfish	2	3.4
Beef	3	5.2	Ice cream	1	1.7
Chicken	1	1.7	Other	8	13.8
Turkey	1	1.7	Unknown	9	15.5
Vegetables	3	5.2			

G. Prevention

Among the predominant contributing factors associated with *Bac. cereus* gastroenteritis, in order of relative importance, are improper holding temperature, contaminated equipment, inadequate cooking, and poor personal hygiene. The heat treatment normally used in food preparation, except for pressure cooking, may not destroy *Bac. cereus* spores. The most important control measure will be to keep food at a temperature where the spores do not germinate and cells do not grow. This can be achieved by uniform quick chilling of the food to about 4 to 5°C or holding the food above 60°C. Quick chilling can be best accomplished by storing a food in a shallow container, no more than 2 to 3 in. thick. As *Bac. cereus* cells, given sufficient time, can grow and produce toxins at refrigerated temperature (≥4°C), a food should not be stored at low temperatures for long periods of time. This means that preparation of a food well in advance should be avoided. Since cells can get in a food through cross-contamination, proper sanitary measures should be adopted in the handling of a food. Finally, as live cells are necessary for the symptoms, uniform reheating of a suspected food to above 75°C before serving should be practiced. However, heating may not destroy the heat-stable toxins associated with emetic symptoms.[3,6]

H. Detection Methods

Bac. cereus can be enumerated by surface plating on an agar medium containing mannitol, egg yolk, and polymyxin B, with polymyxin B as the selective agent. Colonies surrounded by precipitation resulting from lecithinase of the cells are indicative of *Bac. cereus*.

IV. CHOLERA

A. Importance

Cholera is caused by *Vibrio cholerae* 01.[1,2,8,9] It is a non-contagious disease but can cause large epidemics with high mortality. In the 19th century, both epidemic and isolated cholera cases were recorded in the United States. Since 1911, the disease was thought to be eradicated. Most of the outbreaks in the 20th century were recorded in Asian countries. However, in 1973, a cholera case was recorded in Texas. Between 1973 and 1987, a total of 6 outbreaks, involving 916 cases with 12 deaths were recorded, mostly in the coastal states of the United States. Contaminated seafood (cooked crab, fin fish, and raw oysters) was involved in these cases. Between 1991 and 1992, a large cholera epidemic that started in Peru spread to many South American countries and affected about 640,000 people, with 5600 deaths. It was also introduced into the United States by travelers from these countries who either ate contaminated foods before entering the United States or brought contaminated concealed foods to the United States. In addition to 01 strains, *Vib. cholerae* non-01 strains have also been involved in cholera (also designated as non-01 gastroenteritis) cases both in the United States and other countries. Non-01 strains were previously thought to be incapable of causing large epidemics; but in 1992, a non-01 serotype, non-0139, was involved in large epidemics in Bangladesh and India.

B. Characteristics

Vib. cholerae, like other vibrios, is a Gram-negative, motile, curved rod. The species has many serogroups. Strains in 01 are associated with epidemic cholera. This serotype is further characterized by biotype and serotype. The type currently associated with cholera epidemics worldwide is of El Tor biotype and either Inaba or Ogawa serotype. Non-01 serotypes do not agglutinate with antibody prepared against 01 antigen. Also, non-01 serotypes, similar to 01 serotypes, are not sensitive to trimethoprim-sulfamethoxazole with furazolidone. Both types are sensitive to heat and are killed by temperatures used for cooking. Improper heating (at lower temperature for a shorter time) may not be able to kill all the cells present in a food. The optimum temperature of growth is between 30 and 35°C. The growth rate is very rapid, even at room temperature. The cells do not multiply in contaminated live crabs, oysters, or fish. However, in cooked seafoods, rapid growth can occur at 25 to 35°C. Alkaline foods facilitate rapid growth. Survival of cells is better in cooked foods at 5 to 10°C.[8,9]

C. Habitat

Cholera is a human disease. The disease results from the ingestion of infective doses of *Vib. cholerae* cells through food and water contaminated with feces of humans suffering from the disease. Chronic carriers are rare and may not be important in large epidemics. Marine environments may serve as long-term reservoirs. Both 01, and, especially non-01, serotypes have been isolated from water in the United States Gulf Coast states (Florida, Louisiana, and Texas), in the Chesapeake Bay, and along California coasts. Seafood (crabs, oysters, and fin fish) harvested from these areas can carry contamination and provide the infective dose. Contaminated water can also be the source of the disease.

D. Toxins and Toxin Production

The toxin of 01 serotype is a heat-labile, 85-kDa cytotoxic protein, and has two functional units.[8] The active A subunit stimulates adenyl cyclase in the intestinal epithelial cells, causing massive secretion of water along with chloride, potassium, and bicarbonate in the lining of the intestine. The non-01 serotypes produce a cytotoxin and a hemolysin. Following ingestion of *Vib. cholerae* cells in sufficient numbers, the cells colonize the small intestine and multiply rapidly to produce toxins. When the cells die and lyse, the toxins are released into the intestine.

E. Disease and Symptoms

Vib. cholerae is not contagious. A person must consume a large number of viable cells through contaminated food or water to contract the disease. Fecal-oral infection is also possible. The infective dose for cholera is about 10^6 viable cells per person, but varies with the age and health of the individual. The incubation period ranges from 1 to 5 d but is usually 2 d. The symptoms include the sudden onset of profuse watery diarrhea and vomiting. Loss of fluid results in dehydration. Other symptoms in severe cases are painful muscle cramps and clouded mental status. Many infected persons have no symptoms, or mild to moderate diarrhea. Treatment consists of rapid replacement of fluids along with electrolytes and the administration of proper antibiotics. In addition to diarrhea, the non-01 toxins also cause infection of soft tissues and septicemia.

F. Food Association

Food can serve as a source of *Vib. cholerae* if it is contaminated directly with human feces from the patient or previously contaminated water. The handling of food by a person suffering from the disease can also contaminate food, due to poor personal hygiene. In addition, food originating from natural reservoirs of the causative bacteria can be contaminated and spread cholera. The natural reservoirs include marine and brackish water environments. Testing of samples of water, oysters, crab, and shrimp from U.S. coastal states has indicated the presence of both 01 and non-01 serotypes. Table 25.3 lists the types of contaminated food involved in the disease. In the Gulf Coast incidents consumption of raw sea foods (oysters and shrimp) and partially cooked crabs (not to kill *Vib. cholerae* cells) were mainly involved. These foods were probably contaminated with *Vib. cholerae* cells from the water where they were harvested. Foods from the South American countries were also seafoods, eaten either raw or improperly cooked. These foods either had *Vib. cholerae* when harvested or became contaminated after harvest. One of the most important aspects of this data is to recognize how foods from one country can be involved in a disease in another country.[8]

Table 25.3 Contaminated Food Associated with Some Cholera Outbreaks in the United States

	Food source: Gulf Coast (1973–1992)		Food source: South American countries (1991–1992)	
Food type	**Outbreak**	**Cases**	**Outbreak**	**Cases**
Raw oysters	4	4	2[c]	3
Crab[a]	4	32	4	14
Fish & shrimp[b]	1	2	2	77
Cooked rice	1	16	—	—
Seafood[b]	—	—	2	3

[a] Improperly cooked crab.
[b] Some items were raw.
[c] In one incidence, raw clams were consumed.

G. Prevention

The spread of the disease can be prevented or reduced by adopting proper hygienic measures. These include provision for properly treated municipal water, decontamination of other water by boiling or chemical treatment, and proper disposal of sewage. The infected persons may be treated with antibiotics to enhance recovery, along with replacement of body fluids. Unexposed people can be given vaccines to protect them from the disease. Seafood should not be harvested from polluted water nor from water found to harbor *Vib. cholerae*. Finally, seafood should not be eaten raw. The time-temperature of heat treatment of a suspected food should be enough to kill the pathogen. An example of this is the survival of *Vib. cholerae* in contaminated crabs boiled for about 8 min; probably, a 10-min boiling time is necessary to assure killing.

H. Detection

Isolation of *Vib. cholerae* from a sample is achieved by an initial preenrichment in alkaline peptone water, followed by streaking on a selective agar medium plate (such as thiosulfate citrate bile salt sucrose agar). Suspected colonies (yellow) are biochemically and serologically tested for confirmation. Toxin is detected by immunoassay or bioassay. The toxin gene is recognized by the PCR technique.

I. Analysis of an Outbreak

Between March 31 and April 3, 1991, eight people developed profuse watery diarrhea after eating crabmeat transported from Ecuador; of these, five also had vomiting and three had severe leg cramps.[10] The symptoms developed 1 to 6 d after consumption of the crabmeat. The stool samples of some patients yielded *Vib. cholerae* 01, biotype El Tor, serotype Inaba, the same serotype involved in the cholera epidemic in South American countries, including Ecuador. No crabmeat samples were available for analysis.

Investigations revealed that the crabs were purchased at a fish market in Ecuador, then boiled, shelled, wrapped in foil, and transported by air, unrefrigerated in a plastic bag, to the United States on March 30. The meat was delivered to a private residence, refrigerated overnight, and served as a salad on March 31 and April 1.

The sequence of events were as follows. The live crabs were contaminated with *Vib. cholerae* harvested from contaminated water, and cooking was not sufficient to kill all the cells. The viable cells subsequently grew in the crabmeat at nonrefrigerated temperature to reach the high populations necessary to cause the disease. Other possibilities include postheat contamination of the meat during shelling by one or more people with the disease in mild or asymptomatic form, the contamination of equipment, or the water used. Consumption of the crabmeat caused the disease. Preventing harvest of crabs from contaminated water, proper time-temperature of cooking the crabs, proper sanitation in preparing the crabmeat, refrigeration of crabmeat following picking, heat treatment prior to the preparation of salad: any one or more of these steps could have been used to avoid this incident. (Note: It is illegal to bring food into the United States from other countries without the prior permission of regulatory agencies).

V. *ESCHERICHIA COLI* GASTROENTERITIS

A. Importance

The two (of the four) enteropathogenic *Esc. coli* subgroups that correlate well with toxicoinfection probably belong to enteropathogenic and enterotoxigenic *Esc. coli* (EPEC and ETEC, respectively) types. They produce diarrheal diseases when ingested in large numbers through contaminated foods and water. The symptoms are more like cholera. The incidence is high in many developing countries and is directly related to poor sanitation.

B. Characteristics

Many serotypes in both subgroups are involved in human gastroenteritis. They are Gram-negative, small curved rods, nonsporulating and motile (nonmotile strains can be present). The strains are facultative anaerobes and can grow effectively in both simple and complex media and many foods. Growth occurs between 10 and 50°C, with the optimum at 30 to 37°C. There are strains that can grow below 10°C. Rapid growth occurs under optimum conditions. Growth-limiting factors are low pH (below 5.0) and low A_w (below 0.93). The cells are sensitive to low heat treatment, such as pasteurization.

C. Habitat

All strains in both subgroups can establish in the small intestine of humans without producing symptoms. The carriers can shed the organisms in feces and can contaminate food and water directly or indirectly. Many animals, including domestic animals, can also harbor different serotypes of both subgroups and contaminate soil, water, and food. In the animals, some serotypes may not produce the disease symptoms.

D. Toxins and Toxin Production

The strains in the ETEC subgroup produce two types of enterotoxins: one is heat labile (LT) and the other is heat stable (ST). A strain can produce either LT or ST, or both. LT toxin is an antigenic protein, similar to the cholera toxin produced by *Vib. cholerae*, and induces fluid secretion by the epithelial cells of the small intestine. ST is a heat-stable protein, smaller in molecular weight than the LT, and is nonantigenic. It also increases fluid secretion by the intestinal cells but through a different mode of action. The strains in EPEC subgroups were previously thought not to produce enterotoxins like ETEC serotypes. However, some studies have shown that several serotypes produce LT toxin, while several others produce toxin(s) different from LT and ST of ETEC serotypes. In addition to LT and ST enterotoxins, ETEC serotypes are also capable of producing additional factors that enable the cells to colonize, multiply, and initiate infection. The genetic determinants of the enterotoxins in ETEC are plasmid linked and can be transferred to other *Esc. coli* strains.[3,11]

The production of enterotoxins by the ETEC strains is influenced by media composition, culture age, and alteration of a culture during growth. Optimum production occurs in a nutritionally rich medium at pH 8.5. Aeration of broth facilitates good toxin production. The toxins are generally detected in a growing culture within 24 h at 35°C. However, toxins can be produced by the cells growing at 25 to 40°C.[3,11]

E. Disease and Symptoms

EPEC strains were initially thought to be associated with infant diarrhea in many tropical and developing countries, causing high mortality. In contrast, ETEC strains are regarded as the cause of nonfatal diarrheal disease, called "travelers diarrhea."

However, there are some who consider that various EPEC strains can also cause travelers diarrhea. Ingestion of a high level (10^6 to 10^9 cells) of viable cells of the organisms by adults is necessary for the symptoms to occur within 24 to 72 h (by adults, within 24 to 30 h). The symptoms include mild to severe diarrhea. In severe cases, dehydration, prostration, and shock may accompany the diarrhea. Not all individuals will show symptoms, and those who develop the symptoms may shed the organisms in feces after recovery.[3,11]

F. Food Association

Many types of foods, including meat products, fish, milk and dairy products, vegetables, baked products, and water have been associated with gastroenteritis of *Esc. coli* origin in many countries.[3,11] These include serotypes from both EPEC and ETEC subgroups. Direct or indirect contamination of these foods (and water) with fecal materials, along with improper storage temperature and inadequate heat treatment, were involved in these incidences.

G. Prevention

The most important factor in the prevention of gastroenteritis in humans by pathogenic *Esc. coli* is to prevent contamination of food and water, directly or indirectly, by fecal matters. This can be achieved by developing effective sanitation in water supplies, and treatment and disposal of sewage. The other factor to consider is the prevention of contamination of food due to poor personal hygiene by people who are shedding the pathogen. Finally, one needs to recognize that even if the pathogen is present in very small initial numbers in a food, temperature abuse of the food can facilitate multiplication of cells to high levels necessary for disease symptoms. Thus food should be refrigerated or eaten quickly, preferably after reheating.[3,11]

H. Detection Methods

The detection methods used involve selective enrichment of sample (food, water, and feces), isolation of pathogens on selective agar medium, and biochemical characterization of the suspected isolates. Confirmation tests to detect toxins involve one or more serological tests (ELISA). Other methods to detect toxins include injection of test material into the ligated ileal loop of infant mouse and/or exposing Y-1 adrenal cells to toxin (for LT).

REFERENCES

1. Bean, N. H. and Griffin, P. M., Foodborne disease outbreaks in the United States, 1973–1987, *J. Food Prot.*, 53, 804, 1990.
2. Bean, N. H., Griffin, P. M., Goulding, J. S., and Ivey, C. B., Foodborne disease outbreaks, 5 year summary, 1983–1987, *J. Food Prot.*, 53, 711, 1990.

3. Garvani, R. B., Food science facts, *Dairy Food Environ. Sanitation,* 7, 20, 1987.
4. Labbe, R. G., *Clostridium perfringens, Food Technol.,* 42(4), 195, 1988.
5 Labbe, R., *Clostridium perfringens,* in *Foodborne Bacterial Pathogens,* Doyle M. P., Ed., Marcel Dekker, New York, 1989.
6. Johnson, K. M., *Bacillus cereus* foodborne illness — an update, *J. Food Prot.,* 47, 145, 1984.
7. Bryan, F. L., Bartleson, C. A., and Christopherson, N., Hazard analyses, in reference to *Bacillus cereus*, of boiled and fried rice in Cantonese-style restaurants, *J. Food Prot.,* 44, 500, 1981.
8. Popovic, T., Olsvik, O., Blake, P. A., and Wachsmuth, K., Cholera in the Americas: foodborne aspects, *J. Food Prot.,* 56, 811, 1993.
9. Anonymous, Imported cholera associated with a newly described toxigenic *Vibrio cholera* 0139 strain, California, 1993, *Dairy Food Environ. Sanitation,* 14, 48, 1994.
10. Communicable Disease Center, Cholera — New Jersey and Florida, *Morbidity Mortality Weekly Rep.,* 40, 287, 1991.
11. Kornacki, J. L. and Marth, E. H., Foodborne illness caused by *Escherichia coli*: a review, *J. Food Prot.,* 45, 1051, 1982.

CHAPTER 25 QUESTIONS

1. List and discuss five important factors that differentiate foodborne toxicoinfection from food intoxication.

2. List the important characteristics of *Clostridium perfringens* gastroenteritis (symptoms, suspected foods, and toxin production).

3. Discuss the methods to be adopted to prevent foodborne *Clo. perfringens* gastroenteritis.

4. What are the differences in the two types of toxins associated with *Bac. cereus* gastroenteritis? What is the sequence of events that leads to this disease?

5. Discuss the importance of cholera as a foodborne disease.

6. Discuss the factors and foods involved in cholera outbreaks in the United States since the 1970s.

7. Discuss the methods to be implemented to prevent foodborne *Esc. coli* gastroenteritis.

8. Explain the nature of toxins involved in foodborne toxicoinfection by: *Clo. perfringens, Bac. cereus, Vib. cholerae* 01 and non-01, EPEC, and ETEC.

9. Explain the following terms in relation to foodborne diseases: "Banquet disease," "Chinese rice disease," "seafood disease," "travelers diarrhea."

10. From an outbreak incidence of cholera (instructor will provide), discuss the sequence of events that caused the disease. What could have been done to prevent the incidence?

CHAPTER 26

Opportunistic Pathogens, Parasites, and Algal Toxins

CONTENTS

I. INTRODUCTION

Besides the foodborne bacterial pathogens discussed in Chapters 23, 24, and 25, there are several other bacterial species suspected of having the potential of causing foodborne illness. Normally, they are not pathogenic to humans. But there are strains of these species that have been known to produce toxins. Thus consumption of foods contaminated with these bacterial species and strains may cause illness, especially under certain circumstances, such as if they are consumed in extremely high numbers, and the individuals are either very young or not in normal physical condition. These bacterial species-strains are considered opportunistic pathogens. A brief discussion of the characteristics of some of these species, their association with foods, and their disease-producing potential are discussed in this chapter. A brief discussion is also presented on foodborne illnesses caused by several nonmicrobial components. These include biogenic amines (which can be produced by bacterial metabolism in food), algal toxins, and important parasites.

II. OPPORTUNISTIC BACTERIA

A. *Aeromonas Hydrophila*

1. *Characteristics*

The genus includes several species: *Aer. hydrophila, Aer. carvie*, and *Aer. sobria*. *Aer. hydrophila* cells are Gram-negative motile rods. The strains are found in both salt- and freshwater environments and are known to be pathogenic to fish. They are also found in the intestinal contents of humans and animals. Their growth temperature ranges between 3 and 42°C with an optimum between 15 and 20°C; a few strains can grow at 1°C. They are facultative anaerobes, but grow better in an aerobic environment. Pasteurization effectively kills the cells. pH (below 4.5), NaCl (above 4%), low temperature (below 3°C), and other factors can reduce their growth.[1,2]

2. *Food Association*

Due to the nature of their normal habitat, *Aer. hydrophila* are found in many foods, especially foods of animal origin. They may be isolated from milk, fin fish, seafood, red meat, and poultry; in some foods they occur at a level of 10^5 cells per gram or milliliter. The strains have been implicated in the spoilage of foods. Because of the psychrotrophic nature, they can grow in foods at refrigerated temperature and, even from a low initial load, can reach a high population with time during storage. Their presence in high numbers in food can be controlled by heat treatment, preventing postheat contamination, and using one or more of the growth-limiting parameters, namely low pH and low A_w.[1,2]

3. *Disease-Causing Potentiality*

Aer. hydrophila strains have been suspected to cause gastroenteritis in humans, especially when consumed in large numbers and the individuals have impaired health. No confirmed cases of foodborne illness due to *Aer. hydrophila* have been reported. Several incidences of gastroenteritis from the consumption of food and water contaminated with this species were reported. Many strains of *Aer. hydrophila*, especially those isolated from foods, were found to produce cytotoxins and hemolysins. It is not definitively known if these toxins are capable of causing gastroenteritis in humans.[1,2]

B. *Plesiomonas Shigelloides*

1. *Characteristics*

Ple. shigelloides is a Gram-negative, facultative anaerobic, motile, nonsporulating rod. The species has many characteristics similar to *Aeromonas* spp. and was previously classified as *Aeromonas shigelloides*. The organism is isolated from the intestinal contents of humans and warm- and cold-blooded animals. It is found in fresh and brackish water, and in fish and oysters harvested from water. Most strains grow between 8 and 45°C with optimum growth at 25 to 35°C. The cells are killed by pasteurization. Low temperature (below 10°C), low pH (below 4.5), and NaCl (above 5%) can be used, especially in combinations of two or more, to reduce growth.[3,4]

2. *Food Association*

Ple. shigelloides strains are isolated from foods of aquatic origin, such as fish and shellfish. They are present in higher frequencies and levels in oysters collected during warmer months and from muddy beds. Also, due to fecal contamination, they can be present in raw foods of animal, bird, and plant origin. In heat-treated foods, their presence indicates either improper heating or postheat contamination, or both. They can grow rapidly in most foods under optimum growth conditions. They are not expected to grow in foods held in refrigeration (3 to 4°C or below), even for a long time. Foods with low pH (4.5), low A_w (0.95), and 5% salt will discourage their growth.[3,4]

3. *Disease-Causing Potentiality*

Ple. shigelloides strains were implicated in many human gastroenteritis outbreaks associated with contaminated drinking water. In many cases, the organisms were isolated from both the implicated water and stools of the affected people. The typical symptoms are diarrhea, nausea, and abdominal pain; many can also have vomiting, fever, and chills. The incubation period ranged between 24 and 50 h, and the symptoms lasted for 1 to 9 d. Several incidences of gastroenteritis from the consumption of seafood, oysters, crabs, and fish were suspected to be due to *Ple. shigelloides*. In some instances, the organisms were isolated both from foods and stool of the patients. In these cases, foods were eaten either raw or after being improperly cooked or stored.

It is difficult to confirm a direct involvement of *Ple. shigelloides* in human foodborne gastroenteritis. *Ple. shigelloides* was isolated in stools, from both unaffected people and the individuals suffering from gastroenteritis. Thus these strains are considered opportunistic pathogens affecting individuals with less resistance, such as the young, old, and sick. In addition to gastroenteritis, several strains are associated with bacteremia and septicemia. Several studies demonstrated that *Ple. shigelloides* strains have heat-stable toxins that may have enterotoxin properties.[3,4]

C. Non-*Escherichia Coli* Coliforms

Coliform groups include species from genera *Escherichia, Klebsiella, Enterobacter,* and *Citrobacter*, all of which belong to the family *Enterobacteriaceae* and thus share some common characteristics. Previously, *Esc. coli* strains (both nonpathogenic and pathogenic strains) were thought to inhabit mainly the intestinal tract of humans and warm-blooded animals and birds, and most species in the other three genera were thought to be mainly of nonintestinal origin. However, different studies have shown that species and strains from *Klebsiella, Enterobacter,* and *Citrobacter* (together referred to as non-*Esc. coli* coliforms) are capable of colonizing in the human gut and producing potent enterotoxins. In several acute and chronic cases of diarrhea, they were isolated from stool and intestinal tract. Some isolates of *Ent. cloacae, Kle. pneumoniae,* and *Citrobacter* spp. were found to produce enterotoxins similar to heat-labile or heat-stable toxins of enterotoxigenic *Esc. coli*. In enterotoxigenic *Esc. coli,* these traits are plasmid linked. The ability of the non-*Esc. coli* coliforms to produce enterotoxins similar to the pathogenic *Esc. coli* strains probably results from the intergeneric transfer of plasmids encoding these phenotypes.[5]

The non-*Esc. coli* coliforms are normally present in raw food materials, as well as in some pasteurized foods due to postheat contamination. They can grow in many foods if the growth parameters are not limiting. Some strains can grow at refrigerated temperature. Temperature abuse during storage can also facilitate their rapid growth in a food. The significance of their presence in a food may need to be reevaluated. The ability of food isolates to produce enterotoxins may be included in the test protocols.[5]

D. Toxigenic Psychrotrophic *Bacillus* spp.

Cells and spores of many *Bacillus* species can be present in non-heat-treated foods. Also, most heat-treated foods can contain their spores. In refrigerated foods, the spores of the psychrotrophic *Bacillus* spp. can germinate and outgrow, and the vegetative cells can multiply. With time, they can be present in high numbers.[6]

Spoilage of some refrigerated foods, such as pasteurized milk, due to growth of several *Bacillus* spp. has been recognized. The psychrotrophic *Bacillus* spp. isolated in one study includes strains of *Bac. cereus, Bac. mycoides, Bac. circulans, Bac. lentus, Bac. polymyxa,* and *Bac. pumilus*. In this study, an isolate was considered a psychrotroph if it grew at 6°C. Many isolates of these species produced toxins. However, the capability of these toxins to cause gastroenteritis was not examined. This aspect may need further consideration as more and more low heat-processed foods are stored at refrigeration temperature for long times.[6]

III. BIOGENIC AMINES

Different types of amines can form in many protein-rich foods due to the decarboxylation of amino acids by the specific amino acid decarboxylases of microorganisms.[7-9] Some of these amines are biologically active and, when consumed in considerable amounts, cause illness. Examples of biogenic amines are histamine (produced from L-histidine) and tyramine (produced from L-tyrosine). Many bacteria found in foods can produce very active histidine or tyrosine decarboxylases, and, if free L-histidine and L-tyrosine are present, they can convert them to histamine and tyramine, respectively. Some food spoilage bacteria and even several lactobacilli can have very active histidine and tyrosine decarboxylases. The protein-rich foods associated with biogenic amine poisoning include cheeses ripened for a long time, fermented sausages stored for a long time, and different types of fish. Histamine poisoning of fish origin is further discussed here.[7-9]

A. Histamine (Scombroid) Poisoning

Histamine poisoning is called "scromboid poisoning," as the disease was initially recognized to occur from the ingestion of spoiled fish of scromboid groups (tuna, mackerel, and bonito).[8] However, it is now recognized that the illness can occur from the consumption of other types of spoiled fish, such as mahi mahi, bluefish, yellowtail, amberjack, herring, sardines, and anchovies. Most of these fish have a high level of free L-histamine in their tissues. In addition, the breakdown of muscles by autolytic proteases and by bacterial proteases produced during their growth in fish release L-histidine. Histidine decarboxylase produced by bacteria growing in fish catalyzes L-histidine to histamine. The most important bacterial species associated with histamine production in fish is *Morganella morganii*. Several other species are also found, namely *Klebsiella pneumoniae, Proteus* spp. *Enterobacter aerogenes, Hafnia alvei,* and *Vibrio algenolyticus*. In fermented fish products, several *Lactobacillus* spp. have been implicated in histamine production.[8]

The symptoms of histamine poisoning include gastrointestinal (nausea, vomiting, abdominal cramps, and diarrhea), neurological (tingling, flushing, palpitations, headache, burning, and itching), cutaneous (hives, rash), and hypotension. Generally, high levels of histamine in fish are necessary to produce the symptoms. The FDA hazard action level is currently set at 500 ppm in fish. Depending upon the histamine concentrations and susceptibility of an individual, the symptoms can occur very quickly, generally within 1 h. It is not fatal.[8]

Histamine poisoning from fish is a worldwide problem. The frequency of incidence is probably directly related to the popularity of fish and the methods used in preservation from harvest to consumption. In the United States between 1973 and 1987, there were 202 incidents affecting 1216 people. The frequency of incidence seemed to be increasing, as there were 83 incidents between 1983 and 1987. Fin fish were involved in 81 of 83 incidents, and 62.6% occurred with fish served in restaurants while 26.5% occurred at home. The most important factor was improper holding temperature of fish (which caused the fish to decompose). Seasonal variation is not apparent, probably because fish is eaten in the United States throughout the year.[8]

As the consumption of fish increases, the incidence of histamine poisoning is expected to rise. The most important control measure will be to store fish by freezing or by refrigeration at <1°C for a limited time. In addition, proper sanitation to reduce contamination with histamine-producing bacteria will be important. Cleaning of fish soon after catching can also reduce the autolytic effect of muscle tissues. Histamine is heat stable; once formed, cooking is not going to inactivate it. Thus, if there is a doubt that a fish is spoiled, the best way to avoid histamine poisoning is not to eat it.[8]

IV. ALGAL TOXINS

Fin fish and shellfish feeding on several species of toxic algae can accumulate toxins. Consumption of these fish and shellfish can produce disease symptoms.[7-9]

A. Ciguatera Poisoning

Fin fish, especially those feeding on reef and mud, such as sea bass, grouper, and snapper while feeding on toxic algae, *Gambierdiscus toxicus,* and several others, accumulate toxin in the muscle tissues. Liver, intestines, and roe accumulate more toxin than the muscle tissues. Consumption of these fish results in disease with gastrointestinal and neurological symptoms developing within a few hours. The gastrointestinal symptoms are nausea, vomiting, cramps, and diarrhea that last for a short time. Neurological symptoms include tingling and numbness of lips and tongue, dryness of mouth, chills, sweating, blurred vision, and paralysis. The disease can be fatal under severe conditions.

Foodborne disease due to ciguatoxin occurs in many countries, especially where reef fishes are consumed. In the United States between 1973 and 1987, there were 234 confirmed incidences affecting 1052 individuals without any fatalities; 232 were from the consumption of fin fish harvested from unsafe places. Among the 87 incidences between 1983 and 1987, 69 occurred at home and only 4 occurred in food service establishments.

No method exists at the present time for the detection of ciguatoxin, so monitoring of fish for the presence of this toxin is not possible. Incidence of the disease can be reduced by avoiding consumption of large fish from the reef, particularly their inner organs.[7-9]

B. Shellfish Poisoning

1. *Paralytic Shellfish Poisoning*

Scallops, clams, and mussels feeding on *Gonyaulax catenella* and related toxic species of algae accumulate heat-stable toxin in the tissues. Consumption of these shellfish produces paralytic poisoning. The common neurological symptoms include tingling and numbness of lips and fingertips, drowsiness, poor coordination, incoherent speech, and dryness of the throat. In extreme cases, respiratory failure and death can result. The symptoms appear within 1 h and subside within a few days.

Occurrence of paralytic shellfish poisoning is sporadic and depends upon the growth (bloom) of the toxic algae, as influenced by water temperature (about 8°C), salinity, run-off and presence of nutrients, and others. Shellfish growing in shallow water tend to accumulate more toxins. In the United States, the incidence of paralytic shellfish poisoning is quite low. Between 1973 and 1987, there were only 21 incidents affecting 160 people. Between 1983 and 1987, only two incidents were reported. This has been possible due to monitoring systems of shellfish for toxins and closing of contaminated shellfish beds for harvesting.[7-9]

2. *Neurotoxic Shellfish Poisoning*

Growth of toxic algae *Ptychodiscus brevis* produces red tide and feeding of the algae by the shellfish causes them to accumulate toxin in the muscle. Consumption of these shellfish produces neurological symptoms as described for paralytic shellfish poisoning, but less severe. The symptoms appear very quickly and generally subside in a few hours. The incidence is very low in the United States due to effective monitoring systems.[7-9]

3. *Diarrhetic Shellfish Poisoning*

Gastrointestinal disorders from the consumption of scallops, mussels, and clams that feed on toxic algae *Dinophysis fortii* was reported in Japan and the Netherlands. The duration of symptoms is short. The toxin is heat stable; thus cooking does not eliminate it. The incidence of the disease can be reduced by removing the digestive organs in the shellfish since they tend to accumulate the majority of the toxins.[7-9]

V. PARASITES

Included in this group are several roundworms, flatworms, tapeworms, and protozoa that are known to cause human illness and a food association is either confirmed or suspected. Several of these organisms and the disease they produce are briefly described here.

A. Trichinosis by *Trichinella Spiralis*

Trichinella spiralis is a roundworm and can be present in high frequencies in pigs feeding on garbage. It is also found quite frequently among game animals. A person gets infected from the consumption of raw or insufficiently cooked meat of infected animals. The infected meat contains encysted larvae. Once the meat is consumed, the cysts dissolve in the gastrointestinal (GI) tract, releasing the larvae, which then infect the GI tract epithelium. Here, the parasites mate and deposit larvae in the lymphatic system. Through lymphatic circulation, the larvae infect other body tissues. In the United States between 1973 and 1987, a total of 128 incidents of trichinosis infecting 843 people were recorded. Although many types of foods were involved, the highest incidence (55) was from pork. Almost all incidents were due to improper cooking.

The symptoms of trichinosis appear between 2 and 28 d following ingestion. Initial symptoms are nausea, vomiting, and diarrhea, followed by fever, swelling of eyes, muscular pain, and respiratory difficulties. In extreme cases of infection, toxemia and death may result.

The prevention methods consist of not raising pigs on garbage, and cooking meat thoroughly (140°F or 60°C and above). Microwave cooking may not be effective in killing all the larvae. Freezing meat for 20 d at –20°C (home freezer) or in liquid nitrogen for several seconds will also kill the parasite.[8,10]

B. Anisakiasis by *Anisakis Simplex*

The nematode *Anisakis simplex* is a parasite in many fish, particularly of marine origin. Human infection results from the consumption of raw infected fish. Inadequately cooked, brined, or smoked fish have also been implicated in some cases. The symptoms appear after a few days, and generally include irritation of throat and digestive tract. Anisakiasis is more predominant in countries where raw fish is consumed. Proper cooking, salting, or freezing at –20°C for 3 d can be used as a preventative measure.[11]

C. Taeniasis by *Taenia* spp.

Taeniasis is a tapeworm disease caused by *Taenia solium* from pork and by *Taenia saginata* from beef. The illness is associated from the ingestion of raw or improperly cooked meat contaminated with the larvae of the tapeworms. The symptoms appear within a few weeks and consist of digestive disorders. If an organ is infected by the larvae, especially a vital organ, the consequences can be severe.[12]

D. Toxoplasmosis by *Toxoplasma Gondii*

Toxoplasmosis is a protozoan disease transmitted to humans from the consumption of undercooked and raw meat and raw milk. Contamination of food with feces can also be involved. In many people, it does not cause any problems. However, in individuals with low resistance, the symptoms are generally flulike. Preventative measures include cooking meat to a minimal internal temperature of 70°C or freezing at –20°C.[13]

E. Giardiasis by *Giardia Lamblia*

Giardiasis is considered a food- and waterborne disease in many parts of the world. In the United States between 1973 and 1987, five incidents infecting 131 individuals were confirmed. The incidence is much higher in countries with inadequate sanitation facilities and improper water supplies. Contaminated raw vegetables, foods contaminated with water containing the causative agent, and poor personal hygiene are considered the major causes of the disease. The main symptoms are diarrhea and abdominal pain. Improved sanitary conditions and personal hygiene are important for reducing the incidence.[8,14]

REFERENCES

1. Buchanan, R. L. and Palumbo, S. A., *Aeromonas hydrophila* and *Aeromonas sorbia* as potential food poisoning species: a review, *J. Food Safety*, 7, 15, 1985.
2. Abeyta, C., Kaysner, C. A., Wekell, M. M., Sullivan, J. J., and Stelma, G. N., Recovery of *Aeromonas hydrophila* from oysters implicated in an outbreak of foodborne illness, *J. Food Prot.*, 49, 643, 1986.
3. Miller, M. L. and Korburger, J. A., *Plesiomonas shigelloides*: an opportunistic food and waterborne pathogen, *J. Food Prot.*, 48, 449, 1985.
4. Miller, M. L. and Korburger, J. A., Tolerance of *Plesiomonas shigelloides* to pH, NaCl and temperature, *J. Food Prot.*, 49, 877, 1986.
5. Twedt, R. M. and Boutin, B. K., Potential public health significance of non-*Escherichia coli* coliforms in food, *J. Food Prot.*, 42, 161, 1979.
6. Griffiths, M. W., Toxin production by psychrotrophic *Bacillus* spp. present in milk, *J. Food Prot.*, 53, 790, 1970.
7. Taylor, S. L., Marine toxins of microbial origin, *Food Technol.*, 42(3), 94, 1988.
8. Bean, N. H. and Griffin, P. M., Foodborne disease outbreaks in the United States, 1973–1987, *J. Food Prot.*, 53, 804, 1990.
9. Bean, N. H., Griffin, P. M., Goulding, J. S., and Ivey, C. B., Foodborne disease outbreaks, 5 year summary, 1983–1987, *J. Food Prot.*, 53, 711, 1990.
10. Murrell, K. D., Strategies for the control of human trichinosis transmitted by pork, *Food Technol.*, 39(3), 57, 1985.
11. Higashi, G. I., Foodborne parasites transmitted to man from fish and other foods, *Food Technol.*, 39(3), 69, 1985.
12. Hird, D. W. and Pullen, M. M., Tapeworms, meat and man: a brief review and update of cysticercosis caused by *Taenia saginata* and *Taenia solium*, *J. Food Prot.*, 42, 58, 1979.
13. Smith, J. L., Documented outbreaks of toxoplasmosis: transmission of *Toxoplasma gondii* to humans, *J. Food Prot.*, 56, 630, 1993.
14. Smith, J. L., *Cryptosporidium* and *Giardia* as agents of foodborne disease, *J. Food Prot.*, 56, 451, 1993.

CHAPTER 26 QUESTIONS

1. Define an opportunistic pathogen. How can *Aeromonas hydrophila* be considered an opportunistic pathogen?

2. Briefly discuss the possibility of food association in gastroenteritis caused by *Plesiomonas shigelloides*.

3. Name three non-coliforms that may be associated with human foodborne gastroenteritis. Why are they considered to be the causative agents for gastroenteritis?

4. Discuss the importance of psychrotrophic *Bacillus* spp. as a possible causative agent for foodborne gastroenteritis.

5. Discuss the role of microorganisms in histamine poisoning from the consumption of some fish. Name the important microbial species. List the preventative methods.

6. Why is ciguatera poisoning more prevalent with foods served at home? How can one prevent the incidence?

7. List the types of shellfish poisoning. What are their major symptoms?

8. List the major cause(s) of trichinosis in humans. How can the incidence be reduced?

9. Discuss the reasons of foodborne anisakiasis, taeniasis, toxoplasmosis, and giardiasis in humans.

CHAPTER 27

New and Emerging Foodborne Pathogens

CONTENTS

I. INTRODUCTION

A comparison of the lists of bacteria and viruses that were known to be foodborne pathogens in the past and those confirmed as foodborne pathogens today reveals quite an astonishing picture. Many that were not considered or known to be foodborne pathogens in the past have later been implicated in foodborne illnesses. These pathogens, following their recognition, are generally designated as "new or emerging foodborne pathogens." A list of pathogens recognized as new foodborne pathogens during the last 30 years in the United States is presented in Table 27.1. Prior to 1959, four bacterial species were considered as foodborne pathogens: *Staphylococcus aureus, Salmonella* spp. including *typhi* and *paratyphi, Clostridium botulinum* types A and B, and *Shigella* spp. In the 1960s, *Vibrio cholerae* non-01 and *Clostridium botulinum* type E were added to the list of bacteria, along with hepatitis A virus. Between 1971 and 1980, *Vibrio parahaemolyticus*, enteropathogenic *Esc. coli* (0124:B17), *Yersinia enterocolitica, Vibrio cholerae* 01, *Campylobacter jejuni,* and *Vibrio vulnificus* were confirmed either with foodborne disease outbreaks or sporadic cases. The food association of *Clostridium botulinum* with infant botulism was also recognized. In the 1980s, *Listeria monocytogenes*, enterohaemorrhagic *Esc. coli*

Table 27.1 New Pathogenic Bacteria and Viruses Associated with Confirmed Foodborne Diseases in the United States between 1959 and 1990

Prior to 1959	Foodborne pathogens recognized during (year)[a]		
	1959 to 1970	1971 to 1980	1981 to 1990
Sta. aureus	*Clo. perfringens* (1959)	*Vib. parahaemolyticus* (1971)	*Esc. coli* 0157:H7 (1982)
Salmonella spp. (*typhi* and *paratyphi* included)	*Bac. cereus* (1959)	Enteropathogenic *Esc. coli* (0124:B17; 1971)	*Lis. monocytogenes* (1983)
Clo. botulinum (Type A, B)	*Vib. cholerae* (non-01; 1965)	*Clo. botulinum* (Infant botulism; 1976)	*Esc. coli* 0127:H20 (1983)
Shigella spp.	*Clo. botulinum* (Type E; mid 1960s) Hepatitis A (1962)	*Vib. vulnificus* (1976) *Yer. enterocolitica* (1976) *Vib. cholerae* (01; 1978) *Cam. jejuni* (1979)	Norwalk viruses (1982) *Sal. enteritidis*[b]

[a] Year first confirmed in foodborne outbreak.
[b] Recognized as major cause of salmonellosis of egg origin.

0157:H7, enterotoxigenic *Esc. coli* 027:H20, and Norwalk viruses became the new pathogens. *Salmonella enteritidis*, which was involved in foodborne outbreaks before in relatively low frequency, became a major causative agent of foodborne salmonellosis. Similar situations, the sudden emergence of new foodborne pathogens with time, also exist in other countries.[1-4]

Are they really "new" foodborne pathogens? If it means that a pathogen used a food as a vehicle for its transmission to humans for the first time and caused the specific illness, then probably it is not new. Rather, it is appropriate to recognize that it probably has caused foodborne illnesses before but had not been confirmed. Due to changes in several situations, its direct involvement in a foodborne illness has probably been identified for the first time. In this section, the changes in several situations that probably led to the discovery of a new foodborne pathogen are discussed.

A critical analysis of the data in Table 27.1 reveals three important observations: (1) a pathogen that has not been confirmed in the past or at present can emerge as a foodborne pathogen in the future; (2) a pathogen that currently is involved in a few sporadic foodborne illnesses or outbreaks may, in the future, become a major cause of foodborne outbreaks (such as in *Sal. enteritidis*); and (3) conversely, a pathogen considered to be the major cause of foodborne disease in the past or at present can in the future, become associated with fewer incidences (in the United States, there were 26 staphylococcal food intoxication outbreaks in 1976 affecting 930 people; in 1987, there was only one outbreak with 100 cases). The reasons for these changes can be attributed to many factors, some of which are discussed here.

II. BETTER KNOWLEDGE OF THE PATHOGENS

The etiological agents in foodborne disease outbreaks (and in sporadic incidences) are not always confirmed. Epidemiological data of foodborne diseases (bacterial, viral, parasitic, and chemical) from four countries, namely, the United States (between 1960 and 1987), Canada (between 1975 and 1984), the Netherlands (between 1979 and 1982), and Croatia (between 1986 and 1992), are presented in Table 27.2. It is surprising that in the United States, even in the 1980s, the etiological agents for 62% of the outbreaks were not identified. Similar situations exist in the other three countries. In general, among etiological agents, bacterial and viral pathogens are the major causes of foodborne illnesses; many of these with unknown etiology are considered to be of bacterial and viral origin. In a reported outbreak, when the food and environmental samples and samples from the patients are available, they are examined for the presence of the most likely or most common pathogens. Thus, in many incidences, the samples will not be tested for pathogens that are not considered at that time to be a foodborne pathogen. In case of a large outbreak or if a similar type of outbreak occurs frequently, or if the incidence results in death or severe consequences, the samples are generally tested for other suspected pathogens along with the common foodborne pathogens. Most of the new foodborne pathogens were discovered in this way. In the United States, the association of *Yer. enterocolitica, Cam. jejuni, Lis. monocytogenes,* and *Esc. coli* 0157:H7 in foodborne outbreaks has been confirmed this way. In the future, other pathogens will be recognized as "new pathogens" in the same way. This is probably true for several pathogens, such as *Aeromonas hydrophila, Plesiomonas shigelloides,* and other opportunistic pathogens, which have been suspected from the circumstantial evidence to be foodborne pathogens, but have not been directly confirmed as yet.

Table 27.2 Total Foodborne Disease Outbreaks of Unknown Etiology in Several Countries

Period (country)	Outbreaks[a]		Unknown etiology (%)		
	Average/year	Range	Average/year	Range	Ref.
1960 to 1969[b] (U.S.)	240	91–369	44	13–60	1
1970 to 1979 (U.S.)	409	301–497	63	55–71	3
1980 to 1987 (U.S.)	531	387–656	62	54–66	3, 5, 6
1975 to 1984 (Canada)	867	647–1180	78	61–84	8
1979 to 1982 (The Netherlands)	292	163–415	74	71–77	9, 10
1986 to 1992 (Croatia)	49	37–72	87[c]	not available	11

[a] Includes outbreaks of bacterial, viral, parasitic, and chemical origin.
[b] May include a few that were known. In 8 out of 10 years, % unknown ranged between 46 and 60.
[c] For total number of cases.

One of the major reasons for testing available food samples for new pathogens, other than those tested routinely, is because of our current expanded knowledge about the characteristics of many pathogens. Much information about different pathogens with respect to their physiological, biochemical, immunological, and genetic characteristics, and their pathogenicity, habitat, and mode of disease transmission, are available. Recent advances in molecular biology techniques have greatly aided these studies. This information has helped to develop new, effective, and specific methods for the isolation and identification of the foodborne pathogens from the samples. Many of the identification techniques are specific, rapid, and less involved, even in the presence of large numbers of associated bacteria. This aids in testing, economically and effectively, a large number of samples in a relatively short span of time for many pathogens, which includes those routinely tested and those that are probably suspected.

III. IMPROVEMENT IN REGULATORY ACTIONS

In recent years, at least in most developed countries, the local, state, and federal regulatory agencies have been highly active in reporting a foodborne outbreak and taking quick action to identify the etiological agent(s).[4] The number of reported outbreaks has increased in the United States from an average of 240 per year in the 1960s to an average of 530 per year in the 1980s. This increase is due in large part to an increase in reporting. Once a new foodborne pathogen is identified, its frequency of occurrence in foods, mode of transmission in foods, and growth and survival in foods under conditions of processing, storage, and handling at different stages from production to consumption are determined. After this information becomes available, the regulatory agencies develop effective methods to either prevent or reduce its presence in the food ready for consumption. This is accomplished through the development of proper methods of sanitation during processing and handling of foods and preservation. For some pathogens, methods to reduce their presence in raw foods or even in live animals (such as for *Esc. coli* 0157:H7) are being studied. Finally, some improved efforts have been and are being made by the regulatory agencies to educate consumers (especially at home) and food handlers (especially at food service facilities) as to the means by which one can reduce the incidence of contamination (such as for *Salmonella* spp. from uncooked chicken) due to cross-contamination and improper cooking or cooling of foods. This area (educating people) will need increased focus in order to reduce the incidence of foodborne diseases by pathogens, including those that are and will be newly emerging. Since foods served at home and at food service facilities have been implicated in high frequency in foodborne disease outbreaks, including those with unknown etiology, people involved in handling food in these two places need to be aware of the means by which the incidences can be prevented or reduced.

The efforts by the regulatory agencies have probably paid off for several pathogens in the United States. In the 1980s, as compared to the 1970s, foodborne disease outbreaks from *Staphylococcus aureus* and *Clostridium perfringens* were greatly reduced (Table 27.3). In both Croatia and Canada, the incidence of staphylococcal food intoxication is still quite high. The high incidence of outbreaks from *Cam. jejuni* in the United States and the Netherlands (also *Yer. enterocolitica*) could be due to testing foods for the two pathogens on a regular basis in these two countries as compared to in Croatia and probably in Canada. In all four countries, salmonellosis is the most prevalent cause of foodborne disease. This is probably due to the presence of *Salmonella* in high frequency in foods of animal origin and abuse and improper handling of foods, at home, food service establishments, and institutionalized feeding places. The high incidence of shigellosis in the United States is probably due to the same reasons. In the United States, the incidence of botulism is quite high, although the total number of cases is low. Improper home canning of vegetables and foods of animal origin have been the major cause of botulism outbreaks.

Proper education of consumers and food handlers by the regulatory agencies will be an important factor in reducing the incidence of foodborne diseases by currently known pathogens and the new pathogens of the future.

IV. CHANGE IN LIFESTYLE

In this category, several factors can be included that have contributed to the emergence of new foodborne pathogens. One of the most important factors is probably the increase in traveling, especially international traveling. A person arriving at a separate country or returning from a separate country can bring a new foodborne pathogen into a country where it was not recognized previously. *Esc. coli* associated with travelers diarrhea, and *Vib. cholerae* non-01 (new serotype) were probably introduced to the United States in this way.[1-4,15]

Another important factor is the change in food habits. In the United States, increased consumption of seafoods, some of which are eaten raw, have resulted in an increase in foodborne disease outbreaks by *Vib. parahaemolyticus, Vib. vulnificus, Vib. cholerae,* and hepatitis A, all of which were not always recognized as foodborne pathogens. Similarly, the consumer's preference for low heat-processed foods with long shelf-life at refrigerated temperatures has enhanced the chances of psychrotrophic pathogens (e.g., *Lis. monocytogenes* and *Yer. enterocolitica*) to become important foodborne pathogens. Several food preferences, such as consumption of raw milk and undercooked hamburgers, provided the right consequences for *Cam. jejuni* and *Esc. coli* 0157:H7, respectively, to cause foodborne disease outbreaks. Similarly, preference for some imported foods, especially the ready-to-eat type, can introduce a new pathogen. An example is the incidence of enterotoxigenic *Esc. coli* 027:H20 from the consumption of a variety of imported, contaminated cheese in the United States.[1-4,15]

V. NEW FOOD PROCESSING TECHNOLOGY

Due to economic reasons, one of the objectives of food processors is to produce a product in large volume and at a faster rate in a centralized plant. Handling a food in large volumes has its disadvantages; accidental contamination of the product by

Table 27.3 Ranking of Major Foodborne Pathogens in Four Countries

Rank (# outbreaks)	United States (1970s)	United States (1983–1987)	The Netherlands (1979–1982)	Canada (1975–1984)	Croatia (1986 to 1992)
1	*Salmonella* spp.	*Salmonella* spp.	*Salmonella* spp.	*Salmonella* spp.	*Salmonella* spp.
2	*Sta. aureus*	*Clo. botulinum*	*Cam. jejuni*[a]	*Sta. aureus*	*Clo. perfringens*
3	*Clo. botulinum*	*Sta. aureus*	*Yer. enterocolitica*[a]	*Clo. perfringens*	*Sta. aureus*
4	*Clo. perfringens*	*Shigella* spp.	*Clo. perfringens*	*Bac. cereus*	
5	*Shigella* spp.	*Cam. jejuni*			

[a] *Campylobacter jejuni* and *Yersinia enterocolitica* (along with *Vibrio* spp. and *Esc. coli*) were reported for the first time.

a pathogen can cause a foodborne disease outbreak among large numbers of people over a large area. If the outbreak is due to a pathogen previously unrecognized (or recognized from sporadic incidences), the chance of its being isolated and identified is much greater. An outbreak affecting 220 people from the consumption of chocolate milk produced by a processor led to an examination for other possible suspects, along with those that are routinely tested, and resulted in the identification of *Yer. enterocolitica* as a new foodborne pathogen.[3,16]

For faster production, equipment may be designed without much prior consideration of possible microbiological problems, particularly with foodborne pathogens. An incidence of listeriosis from the consumption of *Listeria monocytogenes*-contaminated hot dogs can be used as an example. *Lis. monocytogenes* strains were isolated from many types of foods, including ready-to-eat hot dogs, and epidemiological investigations have suspected that sporadic listeriosis in humans could be caused by the consumption of *Lis. monocytogenes* contaminated hot dogs. In 1989, for the first time in the United States, a case of human listeriosis was linked to eating ready-to-eat hot dogs from the isolation of the same serotype (1/2a) *Lis. monocytogenes* from the patient, the remaining refrigerated hot dogs in the opened package, and hot dogs from unopened retail packages produced in the same processing facility. Subsequently, regulatory agencies investigated products, line samples, and environmental samples from the processing facilities to determine the source(s) of contamination (Table 27.4). Of the seven retail samples from seven lots produced during a 35-d period, six had *Lis. monocytogenes* 1/2a serotype. Among the line samples collected at different phases of production, *Lis. monocytogenes* 1/2a was found in high frequencies in products following peeling and before packaging (following peeling, the products traveled via conveyor belt prior to packaging). Among the environmental samples (swabs) tested, the same serotype was isolated from the cooler floor and from the conveyor belt that the heat-treated products came in contact with following peeling. From the analysis of the data, the conveyor belt, attached to the peeler, was suspected of being the main source of product contamination. The porous conveyor belt in use at this processing plant was difficult to clean. In designing conveyor belts, the possibility of microbiological problems needs to be considered. Use of conveyors that do not have porous belts or large numbers of small segments (such as small links) may help to efficiently sanitize and control microbiological problems.[3,16]

VI. MISCELLANEOUS FACTORS

Several other factors may have contributed to the emergence of new pathogens in foodborne disease outbreaks. Some of these remain speculative, while some others have been confirmed. One of the reasons for the increase in foodborne gastroenteritis by pathogenic *Esc. coli* strains is thought to be the ability of the pathogenic *Enterobacteriaceae* to transfer plasmids among themselves that encode for toxin production and colonization in the digestive tract. Such a plasmid transfer in the environment from a pathogenic strain into a nonpathogenic strain will enable the new variant

Table 27.4 Isolation of *Listeria monocytogenes* Serotype 1/2a from the Product of Environmental Samples in a Processing Plant

Sample	No. positive/No. tested
Packaged products: retail	21/42
Line samples during product manufacture:	
During stuffing	0/8
After cooking	1/12
After cooling	0/13
After peeling and prior to packaging	15/21
Environmental samples (swabs):	
Cooking room	0/3
Cooler floor	1/1
Cooler rack and elements	0/2
Peeler area	0/3
Conveyor belt	1/1
Other equipment	0/6

Note: The products were cooked at 75°C internal temperature over a period of 2 to 3 h. *Lis. monocytogenes* is not expected to survive if the schedule is followed properly.

strain to establish in the intestine and produce illness, and a food can be a vehicle through which the pathogen can be consumed.

There are some concerns that the use of unrestricted antibiotics in animals provides the antibiotic resistance pathogens with a better chance of competing with the sensitive microorganisms and of establishing themselves in the environment. In the absence of competition, these pathogens can become predominant and, through contamination of a food, can emerge as a new foodborne pathogen. Such an assumption was made against some sanitary practices used in the processing and handling of foods. Some of the practices are designed to overcome the problems from the pathogenic and spoilage bacteria traditionally suspected to be present in a food. Elimination of these predominant bacteria may allow minor bacteria to become significant in the absence of competition and to cause problems. However, this is probably more important in the case of food spoilage bacteria and has been discussed in Chapter 20.

There is also some speculation that the present methods of raising food animals and birds may have given some pathogens a better chance to become established in animals and birds. Food from these sources has a greater chance of being contaminated with these pathogens. The occurrence of *Esc. coli* 0157:H7 in hamburger meat prepared from culled dairy cows and the high incidence of *Sal. enteritidis* in eggs and chickens may be due, in part, to the way they are raised now.

Finally, people who are immunocompromised can become easily ill from the consumption of a food contaminated with a pathogen, even at a low dose level. People with normal resistance will have no problem with them. This is particularly true for the opportunistic pathogens. Some, like *Aeromonas hydrophila* and *Plesiomonas shigelloides*, are currently implicated indirectly as possible foodborne pathogens. In the future, people with low immunity may be found to become ill from the consumption of a food contaminated with an opportunistic pathogen, which then will become a "new foodborne pathogen."

REFERENCES

1. Bryan, F. L., Emerging foodborne diseases. I. Their surveillance and epidemiology, *J. Milk Food Technol.,* 35, 618, 1972.
2. Buchanan, R. L., The new pathogens: an update of selected examples, *Assoc. Food Drug Q. Rept.,* 48, 142, 1984.
3. Doyle, M. P., Foodborne pathogens of recent concern, *Annu. Rev. Nutr.,* 5, 25, 1984.
4. Anonymous, New bacteria in the news: a special symposium, *Food Technol.,* 40(8), 16, 1986.
5. Bean, N. H., Griffin, P. M., Goulding, J. S., and Ivey, C. B., Foodborne disease outbreaks, 5-year summary; 1983–1987, *J. Food Prot.,* 53, 766, 1990.
6. Anonymous, Foodborne disease outbreaks annual summary, 1982, *J. Food Prot.,* 49, 933, 1986.
7. Bean, N. H. and Griffin, P. M., Foodborne disease outbreaks in the U.S. 1973–1987: pathogens, vehicles and trends, *J. Food Prot.,* 53, 804, 1990.
8. Todd, E. C. D., Foodborne disease in Canada — a 10 year summary from 1975 to 1984, *J. Food Prot.,* 55, 123, 1992.
9. Beckers, H. J., Incidence of foodborne disease in the Netherlands: annual summary — 1980, *J. Food Prot.,* 48, 181, 1985.
10. Beckers, H. J., Incidence of foodborne disease in the Netherlands: annual summary — 1981, *J. Food Prot.,* 49, 924, 1986.
11. Razem, D. and Katusin-Razem, B., The incidence and costs of foodborne diseases in Croatia, *J. Food Prot.,* 57, 746, 1994.
12. Tauxe, R. V., *Salmonella*: a postmodern pathogen, *J. Food Prot.,* 54, 563, 1984.
13. Palumbo, S. A., Is refrigeration enough to restrain foodborne pathogens?, *J. Food Prot.,* 49, 1003, 1986.
14. Wenger, J. D., Swaminathan, B., Hayes, P. S., Green, S. S., Pratt, M., Pinner, R. W., Schuchat, A., and Broome, C. V., *Listeria monocytogenes* contamination of turkey franks: evaluation of a production facility, *J. Food Prot.*, 53, 1015, 1990.
15. Anonymous, Imported cholera associated with a newly described toxigenic *Vibrio cholerae* 0139 strain in California, *Dairy Food Environ. Sanitation,* 14, 218, 1994.

CHAPTER 27 QUESTIONS

1. Define a "new foodborne pathogen."

2. List the new foodborne pathogens that have emerged in the United States in the last decade.

3. List the possible reasons that lead to the identification of a new foodborne pathogen.

4. Discuss the factors that could be involved in the emergence of a new foodborne pathogen.

5. Discuss how international travel could be a factor in the emergence of a new foodborne pathogen. Give an example.

6. Discuss how new technology can be a factor in increasing the chances of identifying a new foodborne pathogen. Give an example.

7. Why are some pathogens that are recognized as major causes of foodborne illnesses in one country no longer found to be the same in another country? Explain with appropriate examples.

8. "Some opportunistic pathogens of today may be confirmed as new foodborne pathogens of tomorrow." Justify the validity of the statement with an example.

CHAPTER 28

Indicators of Bacterial Pathogens

CONTENTS

I. INTRODUCTION

All pathogenic microorganisms previously implicated in foodborne diseases are, with the exception of *Staphylococcus aureus, Bacillus cereus, Clostridium botulinum* (except in the case of infant botulism), *Clo. perfringens* and toxicogenic molds, considered enteric pathogens. This means they can survive and multiply or establish in the gastrointestinal tract of humans, food animals, and birds. A food contaminated directly or indirectly with fecal materials from these sources may theoretically contain one or more of these pathogens and thus can be potentially hazardous to consumers. To implement regulatory requirements and to ensure the safety of the consumers, it is necessary to know that a food is either free of some enteric pathogens, such as *Salmonella* spp. and *Esc. coli* 0157:H7, or contains low levels of some other enteric pathogens, such as *Yersinia enterocolitica* and *Vibrio parahaemolyticus*. The procedures used in the isolation and confirmation of a pathogen from a food involve several steps, take a relatively long time, and are costly. Some of the new tests involving molecular biology techniques require high initial investment and highly skilled technicians. In a modernized, large commercial operation, involving procurement of ingredients from many different countries, processing of different products, warehousing, distribution over a large area, and retail marketing, it is not practical or economical to test the required number of product samples from each batch for all the pathogens or even those that are suspected of being present in a particular product. Instead, the food samples are examined for the number (or level) of groups or a species of bacteria that are of fecal origin, usually present in higher density than the pathogens, but considered to be nonpathogenic. Their presence is viewed as resulting from direct or indirect contamination of a food with fecal materials and is an indication of the possible presence of enteric pathogen(s) in the food. These bacterial groups or species are termed "indicators of enteric pathogens." Although *Sta. aureus, Clo. botulinum, Clo. perfringens* and *Bac. cereus* can be present in the fecal matters of humans and food animals, they, along with toxigenic molds, are not considered classical enteric pathogens. Their presence in a food is not considered due to fecal contamination, and the indicators of enteric pathogens will not be very effective for the purpose. To determine the presence of these microorganisms and their toxins, specific methods recommended for their detection and identification should be used.

II. CRITERIA FOR THE IDEAL INDICATORS

Several criteria were suggested for selecting a bacterial group or species as indicators of enteric foodborne pathogens.[1,2] Some are listed with brief explanations.

1. The indicator should preferably contain a single species or a few species with some common and identifiable biochemical and other characteristics in order to be able to identify them from the many different types of microorganisms that might be present in a food.
2. The indicator should be of enteric origin, that is, they share the same habitat as the enteric pathogens and will be present when and where the pathogens are likely to be present.
3. The indicator should be nonpathogenic so that their handling in the laboratory does not require safety precautions such as for the pathogens.
4. The indicator should be present in the fecal matter in much higher numbers than the enteric pathogens, so they can be easily detected (enumerated) even when a food is contaminated with small amounts of fecal matter.
5. The indicator should be detected (enumerated) and identified within a short time, easily, and economically so that a product, following processing, can be distributed quickly, and several samples from a batch can be tested.

6. The indicator should be detected using one or more newly developed molecular biology techniques for rapid identification.
7. The indicator should be detected (enumerated) even in the presence of large numbers of associated microorganisms which can be achieved using compounds that will inhibit growth of associated microorganisms but not of indicators.
8. The indicator should have growth and survival rate in a food as the enteric pathogens. They should not grow slower or die off faster than the pathogens in a food. If they die off more rapidly than the pathogen, then theoretically a food can be free of indicators during storage but still can have pathogens.
9. The indicator should not suffer sublethal injury more (in degree) than the pathogens when exposed to physical and chemical stresses. If the indicators are more susceptible to sublethal stresses, they will not be detected by the selective methods used in their enumeration and a food may show no or very low acceptable levels of indicators even when the pathogens are present at higher levels.
10. The indicator should preferably be present when the pathogens are present in a food. Conversely, they should be absent when the enteric pathogens are absent. Unless such correlations exist, the importance of an indicator to indicate the possible presence of a pathogen in a food reduces greatly.
11. The indicator should preferably have a direct relationship between the level of an indicator present and the probability of the presence of an enteric pathogen in a food. This will help set up regulatory standards for an indicator limit for the acceptance or rejection of a food for consumption. For this criterion, it is very important to recognize if the high numbers of an indicator in a food have resulted from a high level of initial contamination (and a greater chance for the presence of a pathogen) or from their growth in the food from a very low initial contamination (in which case a pathogen may not be present even when the indicator is present in high numbers).

It is quite apparent that no single bacterial group or species will be able to meet all the criteria of an ideal indicator. There are several bacterial groups or species that satisfy many of these criteria. The characteristics, advantages, and disadvantages of some of the important and accepted indicator bacterial groups and species (of enteric pathogens) are described here.[1,2]

III. COLIFORM BACTERIA

A. Organisms and Sources

The term *coliform* does not have taxonomic value; rather, it represents a group of species from several genera, namely, *Escherichia, Enterobacter, Klebsiella, Citrobacter*, and probably *Aeromonas* and *Serratia*. The main reason for grouping them together is due to their many common characteristics. They are all Gram-negative, nonsporeforming rods, many are motile, facultative anaerobes, resistant to many surface-active agents, and ferment lactose producing acid and gas within 48 h at 32 or 35°C. Some species can grow at higher temperature (44.5°C), while other species can grow at 4 to 5°C. All are able to grow in foods except in those that are at pH 4.0 or below and A_w 0.92 or below. They are all sensitive to low heat treatments and are killed by pasteurization.

They can be present in the feces of humans and warm-blooded animals and birds. Some can be present in the environment. Thus some *Klebsiella* spp. and *Enterobacter* spp. are found in soil where they can multiply and reach high population levels. Some are found in water and plants.[1,3-5]

B. Occurrence and Significance in Food

Coliforms are expected to be present in many raw foods and food ingredients of animal and plant origin. In some plant foods, they are present in very high numbers due to contamination from soil. Due to their ability to grow in foods, some even at refrigerated temperature, a low initial number can reach a high level during storage. The occurrence of some coliforms of nonfecal origin and their ability to grow in many foods reduce the specificity of coliforms as an indicator of fecal contamination in raw foods. In contrast, in heat-processed (pasteurized) products, their presence is considered postheat treatment contamination due to improper sanitation. In heat-processed food, their presence (even in small numbers) is viewed with caution. Thus, in heat-processed food, their specificity as an indicator is considered favorably (more as an indicator of improper sanitation than fecal contamination). Several selective media have been recommended for the determination of coliform numbers in food samples. These are selective-differential media and differ greatly in their recovery ability of coliforms. The results based on lactose fermentation are available in 1 to 2 d. The presence of sublethally stressed or injured cells can considerably reduce the recovery in the selective media. Several other factors, such as high temperature of melted agar media in pour plating, high acidity of a food, and presence of lysozyme (egg-based products) in a food, can further reduce the enumeration of stressed cells. Even with some disadvantages, they are probably the most useful and most extensively used as indicators.[1,3-5]

IV. FECAL COLIFORM BACTERIA

A. Organisms and Sources

Fecal coliform bacteria also constitute a group of bacteria and include those coliforms whose specificity as fecal contaminants is much higher than coliforms. This group includes mostly *Escherichia coli*, along with some *Klebsiella* and *Enterobacter* spp. The nonfecal coliforms are eliminated due to use of high incubation temperature (44.5 ± 0.2 or 45.0 ± 0.2°C) for 24 h in selective broths containing lactose. Lactose fermentation, with the production of gas, is considered as a presumptive positive test.[1,6]

B. Occurrence and Significance in Food

Some fecal coliforms are present in raw foods of animal origin. They can be present in plant foods from contaminated soil and water. High numbers can be either due to gross contamination or due to growth from a low initial level, probably due to improper storage temperature. Their presence in heat-processed (pasteurized) foods is considered due to improper sanitation after heat treatment.

In raw foods that will be given heat treatment, their presence, even in high numbers (10^3/g or ml) is not viewed gravely; if the numbers go higher, some importance is given to contamination of fecal matter, improper sanitation, and the possible presence of enteric pathogen. A need for corrective measures becomes important. In contrast, in heated products and ready-to-eat products (even raw), their presence, especially above a certain level, is viewed cautiously for possible fecal contamination and presence of enteric pathogens. Acceptance or rejection of a food can be considered on the basis of the numbers present. This group is extensively used as indicators in foods of marine origin (shellfish) and in water and wastewater.[1,6]

V. *ESCHERICHIA COLI*

A. Organisms and Sources

In contrast to either coliform or fecal coliform groups, *Escherichia coli* have taxonomic basis.[1] They include only the *Escherichia* spp. of the coliform and fecal coliform groups. *Esc. coli* strains conform to the general characteristics described for coliform groups. Biochemically, they are differentiated from other coliforms by the IMViC (indole production from tryptone, methyl red reduction due to acid production [red coloration], Voges Proskauer reaction [production of acetylmethyl carbinol from glucose] and citrate utilization as a C-source) reaction patterns. *Esc. coli* type I and type II give the IMViC reaction patterns, respectively, of ++ – – and – +– –. The – +– – reaction pattern of *Esc. coli* type II could also be due to slow or low production of indole from tryptone (or peptone). The IMViC tests are conducted with an isolate obtained following the testing a food sample for coliform group or fecal coliform group. However, there is a concern now about the adequacy of these reaction patterns to identify *Esc. coli* types.

Initially, *Esc. coli* types were used as indicators of fecal contamination and possible presence of enteric pathogens (in food) with the considerations that they are nonpathogenic and occur in the GI tract of humans, animals, and birds in high numbers. However, it is now known that some variants and strains of *Esc. coli* are pathogenic. None of the methods mentioned above are able to differentiate the pathogenic and nonpathogenic *Esc. coli* strains; that can be achieved only through specific tests designed to identify different pathogenic *Esc. coli* strains. In this discussion, emphasis is given to the value of *Esc. coli* as indicators.[1] The significance and importance of pathogenic *Esc. coli* have been discussed previously.

B. Occurrence and Significance in Food

Esc. coli are present in the lower intestinal tract of humans and warm-blooded animals and birds. Their presence in raw foods is considered an indication of direct or indirect fecal contamination. Direct fecal contamination occurs during the processing of raw foods of animal origin and poor personal hygiene of food handlers. Indirect contamination can occur through sewage and polluted water. In heat-processed (pasteurized) foods, their presence is viewed with great concern. Their value as an indicator of fecal contamination and the possible presence of enteric pathogens is much greater than the coliform and fecal coliform groups. However, the time to complete the tests (IMViC) takes about 5 d. Currently, some direct plating methods have been developed that give indication for *Esc. coli* in a shorter time. There are several other inadequacies of *Esc. coli* as an indicator. They may die at a faster rate in dried, frozen, and low pH products than some enteric pathogens, and some enteric pathogens are capable of growing at low temperatures (0 to 2°C) temperature at which *Esc. coli* strains can die. In addition, *Esc. coli* strains can be injured by sublethal stresses in higher degrees than some enteric pathogens, and may not be effectively detected by the recommended selective media unless a prior resuscitation (repair) step is included.

VI. *ENTEROBACTERIACEAE* GROUP

The methods recommended for the detection of coliform and fecal coliform groups and *Esc. coli* are based on the ability of these bacterial species to ferment lactose to produce gas and acid. In contrast, some enteric pathogens do not ferment lactose, such as most *Salmonella* spp. Thus, instead of only enumerating coliforms or fecal coliforms in a food, enumeration of all the genera and species in the *Enterobacteriaceae*[1,1a] family is advocated. As this family includes many genera and species that are enteric pathogens, enumeration of the whole group could be used as a better indicator of the level of sanitation, possible fecal contamination, and possible presence of enteric pathogens (Table 28.1). In European countries, this concept has been used to a certain degree. The method includes the enumeration of the organisms from colony-forming units in a selective agar medium containing glucose instead of lactose.[1]

The criticisms against this idea are that many species in the *Enterobacteriaceae* are not of fecal origin, many are found naturally in the environment, including plants, and those forming typical colonies due to glucose fermentation in the selective medium do not all belong to this family. However, in heat-processed foods (all are sensitive to pasteurization) and in ready-to-eat foods, their presence in high numbers should have public health significance.

Table 28.1 Genera, Habitat, and Association to Foodborne Illnesses of Some *Enterobacteriaceae*

Genera	Main habitat[a]	Associated with foodborne illness[b]
Escherichia	Lower intestine of humans and warm-blooded animals and birds	Only the pathogenic strains
Shigella	Intestine of humans and primates	All strains
Salmonella	Intestine of humans, animals, birds, and insects	considered to be potential
Citrobacter	Intestine of humans, animals, birds; also soil, water, and sewage	Can be opportunistic
Klebsiella	Intestine of humans, animals, birds; also soil, water, and grain	Can be opportunistic
Enterobacter	Intestine of humans, animals, birds; widely distributed in nature, mostly plants	Can be opportunistic
Erwinia	Mostly in plants	No association
Serratia	Soil, water, plants, and rodents	Can be opportunistic
Hafnia	Intestine of humans, animals, birds; also soil, water, and sewage	No association
Edwardsiella	Cold-blooded animals and water	No association
Proteus	Intestine of humans, animals, birds; soil and polluted water	Can be opportunistic
Providencia	Intestine of humans and animals	Can be opportunistic
Morganella	Intestine of humans, animals, and reptiles	Can be opportunistic
Yersinia	Intestine of humans and animals; environment	Some species/strains are pathogenic
Obesumbacterium	Brewery contaminant	No association
Xenorhabdus	Nematodes	No association
Kluyvera	Soil, sewage, and water	Can be opportunistic
Rahnella	Fresh water	No association
Tatumella	Human respiratory tract	No association

[a] Some are of non-fecal origin.
[b] The pathogenic species and strains in the indicated genera are confirmed with foodborne and waterborne illnesses. Some species and strains in some genera, indicated as opportunistic, are suspected of being foodborne pathogens.

VII. *ENTEROCOCCUS* GROUP

A. Characteristics and Habitat

The genus *Enterococcus* is relatively new and includes many species that were previously grouped as fecal streptococci and other streptococci.[7-10] They are Gram-positive, nonsporeforming, nonmotile cocci or coccobacilli, catalase negative, and facultative anaerobic. They can grow between 10 and 45°C, and some species can grow at 50°C. Some require B vitamins and amino acids for growth. Some can survive pasteurization temperature. In general, they are more resistant than most coliforms to refrigeration, freezing, drying, low pH, and NaCl. They are found in the intestinal tracts of humans and warm- and cold-blooded animals, birds, and insects. Some can be species specific while others can be present in humans, warm-blooded animals, and/or birds. Among the currently recognized species, several are found in the intestine of humans and food animals and birds and include *Enterococcus faecalis, Ent. faecium, Ent. durans, Ent. gallinarum, Ent. avium,* and *Ent. hirae*. Many are found in vegetation, processing equipment, and processing environments. Once established, they can continue to multiply in the equipment, and environment and are often difficult to completely remove. They are found in sewage and water, especially polluted water and mud. They probably do not multiply in water, but can survive longer than many coliforms. They can grow in most foods.

B. Occurrence and Significance in Food

Enterococcus can get into different foods, either through fecal contamination or through water, vegetation, or equipment and processing environments, and may not be of fecal origin. In this respect, their value as indicators of fecal contamination and possible presence of enteric pathogens in food is questionable. Also, the ability of some strains to survive pasteurization temperature reduces their value as an indicator. On the other hand, their better survival ability in dried, frozen, refrigerated, and low pH foods can place them in a favorable position as indicators. At present, their presence in high numbers, especially in heat-processed foods, could be used to indicate their possible presence in high numbers in raw materials and improper sanitation of the processing equipment and environment. They are used to determine the sanitary quality of water in shellfish beds and are considered to be better as indicators than coliforms for shellfish. Some strains are also associated with food-borne gastroenteritis, probably as opportunistic pathogens.[7-10]

REFERENCES

1. Hitchins, A. D., Hartman, P. A., and Todd, E. C. D., Coliforms — *Escherichia coli* and its toxins, in *Compendium of Methods for the Microbiological Examination of Foods,* 3rd ed., Vanderzant, C. and Splittstoesser, D. F., Eds., American Public Health Association, Washington, DC, 1992, 325.

1a. Krieg, N. R. and Holt, J. G., Eds., *Bergey's Manual of Systematic Bacteriology*, vol. 1, Williams & Wilkins, Baltimore, 1984, 408.

2. Foegeding, P. M. and Ray, B., Repair and detection of injured microorganisms, *Compendium of Methods for the Microbiological Examination of Food*, 3rd ed., Vanderzant, C. and Splittstoesser, D. F., Eds., American Public Health Association, Washington, DC, 1992, 121.

3. Splittstoesser, D. F., Indicator organisms of frozen blanded vegetables, *Food Technol.*, 37(6), 105, 1983.

4. Tompkin, R. B., Indicator organisms in meat and poultry products, *Food Technol.*, 37(6), 107, 1983.

5. Reinbold, G. W., Indicator organisms in dairy products, *Food Technol.*, 37, 111, 1983.

6. Matches, J. R. and Abeyta, C., Indicator organisms in fish and shellfish, *Food Technol.*, 37(6), 114, 1983.
7. Hartman, P. A., Deibel, R. H., and Sieverding, L. M., *Enterococci*, in *Compendium of Methods for the Microbiological Examination of Foods,* 3rd ed., Vanderzant, C. and Splittstoesser, D. F., Eds., American Public Health Association, Washington, DC, 1992, 523.
8. Schleifer, K. H. and Kilpper-Bälz, R., Molecular and chemotoxonomic approaches to the classification of streptococci, enterococci and lactococci: a review, *Syst. Appl. Microbiol.,* 10, 1, 1987.
9. Hackney, C. R., Ray, B., and Speck, M., Repair detection procedure for enumeration of fecal coliforms and enterococci from seafoods and marine environments, *Appl. Environ. Microbiol.,* 37, 947, 1979.
10. Hartman, P. A., Reinbold, G. W., and Saraswat, D. S., Indicator organisms: a review. II. The role of enterococci in food poisoning, *J. Milk Food Technol.,* 28, 344, 1966.

CHAPTER 28 QUESTIONS

1. Discuss the need for using indicator bacteria for enteric pathogens in food.

2. List six criteria that should be considered in selecting an indicator of enteric pathogens.

3. Which foodborne pathogens are generally not considered enteric? Why? What could be done to determine their possible presence in a food?

4. What are some similarities and differences in characteristics between coliforms and fecal coliforms? What is the significance of their presence in high numbers $\geq 10^4$/g or ml) and low numbers ($\leq 10^1$/g or ml) in a raw food, in a pasteurized food, and in a ready-to-eat food?

5. What are the advantages and disadvantages of using *Esc. coli* as an indicator of enteric pathogens in food? "An indicator should be present when an enteric pathogen is present in a food." Why may this statement not be valid for *Esc. coli* as an indicator?

6. List the conditions for which coliforms, fecal coliforms, and *Esc. coli* have questionable value as indicators of enteric pathogens.

7. What are the justifications of using the *Enterobacteriaceae* family as indicators? What are the disadvantages of using this group as indicators of enteric pathogens?

8. List some characteristics of enterococci that put them in an advantageous position over coliforms as indicators of enteric pathogens. What are the disadvantages of using enterococci as indicators? Why can they be used as a better indicator for some seafoods?

SECTION VI

CONTROL OF MICROORGANISMS IN FOODS

It is apparent from the materials presented in the previous chapters that although some microorganisms are desirable for the production of bioprocessed food, many are undesirable due to their ability to cause food spoilage and foodborne diseases. For efficient production of bioprocessed food, our objectives are to stimulate growth and increase the viability of desirable microorganisms. In contrast, with respect to spoilage and pathogenic microorganisms, our objective is to minimize their numbers in food. Several methods, individually or in combinations, are used to achieve this goal by: (1) controlling access of the microorganisms in the foods, (2) physically removing the microorganisms present in foods, (3) preventing or reducing the growth of microorganisms and germination of spores present in foods, and (4) killing microbial cells and spores present in foods. The influence of intrinsic and extrinsic factors necessary for optimum microbial growth and the range in which each factor will support microbial growth have been discussed before (Chapter 6). To control growth and kill microorganisms in a food, some of these factors are used, but only beyond the range that will support growth and induce germination of spores. Some of these controlling methods are discussed here. Irrespective of the method(s) used, it is important to recognize that a control method is more effective when food has fewer microbial cells and when the cells are at exponential growth phase. Also, spores are more resistant than the vegetative cells and Gram-negative are generally more susceptible than Gram-positive cells to many control methods. Finally, bacteria, yeasts, molds, phages, and viruses differ in resistance to the methods used to control them. The following control or preservation methods are discussed in this section.

Chapter 36: Control by Irradiation
Chapter 37: Control by Antimicrobial Preservatives
Chapter 38: Control by New Nonthermal Methods
Chapter 39: Control by Combination of Methods (Hurdle Concept)

CHAPTER 29

Control of Access (Cleaning and Sanitation)

CONTENTS

I. HISTORY

The internal tissues of plants and animals used as foods are essentially sterile. However, many types of microorganisms capable of causing food spoilage and foodborne diseases enter foods from different sources. This is discussed in Chapter 3. It is impossible to prevent access to food of microorganisms from these sources. However, it is possible to control their access to food so as to reduce initial load and minimize microbial spoilage and health hazard. This is what the regulatory agencies advocate and food processors try to achieve through sanitation.

When and how sanitation was introduced in food handling operations is not clearly known. However, the consequences of changes in food consumption and production patterns during the late 19th and early 20th centuries, and the understanding of the scientific basis of food spoilage and foodborne diseases, may have helped to enhance food sanitation. One of the outcomes of the Industrial Revolution was the increase in population in the cities and urban areas that needed processed foods for convenience and stable supply. The foods produced by the available processing techniques were not safe. With the knowledge of microbial association with food spoilage and foodborne diseases, food processing techniques to reduce spoilage and ensure the safety of foods were studied. It was recognized during this time that microorganisms can get into food from various sources, but that proper cleaning and sanitation during handling of foods can reduce their level; and a food with lower levels of microorganisms could be processed and preserved more effectively to ensure stability and safety than a food with high initial microbial load. Thus sanitation became an integral part of food processing operations.

In recent years, more foods than ever are being processed in both developed and developing countries. In addition, particularly in some countries like the United States, many centralized processing facilities are producing foods in large volume. In these centralized plants, many types of raw materials and finished products are handled at a rapid rate. This has been possible due to the availability of needed processing technologies capable of handling large volumes. Some of the food contact machineries are extremely complex and automated and require special methods for effective sanitation.

Although our understanding of the mechanisms by which microorganisms contaminate foods and the means by which that can be intervened have increased, the volume of foods that are spoiled and the incidence of foodborne diseases remain high. This indicates the need for more effective methods to control microbial contamination of foods through efficient sanitation.

II. OBJECTIVES

The main objective of sanitation is to minimize the access of microorganisms in food from various sources at all stages of handling.[1,2] As the microbial sources and level of handling vary with each food of plant and animal origin and fabricated foods, the methods by which microorganisms contaminate foods differ.

Proper sanitation helps reduce the microbial load to desired levels in further processed food. An example of this is that a low microbial level in raw milk produced through effective sanitation makes it easier to produce pasteurized milk that meets the microbial standard. Also, proper sanitation helps produce food that, when properly handled and stored, will have a long shelf life. Finally, proper sanitation helps to reduce the incidence of foodborne diseases.[1,2]

III. FACTORS TO CONSIDER

In order to minimize the access of microorganisms in foods, the microbiological quality of the environment to which a food is exposed (food contact surfaces) and the ingredients that are added to a food should be of good microbiological quality. To achieve these goals, several factors need to be considered; these are briefly discussed here.[1-4]

A. Plant Design

At the initial design stage of a food processing plant, an efficient sanitary program must be integrated in order to provide maximum protection against microbial contamination of foods. This includes both the outside and the inside of the plant. Some of these are specific floor plan, approved materials used in construction, adequate light, air ventilation, direction of air flow, separation of processing areas for the raw and finished products, sufficient space for operation and movement, approved plumbing, water supply, sewage disposal system, waste treatment facilities, drainage, soil conditions, and surrounding environment. Regulatory agencies have specifications for many of these requirements and can be consulted at the initial stage of planning to avoid costly modifications.[1-4]

B. Quality of Water, Ice, Brine, and Curing Solution

Water is used as an ingredient in many foods and is also used in some products after heat treatment. The microbiological quality of this water, especially if the foods are ready-to-eat types, should not only be free from pathogens (like drinking water), but also be low (if not free) in spoilage bacteria, such as *Pseudomonas* spp. This is particularly important for foods that are kept at low temperature for extended shelf life. The ice used for chilling unpackaged foods should also not contaminate a food with pathogenic and spoilage bacteria. Water used for chilling products, such as chicken at the final stage of processing, can be a source of cross-contamination of a large number of birds from a single bird contaminated with an enteric pathogen. Similarly, the warm water used in defeathering chickens can be a source of thermoduric bacteria.

Brine and curing solutions used in products such as ham, bacon, turkey-ham, and cured beef brisket can be the source of microbial contamination. To reduce this, brine and curing solutions should be made fresh and used daily. Storing brine for extended periods before use may reduce the concentration of nitrite through formation and dissipation of nitrous oxide and may reduce the shelf life of the products.[1-4]

C. Quality of Air

Some food processing operations, such as spray drying of nonfat dry milk, require large volumes of air that come into direct contact with the food. Although the air is heated, it does not kill all the microorganisms present in the dust of the air and thus can be a source of microbial contamination of foods. The installation of air inlets to obtain dry air with the least amount of dust and filtration of the air are important to reduce microbial contamination from this source.[1-4]

D. Training of Personnel

A processing plant should have an active program to teach the plant personnel the importance of sanitation and personal hygiene in order to ensure product safety and stability. The program will not only teach how to achieve good sanitation and personal hygiene, but also monitor the implementation of the program. People with an illness or infection should be kept away from handling the products. Some kind of incentive might help make the program efficient.[1-4]

E. Equipment

The most important microbiological criterion that should be considered during the design of food processing equipment is that it should protect a food from microbial contamination. This can be achieved if the equipment does not contain dead spots where microorganisms harbor and grow, and cannot be easily and readily cleaned in place or by disassembling. Some of the equipment, such as meat grinders, choppers, or slicers and several types of conveyor systems, may not be cleaned and sanitized very effectively and therefore serve as a source of contamination to a large volume of product. This is particularly important for products that come into contact with equipment surfaces after heat treatment and prior to packaging.[1-4]

F. Cleaning of Processing Facilities

Cleaning is used to remove visible and invisible soil and dirt from the food processing surroundings and equipment. The nature of soil varies greatly with the type of food being processed, but chemically it consists of lipids, proteins, carbohydrates, and some minerals. Although water is used for some cleaning, in order to increase the efficiency of cleaning, chemical agents or detergents are used with water. In addition, some form of energy, such as spraying, scrubbing, or turbulent flow, with the liquid is used for better cleaning.

Many types of detergents are available and their selection is made on the basis of a specific need. The effectiveness of a cleaning agent to remove soil from surfaces is dependent upon several characteristics. Some of these are efficiency of emulsifying lipids, dissolving proteins, and solubilizing or suspending carbohydrates and minerals. In addition, a detergent should be noncorrosive, safe, rinsed easily, and compatible, when required, with other chemical agents. The detergents frequently used in food processing facilities are synthetic, which could be anionic, cationic, or nonionic. Among these, anionic detergents are used with higher frequency. Examples of anionic detergents include sodium lauryl sulfate and different alkyl benzene sulfonates and alkyl sulfonates. Each molecule has a hydrophobic or lipophilic (nonpolar) segment and a hydrophilic or lipophobic (polar) segment. The ability of a detergent to remove dirt from a surface is attributed to the hydrophobic segment of a molecule. They dissolve the lipid materials of the soil on the surface by forming micelles with the polar segments protruding outside in the water. The concentration of a detergent at which micelle formation starts is called "critical micelle concentration" (CMC), which varies with the detergent. The concentration of a detergent is used above its CMC level. Generally, this is about 800 to 900 ppm, but could be 1000 to 3000 ppm if skin contact does not occur (as in the clean-in-place, CIP, method) or where a heavy-duty cleaning is required.

The frequency of cleaning depends upon the products being processed and the commitment of the management to good sanitation. From a microbiological standpoint, prior microbiological evaluation of a product can give an indication as to the frequency of cleaning necessary in a particular facility. Cleaning of the equipment is done either after disassembling the equipment or by the CIP system. Due to efficiency and lower cost, CIP cleaning is becoming popular. The system uses detergent solutions at high pressure. Since microorganisms can grow in some detergent solutions, they preferably should be prepared fresh (not exceeding 48 h).[1-4]

G. Sanitation of Food Processing Equipment

Efficient cleaning can remove some microorganisms along with the soil from the food contact surfaces, but cannot ensure complete removal of pathogens. To achieve this goal, food contact surfaces are subjected to sanitation after cleaning. The methods should effectively destroy pathogenic microorganisms as well as reduce total microbial load. Several physical and chemical methods are used for the sanitation of food processing equipment.

Physical agents used for sanitation of food processing equipment include: hot water, steam, hot air, and UV irradiation. UV irradiation is used to disinfect surfaces and is discussed in Chapter 36. Hot water and steam, although less costly and more efficient for the destruction of vegetative cells, viruses, and spores (especially steam), can be used only in a limited way.

Chemical sanitizers are used more frequently than physical sanitizers. There are several groups of sanitizers approved for use in food processing plants. They vary greatly in their antimicrobial efficiency. Some of the desirable characteristics used in selecting a chemical sanitizer are: effectiveness for a specific need, nontoxicity, noncorrosiveness, no effect on food quality, easy to use and rinse, stability, and cost effectiveness. Important factors for the antimicrobial efficiency are exposure time, temperature, concentrations used, pH, microbial load and type, microbial attachment to surface, and water hardness. Microbial attachment has been discussed separately (see Appendix A). The mechanism of antimicrobial action, and the advantages and disadvantages of some of the sanitizers used in food processing plants, are briefly discussed here. Some sanitizers, designated as detergent-sanitizers, have both cleaning and sanitation capabilities. They can be used in a single operation instead of first using detergent to remove the soil and then using sanitizers to control microorganisms. They are also discussed here.[1-4]

1. Chlorine-Based Sanitizers

Some of the chlorine compounds used as sanitizers are liquid chlorine, hypochlorites, inorganic or organic chloramines, and chlorine dioxide. Chlorine compounds are effective against vegetative cells of bacteria, yeasts, and molds, spores, and viruses. Clostridial spores are more sensitive than bacilli spores. The antimicrobial (germicidal) action of the chlorine compounds is due to the oxidizing effect of chlorine to the –SH group in many enzymes and structural proteins. In addition, damage to membrane, disruption of protein synthesis, reactions with nucleic acids, and interference with metabolisms have been suggested.

The germicidal action of liquid chlorine and hypochlorites is produced by hypochlorous acid (HOCl). It probably enters the cell and reacts with the –SH group of proteins. HOCl is stable at acid pH and thus is more effective; at alkaline pH, it dissociates to H^+ and OCl^- (hypochlorite ions), which reduces its germicidal effectiveness. They are also less effective in the presence of organic matter. Chloramines (inorganic or organic), such as chloramine T, release chlorine slowly but they are less active against bacterial spores and viruses. They are effective to some extent against vegetative cells at alkaline pH. Chlorine dioxide is more effective at alkaline pH and in the presence of organic matter.

Chlorine compounds are fast acting against all types of microorganisms, less costly, and easy to use. However, they are unstable, especially at higher temperatures and with organic matter, corrosive to metals and can oxidize food (color, lipid), and are less active in hard water.[1-4]

2. Iodophores

Iodophores are prepared by combining iodine with surface-active compounds, such as alkylphenoxypolyglycol. Because of the surface-active compounds, they are relatively soluble in water. Iodophores are effective against Gram-positive and Gram-negative bacteria, bacterial spores, viruses, and fungi. Their germicidal property is attributed to elemental iodine (I_2) and hypoiodous acid, which oxidize the –SH group of proteins, including key enzymes. They are more effective at acidic pH and higher temperatures. In the presence of organic matter, they do not lose their germicidal property as rapidly as chlorine. However, their effectiveness is reduced in hard water.

They are fast acting, noncorrosive, easy to use, nonirritating, and stable. However, they are expensive, less effective than hypochlorites against spores and viruses, can cause flavor problems in products, and react with starch.[1-4]

3. Quaternary Ammonium Compounds (QAC)

QACs can be used as detergent-sanitizers as they have cleaning properties along with germicidal abilities. However, they are principally used as sanitizers. They are synthesized by reacting tertiary amines with alkyl halides or benzyl chloride. The general structure is:

$$\left[\begin{array}{c} R_2 \\ \ddot{} \\ R_1 : \underset{\cdot\cdot}{\overset{\cdot\cdot}{N}} : R_3 \\ \\ R_4 \end{array}\right] + Cl^- \text{ or } Br^-$$

where R_1, R_2, R_3, and R_4 represent alkyl and other groups. The cationic group is hydrophobic, while the anionic group is hydrophilic. They can act as bactericides in high concentrations and when used in solution. However, they form a film on the equipment surface, in which state (low concentrations) they are bacteriostatic. They are more effective against Gram-positive bacteria but less effective against many Gram-negative bacteria, bacterial spores, fungi, and viruses. The antimicrobial action is produced by the denaturation of microbial proteins and destabilization of membrane functions. They are more effective against microorganisms at acidic pH and higher temperature. Their effectiveness is not greatly reduced in the presence of organic matter. However, in hard water, they are less effective.

Some of the advantages of QACs as sanitizers are high stability, noncorrosive, nonirritating, residual bacteriostatic effect (however, needs to be rinsed before using the equipment), nontoxic, and detergent effect. The disadvantages are high cost, low activity against many Gram negatives, spores, and viruses, incompatibility with anionic synthetic detergents, and rinsing required due to film formation on equipment surfaces. Some Gram-negative bacteria, such as *Pseudomonas* spp., can grow in diluted QAC solutions.[1-4]

4. H_2O_2

H_2O_2 is a very effective germicide and kills vegetative cells, spores, and viruses. The use of H_2O_2 solutions in food (milk and liquid egg) is discussed in Chapters 15 and 37. Its use as a sanitizer is briefly described here. The U.S. Food and Drug Administration has approved its use for the sanitation of equipment and containers used in aseptic packaging of foods and beverages. Equipment and container surfaces can be sterilized in 15 min with a 30 to 50% solution; the treatment time can be reduced if the temperature of the solution is raised to 150 to 160°F. Use of H_2O_2 in vapor phase can also be effective in killing microorganisms on food contact surfaces. Organic materials greatly reduce the germicidal effect of H_2O_2.[1-4]

H. Microbiological Standards, Specifications, and Guidelines

Microbiological standards, specifications, and guidelines are useful in keeping the microbial load of foods at acceptable levels by various methods, one of which is by controlling their access to foods. Microbiological standards of food are set and enforced by the regulatory agencies to increase consumer safety and product stability. A standard dictates the maximum microbial level that can be accepted in a food.

With proper sanitation and quality control, this level is generally attainable. Some examples are maximum acceptable levels of standard plate counts (SPC) of Grade A raw milk, 100,000/ml; pasteurized Grade A milk, SPC 20,000/ml; and coliforms, 10/ml. However, very few foods have microbiological standards. Instead, many foods and food ingredients have microbiological specifications. A specification indicates maximum permissible microbial load for the acceptance of a food or food ingredient. It should be attainable and agreed upon by the buyers and sellers of the products. It is not set up or enforced by the regulatory agencies. In the United States, the military has microbiological specifications for foods purchased outside for army rations. An example: for dried whole egg, aerobic plate count (APC), 25,000/g; coliforms, 10/g; and *Salmonella*, negative in 25 g. The specifications discourage mixing of a microbiologically poor quality product with a good quality product. Microbiological guidelines are generally set either by the regulatory agencies or by the food processors to help produce products with acceptable microbiological qualities. A guideline is set at a level that can be achieved if a food processing facility is using good cleaning, sanitation, and handling procedures. It also helps detect if a failure has occurred during processing and handling, and thus alerts the processor to take corrective measures.[1-4]

REFERENCES

1. Cords, B. R. and Dychdala, G. R., Sanitizers: halogens, surface active agents and peroxides, *Antimicrobials in Foods,* Davidson, P. M. and Branen, A. L., Eds., Marcel Dekker, New York, 1993, 469.
2. Marriot, N. G., *Principles of Food Sanitation,* Van Nostrand Reinhold, New York, 1989, 71, 101.
3. Troller, J. A., *Sanitation in Food Processing,* Academic Press, New York, 1982, 21.
4. Lewis, K. H., Cleaning, disinfection and hygiene, *Microbial Ecology of Foods,* Vol. I., Silliker, J. H., Ed., Academic Press, New York, 1980, 232.

CHAPTER 29 QUESTIONS

1. Discuss the objectives of food sanitation.

2. List five factors that are important in reducing microbial access to foods.

3. Describe how water, ice, brine, and curing solution can contribute to the microbial load in a food.

4. List the objectives of using cleaning agents in food processing facilities. Describe the properties of a desirable detergent for use in a food processing plant.

5. Discuss the functions of sanitizers for use in food. Describe the advantages and disadvantages of the following agents as sanitizers: hot water, steam, UV light, hypochlorite, iodophores, QAC, and H_2O_2.

CHAPTER 30

Control by Physical Removal

CONTENTS

I. INTRODUCTION

Microorganisms can be physically removed from solid and liquid foods by several methods. In general, these methods can partially remove microorganisms from food, and by doing so they reduce the microbial level and help other antimicrobial step(s) that follow them become more effective. They are generally used with raw foods before further processing.

II. CENTRIFUGATION

Centrifugation[1] is used in some liquid foods, such as milk, fruit juices, and syrups, to remove suspended undesirable particles (dust, leukocytes, and food particles). The process consists of exposing the food in thin layers to a high centrifugal force. The heavier particles move outward and are separated from the lighter liquid mass. Although this is not intended to remove microorganisms, spores, large bacterial rods, yeasts, and molds can be removed, due to their heavier mass. Under high force, as much as 90% of the microbial population can be removed. Following centrifugation, a food will have fewer thermoduric microorganisms (bacterial spores) that otherwise would have survived mild heat treatment, (e.g., milk pasteurization) and increased the microbial load of the pasteurized product.

III. FILTRATION

Filtration[2] is used in some liquid foods, such as soft drinks, fruit juices, beer, wine, and water, to remove undesirable solids and microorganisms and to give a sparkling clear appearance. As heating is avoided or given only at minimum levels, the natural flavor of the products and heat-sensitive nutrients (e.g., vitamin C in citrus juices) are retained to give the products natural characteristics. The filtration process can also be used as a step in the production of concentrated juice with better flavor and higher vitamins. Many types of filtration systems are available. In many filtration processes, coarse filters are initially used to remove the large components; this is followed by ultrafiltration. Ultrafiltration methods, depending upon pore size of the filter materials (0.45 to 0.7 μm), are effective in removing yeasts, molds, and most bacterial cells and spores from the liquid products.

Filtration of air is also used in some food processing operations, such as spray drying of milk, to remove dust from air used for drying. The process also removes some microorganisms with dust, and they reduce the microbial level in food from this source (air).[2]

IV. TRIMMING

Fruits and vegetables showing damage (greater chance of microbial contamination) and spoilage are generally trimmed.[2] In this manner, areas heavily contaminated with microorganisms are removed. Trimming of outside leaves in cabbage used for sauerkraut production also helps reduce microorganisms coming from soil. Trimming is also practiced for the same reason to remove visible mold growth from hard cheeses, fermented sausages, bread, and some low pH products. However, if a mold strain is a mycotoxin producer, trimming will not ensure removal of toxins from the remaining food. Trimming is also used regularly to remove fecal stain marks, unusual growths, and abscesses or small infected areas from carcasses of food animals and birds. Although this method helps remove highly contaminated areas, it does not ensure complete removal of the causative microorganisms. Thus a beef carcass can have an area contaminated with feces along with enteric pathogens. Just removing the visibly tainted area by trimming does not help the removal of the pathogens from the surrounding areas that do not show the taint. This is an important concern in the production of safer foods.

V. WASHING

Washing equipment and work areas is discussed under cleaning and sanitation (Chapter 29). Here only washing of foods will be discussed.[3-6] Fruits and vegetables are washed regularly to reduce temperature (that helps to reduce metabolic rate of a produce and microbial growth) and remove soil. Washing also helps remove the microorganisms present, especially from the soil. It is also used for shell eggs to remove fecal materials and dirt. During the processing of chicken and turkey, the carcasses are exposed to water several times. During defeathering, they are exposed to hot water; and then, following removal of the gut materials, they are given spray washings and finally exposed to cold water in a chilling tank. Although these treatments are expected to reduce microbial load, they can spread contamination of undesirable microorganisms, particularly the enteric pathogens. Thus higher percentages of chicken were demonstrated to be contaminated with salmonellae when coming out of the chill tank than before entering the tank. This aspect is discussed in Chapter 34. Carcasses of food animals, such as beef, pork, and lamb, are washed to remove hair, soil particles, and microorganisms. Instead of hand washing, automated machine washing at a high pressure is currently used in an effort to effectively remove the undesirable materials and microorganisms from the carcass. In addition to high pressure, the effectiveness of hot water, steam, ozonated water and water containing chlorine, acetic and propionic acids, lactic acid, tripolyphosphates, or bacteriocins (nisin and pediocin) of lactic acid bacteria have been studied to determine their effectiveness in removing microorganisms, particularly enteric pathogens such as *Salmonella* spp., *Campylobacter jejuni, Esc. coli* 0157:H7, and *Listeria monocytogenes*, separately or in combination. The results were not consistent. Some of these agents also have bactericidal properties. However, the studies showed that all of these agents are capable of reducing bacterial contamination to a certain level from the carcass surfaces, and that a combination of two or more components may be better. However, the suitability of the combinations as well as their concentrations and duration of application must to be determined. One has to recognize that, with time, microorganisms can form biofilm on the carcass surface (Appendix A). The nature of the biofilm varies with microbial species and strains. Also, after some time, the biofilm becomes more stable and removing microorganisms by washing after the formation of stable biofilm is relatively difficult. This aspect needs to be considered in developing effective methods of carcass washing. Removal of pathogens from carcass surfaces will be an important area to develop a suitable intervention strategy for ensuring the safety of meat and meat products.

REFERENCES

1. Porter, N. N., *Food Science,* 2nd ed., AVI Publishing, Westport, CT, 1973, 352.
2. Koseoglu, S. S., Lawhon, J. T., and Lusas, E. W., Use of membranes in citrus juice processing, *Food Technol.,* 44(12), 90, 1990.
3. Dickson, J. S. and Anderson, M. E., Microbiological decontamination of food animal carcasses by washing and sanitizing systems: a review, *J. Food Prot.,* 55, 133, 1992.
4. El-Khateib, T., Yousef, A. E. and Ockerman, H. W., Inactivation and attachment of *Listeria monocytogenes* on beef muscle treated with lactic acid and selected bacteriocins, *J. Food Prot.,* 56, 29, 1993.
5. Lillard, H. S., Effect of trisodium phosphate on salmonellae attached to chicken skin, *J. Food Prot.*, 57, 465, 1994.

CHAPTER 30 QUESTIONS

1. List the methods used in the removal of microorganisms from foods, and discuss one advantage and one disadvantage of each method.

2. Discuss two specific microbiological disadvantages, with examples, of trimming food for visible microbial growth or microbial contamination.

3. Discuss two specific microbiological disadvantages, with examples, of washing carcasses of animals and birds to reduce microorganisms.

4. List six different agents that were tested for their relative efficiency in removing microorganisms from beef carcasses. From the references, read one article and discuss the efficiency of a particular agent.

CHAPTER 31

Control by Heat

CONTENTS

I. HISTORY

The desirable effect of heat (fire) on the taste of foods of animal and plant origin, especially seeds, tubers, and roots, was probably accidentally discovered, following a natural forest fire, by our ancestors long before civilization. They also possibly recognized that heated foods did not spoil as fast as raw foods. Since then, particularly following the invention of pottery and ovens, heat has been used to roast, boil, bake, and concentrate foods to improve taste and to enhance shelf life (and probably safety). However, it was not until around 1810 that extended shelf life of perishable foods was achieved by a different and specific method, appertization.

Nicolas Appert, for whom the method was named in France, reported that by filling a clean glass jar with a desirable food, heating the contents in boiling water for 6 h or more, and hermetically sealing the container with a cork kept the food unspoiled and safe for a long time. Some of his products (meat stew) prepared in 1824 were found nontoxic when opened in 1938, but were found to contain dormant spores. Heating of foods in cans was developed by Durand in England in 1810, and by 1820 the method was used in the United States. Methods to reduce the heating time to about 30 min were also developed by adding different types of salt in water to raise the boiling point above 212°F (100°C). In 1870, the autoclave was invented, providing the possibility of developing pressure vessels that could be used to heat canned food at a much higher temperature for a relatively shorter period of time in order to retain the quality of the foods as well as to enhance shelf life.

All these developments occurred without clearly knowing why a perishable food, following appertization or canning, did not spoil (or cause foodborne diseases). The role of microorganisms in the spoilage of wine and milk was discovered by Louis Pasteur around 1870. He also showed that a mild heat treatment (pasteurization) killed these microorganisms, and in the absence of recontamination, the products stayed unspoiled. Subsequently, the microbial role in foodborne diseases was also recognized. Following these, studies were conducted to isolate the most heat-resistant microorganisms (spores) that could survive heating in canned foods and cause spoilage and foodborne diseases, and to determine the temperature and time requirements for their destruction. Studies were also conducted to identify the time-temperature relationships for the destruction of less heat-resistant microorganisms (vegetative cells). From these results, mathematical expressions were developed to accurately predict time-temperature relationships to destroy microorganisms (also some enzymes and toxins) by heating foods at different temperatures.

II. OBJECTIVES OF HEAT PRESERVATION

The main objective (microbiological) of heating food is to destroy vegetative cells and spores of microorganisms that include molds, yeasts, bacteria, and viruses (including bacteriophages).[1] Although very drastic heat treatment (sterilization) can be used to kill all the microorganisms present in a food, most foods are heated to destroy specific pathogens and some spoilage microorganisms, which is important in a food. This is necessary in order to retain the acceptance and nutritional qualities of a food. To control the growth of surviving microorganisms in the food, other control methods are used following heat treatment.

Heating of foods also helps to destroy undesirable enzymes (microbial and food) that would otherwise adversely affect the acceptance quality of food. Some microorganisms also produce heat-stable proteinases and lipases in food. Heating a food to a desired temperature for a period of time can help to destroy or reduce the activity of these enzymes. This is especially important in foods that are stored for a long time at room temperature.

Some microorganisms can release toxins in food; also, some foods can have natural toxins. If a toxin is heat sensitive, sufficient heating will destroy it and consumption of such a food will not cause health hazards. It is also important to recognize that microbial heat-stable toxins are not completely destroyed even after a high heat treatment.

Heating (warming) of ready-to-eat foods prior to serving is also usually used to prevent the growth of pathogenic and spoilage microorganisms. A temperature above 50°C, preferably 60°C, is important in controlling microbial growth in such foods.

Finally, heating of raw materials, such as milk, is used prior to adding starter culture bacteria for fermentation to kill undesired microorganisms (including bacteriophages) and to allow for growth of starter cultures without competition.

III. MECHANISM OF ANTIMICROBIAL ACTION OF HEAT

Depending upon the temperature and time of heating, microbial cells and spores can be sublethally injured or dead. The sublethally injured cells and spores are capable of repair and multiplication; this is discussed in Chapter 7.

Results of different studies have shown that following heat injury, bacterial cells show loss of permeability and increased sensitivity to some compounds to which they are normally resistant. Sublethally injured cells seem to suffer injury in the cell membrane, cell wall, DNA (strand break), ribosomal RNA (degradation), and enzymes (denaturation). Death occurs from damages in some vital functional and structural components. Bacterial spores, following heating, were found to lose structural components from spore coat, suffer damage to the structures that are designed to become membrane and wall, and develop an inability to use water for hydration during germination. Death results from the inability of a spore either to germinate or to outgrow (see Chapter 7).

Exposure of microbial cells to about 45 to 50°C for a short time, which can occur while heating a large volume of a food such as a large rare roast, may induce production of heat shock proteins (stress proteins) by the cells. In the presence of these proteins, the microbial cells can develop greater resistance to subsequent heating at higher temperatures. The implication of this phenomenon in the thermal destruction of microbial cells in low heat processed foods will be important to study.[2]

IV. FACTORS TO CONSIDER IN HEAT TREATMENT

The effectiveness of heat in killing microbial cells and spores is dependent on many factors, some of which are related to the inherent nature of the foods, while others are dependent on both the nature of microorganisms and the nature of processing. An understanding of these factors is important in developing and adopting an effective heat processing procedure for a food.[1]

A. Nature of Food

Composition (amount of carbohydrates, proteins, lipids, and solutes), A_w (moisture), pH, and antimicrobial content (natural or added) greatly influence microbial destruction by heat in a food. In general, carbohydrates, proteins, lipids, and solutes provide protection to microorganisms against heat. Greater microbial resistance results with higher concentrations of these components. Microorganisms in liquid food and food containing small-sized particles suspended in a liquid are more susceptible to heat destruction than in a solid food or in a food with large chunks in liquid. Microorganisms are more susceptible to heat damage in foods that have higher A_w and near-neutral pH range. In low pH foods, heating is more lethal to microorganisms in the presence of acetic, propionic, and lactic acids than phosphoric or citric acids at the same pH. In the presence of antimicrobials, not inactivated by heat, microorganisms are destroyed more rapidly; the rate differs with the nature of the antimicrobials.

B. Nature of Microorganisms

Some of the factors that influence microbial sensitivity to heat are inherent resistance, stage of growth, previous exposure to heat, and initial load. In general, vegetative cells of molds, yeasts, and bacteria are more sensitive than spores. Cells of molds, yeasts, and many bacteria (except thermoduric and thermophilic), as well as viruses, are destroyed within 10 min at 65°C. Most thermoduric and thermophilic bacterial cells important in foods are destroyed in 5 to 10 min at 75 to 80°C. Yeast and most mold spores are destroyed at 65 to 70°C in a few minutes, but spores of some molds can survive as high as 90°C for 4 to 5 h. Bacterial spores vary greatly in their sensitivity to heat. Generally, heating at 80 to 85°C for a few minutes does not kill them. Many are destroyed at 100°C in 30 min, but there are bacterial species whose spores are not destroyed even after boiling (100°C) for 24 h. All spores are destroyed at 121°C in 15 min (sterilization temperature and time). Below this temperature (and time), spores of some bacterial species can survive; however, it depends upon the initial number of spores and the suspending medium. Species and strains of bacterial cells and spores also differ in heat sensitivity. This is especially important if a food is heat treated on the basis of results obtained using a heat-sensitive species or strain but contains heat-resistant variants.

Cells at exponential stage of growth are more susceptible to heat than the resting cells (stationary phase). Also, cells previously exposed to low heat become relatively resistant to subsequent heat treatment (due to stress protein synthesis). Finally, the higher the initial microbial load in a food, the longer time at a given temperature it takes to reduce the population to a predetermined level. This is because the rate of heat destruction of microorganisms follows first-order kinetics, and is discussed later. This suggests the importance of lower initial microbial loads (through sanitation and controlling growth) in a food prior to heat treatment.

C. Nature of Process

Microbial destruction in food by heat is expressed in terms of its exposure to a specific temperature for a period of time, and these are inversely related; the higher the temperature, the shorther the period of time required to get the same amount of destruction when other factors are kept constant. As a food is heated by conduction (molecule-to-molecule energy transfer) and convection (movement of heated molecules), a liquid food is heated more rapidly than a solid food, and a container with high conduction (metal) is better. Also, food in a small container is heated more rapidly than in a large container. A product can have a cold point at the center (in a solid food in a can) or near the end (in a liquid food in a can), which may not attain the desired temperature within the given time. Finally, it needs to be emphasized that heating a food at a given temperature for a specific time means that every particle of that food should be heated to the specified temperature (say 71.6°C or 161°F) and stay at that temperature for the specified time (15 s; used in milk pasteurization). This time is also called the "holding time." The time during which milk is heated before the temperature reaches 161°F in this case is not considered.

V. MATHEMATICAL EXPRESSIONS

When a microbial population is heated at a specific temperature, the cells die at a constant rate. This observation helps in expressing the microbial death rate due to heat as a function of time and temperature under a given condition. These expressions are helpful in designing a heat treatment method for a food.

A. Decimal Reduction Time (D Value)

The D value is the time in minutes during which the number of a specific microbial (cells or spores) population exposed to a specific temperature is reduced by 90% or 1 log. It is expressed as D_T = t min, where T is the temperature and t is the time in min for 1 log reduction of the microbial strain used. Thus it is a measure of the heat sensitivity of microorganisms and varies with microbial species and strains, temperature used, and other variables, such as suspending media and age of the culture.

It can be determined using the expression:

$$D_T = \frac{t}{\log_{10} x - \log_{10} y}$$

where x and y represent microbial numbers before and after exposing at temperature T for t min. It also can be determined by plotting $\log_{10}$ survivors against time of exposure in minutes for a specific temperature (Figure 31.1). It is a straight-line graph and is independent of the initial number of a microbial population. It can be extrapolated to $-\log_{10}$ values to obtain very low levels of microbial survivors, such as 1 cell or spore in 10 g, 100 g, or 1000 g of product, and thus can be used to design heat treatment parameters to obtain a desirable low level of a microbial population in a food. It is evident from the plot that, to reach a desirable microbial level, a food with lower initial numbers will require less time (fewer D) than a food with higher initial numbers at a specific temperature. It can also be used to determine

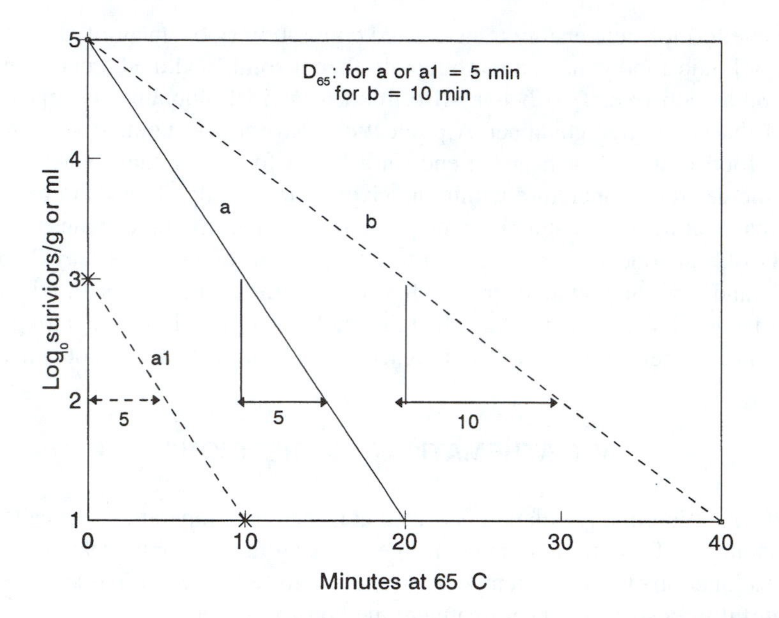

Figure 31.1 Graphical representation of decimal reduction time (D). The graph also illustrates the number of D required with high and low populations of bacteria with the same heat sensitivity (a and a1) to obtain a desired survivor level (say 10^1/g or ml) and different D values for two bacterial species with different heat sensitivity at 65°C (a and b).

the relative sensitivity of two or more microbial species-strains with respect to a specific temperature (Figure 31.1).

The 12D concept is used in heat processing of high pH foods (pH > 4.6, low-acid foods such as corn, beans, and meat) to destroy the most heat-resistant spores of pathogenic bacteria, *Clostridium botulinum*. It means these products are given heat treatment to reduce the population of *Clo. botulinum* spores by 12 log cycles. Hypothetically, if 1 billion (10^9) cans, each containing 10^3 *Clo. botulinum* spores, are given proper heat treatment, only one can will contain one viable spore. This is an extreme processing condition used for a high degree of safety. The 12D value at $D_{121.1°C}$ is about 2.8 to 3.0 min.[1,3]

B. Thermal Death Time (TDT), Z Value, and F Value

TDT is the time in log that is necessary to completely destroy a specific number of microbial cells or spores in a population at a specific temperature. It indicates the relative sensitivity of a microorganism to different temperatures. A TDT curve can be constructed, either by plotting log time of complete destruction against temperature or by plotting log D values against temperature (Figure 31.2; this is called a phantom TDT curve). The slope of the curve is the Z value, which indicates the temperature (°C or °F) required to change the D value (or TDT) to transverse by 1 log. A value of Z = 10 in °C implies that if D value at 100°C is 50 min, at

110°C it will be 5 min, and at 120°C it will be 0.5 min. In developing heat-processing conditions for a food, D and Z values are used to obtain desirable destruction of microorganisms. In addition, a symbol F is used to express the time (min) necessary to completely destroy a specific number of microbial spores or cells at a reference temperature (121.1°C for spores and 60°C for cells). The $F_{121.1°C}$ value for *Clo. botulinum* type A spores was found to be 0.23 min and was used to calculate the 12 D value (12 × 0.23 = 2.78 or 3 min) for heat processing of low-acid canned foods.[1,3]

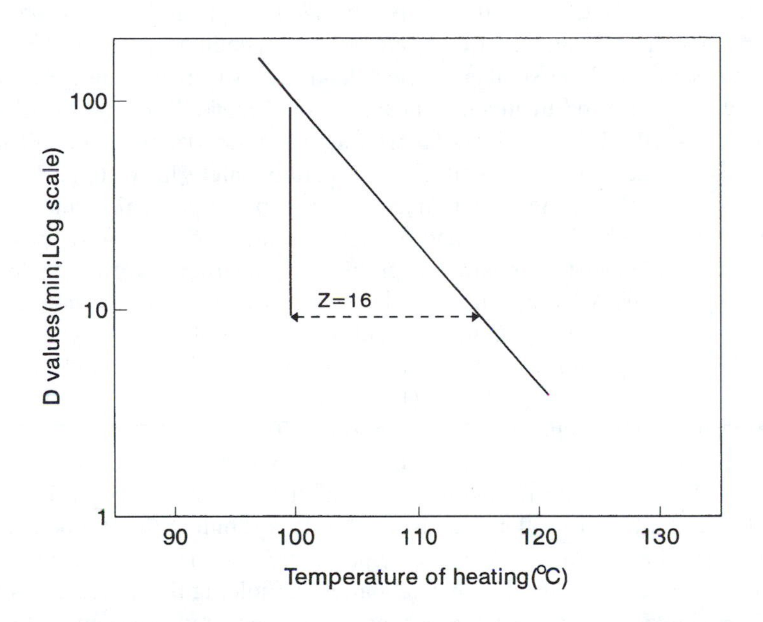

Figure 31.2 Hypothetical thermal death time curve. D, decimal reduction time; Z, °C required for the thermal death time curve to transverse over a $\log_{10}$ cycle.

VI. METHODS OF HEAT PROCESSING

On the basis of the temperature and time of heating the food used to destroy microorganisms, the methods can be broadly divided into low-heat processing and high-heat processing. The low-heat processing is used to destroy mainly the microorganisms relatively sensitive to heat and not generally effective against thermoduric microorganisms. In contrast, high-heat processing is used to destroy the thermodurics, and especially the most heat-resistant spores of spoilage and pathogenic bacteria. Methods, (e.g., baking, broiling, simmering, roasting, and frying) used in cooking foods and blanching used to destroy some natural enzymes in fresh vegetables and fruits are not discussed here. Although they destroy microorganisms, their main use is for food preparation, not destruction of microorganisms in foods.[1,4]

A. Low-Heat Processing or Pasteurization

The temperature used for low-heat processing or pasteurization is below 100°C. The objectives of pasteurization are to destroy all the vegetative cells of the pathogens and a large number (~90%) of associative (spoilage) microorganisms (yeasts, molds, and bacteria). In certain foods, pasteurization also destroys some natural enzymes (e.g., lipases in milk). The temperature and time are set to the lowest level to meet the objectives and to minimize the thermal damage of the food, which otherwise could reduce the acceptance quality (such as heated flavor in milk) or pose processing difficulties (such as coagulation of liquid egg). Depending upon the temperature used, thermoduric cells of spoilage bacteria and spores of pathogenic and spoilage bacteria will survive the treatment. Thus additional method(s) need to be used to control the growth of the survivors (as well as postpasteurization contaminants) of pasteurized products, unless a product has a natural safety factor (e.g., low pH in some acid products). Refrigeration, modified atmosphere packaging, incorporation of preservatives, reduction of A_w, and other techniques are used, when possible in combination, to prevent or retard the problem of microbial growth in low heat processed products. Microbial heat-stable enzymes and toxins are not destroyed unless a food is heated for 30 min or longer at or above 90°C.

Pasteurization of milk has been used for a long time. Two methods, heating at 145°F (62.8°C) for 30 min or 161°F (71.7°C) for 15 s, are used to destroy the most heat-resistant Q fever pathogen, *Coxiella burnetii*. The methods are also designated, respectively, as low temperature long time (LTLT) and high temperature short time (HTST) methods. As indicated before, the regulation requires every particle of milk to be heated at 145°F for 30 s or at 161°F for 15 s (holding time). Foods that are not uniformly heated to the specified temperature and time can be involved in foodborne disease outbreaks. Immediately after the holding time, the milk is cooled to 40°F, packaged, and maintained at that temperature until consumed. Low-heat treatment is used for processing many products using different times and temperatures. Some of these are ice cream mix, 180°F (82.3°C) for 25 s or 160°F (71.2°C) for 30 min; liquid whole egg, 140°F (60°C) for 3.5 min (intended to destroy *Salmonella*); fruit juices, 60 to 70°C for 15 min or 80 to 85°C for 1 min; wine, 82 to 85°C for 1 min; pickles in acid (pH 3.7), 74°C for 15 min; vinegar, 65.6 to 71.1°C for 1 min or 60°C for 30 min (heating of all low pH products is designed to destroy spoilage microorganisms); crabmeat, 70°C for 10 min; low-heat processed meat products, 60 to 70°C internal temperature (depending upon the size, it can take 2 h for the center of a product to reach the desired temperature; heating is not intended to kill *Clo. botulinum* type AB, or nonproteolytic spores). The time and temperature of heating for different foods is also specified by the regulatory agencies.

In addition to use of hot water (e.g., in milk) or moist heat (e.g., in meat products) to heat process some foods, other products, such as dried egg whites and dried coconut, are pasteurized by dry heat. In such a condition, the products are exposed to 50 to 70°C for 5 to 7 d.

In the production of some fermented foods, the raw materials are heated to high temperature in order to destroy vegetative cells of pathogens and spoilage microorganisms, which include thermoduric bacterial cells. Raw milk used for the production of buttermilk, acidophilus milk, and yogurt are given a 30- 60-min heat treatment at about 90°C prior to adding starter culture bacteria. Heating helps the starter culture bacteria grow preferentially, in addition to improving the gelling properties of the milk proteins at low pH.[1,4]

B. High-Heat Processing

The process involves heating foods at or above 100°C for the desired period of time. The temperature and time of heating are selected on the basis of product characteristics and the specific microorganism(s) to be destroyed. Most products are given a commercially sterile treatment to destroy microorganisms growing in a product under normal storage conditions. The low-acid or high-pH (pH > 4.6) products are given 12D treatment to destroy *Clo. botulinum* type A or B spores (the most resistant spores of a pathogen). However, the products can have viable spores of thermophilic spoilage bacteria (e.g., *Bacillus stearothermophilus, Bac. coagulans, Clostridium thermosaccharolyticum,* and *Desulfotomaculum nigrificans*; see Chapter 18). As long as the products are stored at or below 30°C, these spores will not germinate. If the products are temperature abused to 40°C and above, even for a short time, the spores will germinate. Subsequent storage below 40°C will not prevent outgrowth and multiplication of these thermophiles to cause food spoilage. The time and temperature required for commercial sterility of a particular food is determined by actual pack inoculation studies. Generally, *Clo. sporogenes* PA 3679 is used to simulate *Clo. botulinum* as this is a nonpathogenic strain, but the spores have the same heat resistance as *Clo. botulinum* type A or B (nonproteolytic). For spoilage studies, spores of *Bac. stearothermophilus* are used since spores of this species are the most heat resistant.

For high-acid or low-pH (pH ≤ 4.6) products, such as tomato products, fruit products, and acidified foods, a much lower heat treatment is used. Since *Clo. botulinum* spores cannot germinate and outgrow at this low pH, their presence is of little health significance. The sporeformers that can germinate and grow in low-pH products, (e.g., *Bac. coagulans*) and the aciduric nonsporeforming bacteria (e.g., *Lactobacillus* and *Leuconostoc* spp.) yeasts, and molds that can grow at low pH are relatively heat sensitive. These products are generally heated around 100°C for a desirable period of time. High-heat-treated products are either first packed in containers and then heated, or heated first and then packed in sterile containers while still hot (hot pack).

Commercial sterility is also obtained by heating a food at very high temperatures for a short time. This process is designated as ultrahigh temperature (UHT) processing. Milk heated to about 150°C for 2 to 3 s can be stored at room temperature (≤30°C), and the products generally have a 3-month shelf life. However, if microbial heat-stable proteinases or lipases are present in the raw milk, the product can show spoilage. In the UHT process, the milk is heated by injecting steam at high pressure for a rapid temperature increase. Following heat treatment in bulk, the milk is packed in small serving containers. Microbial heat-sensitive toxins will be destroyed, but heat-stable toxins may remain active even after heating for commercial sterility.[1,4]

Under special circumstances, foods are heated to destroy all microorganisms (cells and spores) and to achieve sterility. Sterile foods are necessary for immunosuppressant individuals in order to avoid any complications from the microorganisms that are normally present in heated but nonsterile foods.

C. Microwave Heating

Heating of foods by microwave (or cooking foods) at home has become quite common in developed countries.[5,6] Frozen foods can be thawed and heated very rapidly, in a few minutes, depending upon the size of a product. However, the method has not been well accepted as a source of rapidly generated high heat for commercial operations.

In a microwave oven, the waves change their polarity very quickly. Oppositely charged water molecules in a food rapidly move to align along the waves. The movement of the water molecules generates frictional heat, causing the temperature of the food to rise very rapidly. Depending upon the exposure time and the intensity of the wave, the temperature can be very high. Microwave treatment is quite lethal to microorganisms and the destruction is caused by the high temperature. At present, microwave-heated foods cannot be considered safe from pathogens. Generally, when a food is heated in a microwave oven, it is not heated uniformly and some areas can remain cold. If a food harbors pathogens, there are chances that they will survive in the cold spots.

REFERENCES

1. Olson, J. C., Jr. and Nottingham, P. M., Temperature, in *Microbial Ecology of Foods,* Vol. I., Silliker, J. H., Ed., Academic Press, New York, 1980, 1.
2. Gould, G. W., Heat induced injury and inactivation, in *Mechanism of Action of Food Preservation Procedures,* Gould, G. W., Ed., Elsevier, New York, 1989, 11.
3. Pflug, I. J., Calculating F_t-values for heat preservation of shelf-stable low acid canned foods using the straight-line semilogarithmatic model, *J. Food Prot.,* 50, 608, 1987.
4. Anonymous, Sterilization methods: principles of operation and temperature distribution studies: symposium, *Food Technol.,* 44(12), 100, 1990.
5. Heddleson, R. A. and Doores, S., Factors affecting microwave heating of foods and microwave induced destruction of foodborne pathogens, *J. Food Prot.*, 57, 1026, 1994.
6. Anonymous, Dielectric and ohmic sterilization: symposium, *Food Technol.,* 46(12), 50, 1992.

CHAPTER 31 QUESTIONS

1. Discuss the ideological differences in the use of heating in food preservation before and after 1870 A.D.

2. List the microbiologically related objectives of heating a food.

3. Describe the mechanisms of sublethal and lethal heat injury in bacterial cells and spores. How can stress protein alter this effect in food processed slowly at low heat?

4. List three important factors that one should consider in designing thermal preservation of a food. Discuss the implications of one of the factors.

5. (a) Define D value, Z value, F value, and 12D value. (b) Draw hypothetical plots to show how D and Z values can indicate the relative heat resistance of two microbial species.

6. List the objectives of pasteurization of food. How do these objectives differ from those used in commercial sterilization of foods? Use a food system for each method of treatment to justify your explanations.

7. How does microwave heating differ from conventional heating to ensure food safety?

CHAPTER 32

Control by Low Temperature

CONTENTS

I. HISTORY

The effectiveness of low temperature, especially freezing, in food preservation was probably recognized by our ancestors at least in the last Ice Age, before 40,000 B.C. Natural freezing and thawing could also have been used to preserve food during the very early stages of civilization, about 10,000 to 12,000 B.C. In the colder regions of the world, foods (e.g., meat and fish) are still preserved in natural ice. Ice was used by the wealthy Romans to reduce the temperature of foods. Until about 1800 A.D., ice blocks from frozen lakes were cut, stored, and used to preserve raw foods (e.g., meats, milk, fish, and produce) by lowering the temperature. By 1840, with the help of ammonia compressed refrigerator units, ice blocks were produced commercially and used to reduce the temperature of food for preservation. In 1880, refrigeration was used on ships and trains in Europe to transport meat and fish from other countries. Linde, in Germany, developed the first domestic refrigerator around 1874 and started commercial production before 1890. The popularity of domestic refrigerators was initially low in the United States; in 1930, only about 2 to 3% of households had the units. In the United States methods to freeze fruits and vegetables were developed and commercially used around the 1930s. During this time, retail stores also started using cabinets to display frozen foods.

During World War II, consumer interest in refrigerated and frozen foods increased dramatically and that helped develop the technology necessary for processing, transportation, retailing, and home storage of refrigerated and frozen foods in the United States. The popularity of refrigerated and frozen foods has increased steadily since then. In the 1960s, 1970s, and early 1980s, frozen food consumption increased sharply, mainly because of their long shelf life. Since the mid-1980s, there has been increased interest for the refrigerated and chilled foods that consumers view as natural and healthy. The major drawback of many refrigerated foods is their relatively short shelf life, about 1 to 2 weeks. But in recent years, several technological improvements, such as oxygen-impermeable packaging materials, good vacuum-packaging equipment, innovative packaging systems, and low-temperature refrigeration unit (≤–1°C), have helped to increase the shelf life of refrigerated foods to about 60 d, and for some products, over 90 d. At present, of the total foods consumed, low-temperature preserved foods constitute over 65%, and the trend shows a steady increase in the future. To suit the taste of consumers, a large number of new products are being developed that are low in fat (caloric), high in fiber, phosphates, and other additives, and have low amounts or no preservatives. To achieve the long shelf life and to make these products safe, extra precautions are being introduced for microbiological control. This has helped many new or emerging pathogenic and spoilage bacteria, in the absence of competition from the associated microorganisms, to become predominant. New designs of processing equipment for high production efficiency, centralized production of large volumes of products, transportation of the products for long distances in regional storage facilities, retailing conditions, and consumer handling of the products have facilitated these so-called new pathogenic and spoilage bacteria in gaining prominence. Unless some effective intervention strategies are developed, the new pathogenic and spoilage microorganisms will continue to surface in refrigerated foods stored for a long time.[1a,1b]

II. OBJECTIVES

The main microbiological objective in low-temperature preservation of food is to prevent or reduce growth of microorganisms. Low temperature also reduces or prevents catalytic activity of microbial enzymes, especially heat-stable proteinases and lipases. Germination of spores is also reduced. Low-temperature storage, especially freezing (and thawing), is also lethal to microbial cells, and under specific conditions, 90% or more of the population can die during low-temperature preservation. However, the death rate of microorganisms at low-temperature, as compared to heat treatment, cannot be predicted (as D and Z values in heating). Also, spores are not killed at low temperature. Thus, foods are not preserved at low temperature in order to kill microbial cells. Freezing is also used to preserve starter cultures for use in food bioprocessing. This has been discussed separately (see Chapter 12).

III. MECHANISMS OF MICROBIAL CONTROL

The metabolic activities, enzymatic reactions, and growth rates of microorganisms are maximum at the optimum growth temperature. As the temperature is lowered, microbial activities associated with growth slow down. Normally, the generation time is doubled for every 10°C reduction in temperature. Thus a species dividing every 60 min in a food at 22°C will take 120 min to divide if the temperature is reduced to 12°C. In a lower range, generation time can be even higher than double. This means that if the temperature is reduced from 12 to 2°C, the generation time for the species could be greater than 240 min. The lag and exponential phases and the germination time for some psychrotrophs (mesophilic types) become longer and longer as the temperature is reduced to about 0°C or even to about –1°C. At this temperature, nongrowing cells of some mesophiles (nonpsychrotrophic) and thermophiles may die, especially if they are stored for a long time (weeks) at 2°C or below and the foods have low A_w, low pH, or preservative(s). The rate of catalytic activity of some enzymes also decreases as the temperature of an environment is reduced.[2-5]

Water is present in food as free water and bound (with the hydrated molecules, etc.) water. As the temperature in a food system drops to about –2°C, free water in the food starts freezing and forming ice crystals (pure water freezes at 0°C, but in a food with solutes it freezes below 0°C). As the temperature drops further, and more ice crystals form, the solutes get concentrated in the remaining water, which in turn further depresses the freezing point of the water in the solution. The A_w is also reduced. When the temperature is reduced to about –20°C, almost all of the free water freezes.

As the temperature of a food is reduced below –2°C, free water inside the microbial cells also undergoes similar changes. At slow rates of freezing, as the water molecules in the food start freezing, water molecules from inside the microbial cells migrate outside, causing dehydration of cells and concentration of solutes and ions inside. When the temperature is reduced dramatically (above –20°C), so that the water in the food has frozen, water inside the cell also freezes. However, prior to that, microbial cells are exposed to low pH (concentration of ions) and low A_w (concentration of solutes) inside and outside the cells. This can cause denaturation and destabilization of the structural and functional macromolecules in the microbial cells, whose stability and functions are dependent on three-dimensional structures, and can injure the cells. If the freezing is rapid, the very small ice crystals form rapidly and the cells are not exposed to solution effect.

Microbial cells subjected to freezing and thawing are found to suffer sublethal (repairable) as well as lethal injury. Studies show that different components of the cell wall (or outer membrane) and cell membrane (or inner membrane) are injured. DNA strand break, ribosomal RNA degradation, and activation/inactivation of some enzymes have also been reported in some studies. In sublethally injured cells, the structural and functional injuries are reversible. In lethally injured (or dead) cells, the damages are irreversible.[2-5]

IV. FACTORS TO CONSIDER

The effectiveness of low temperature in controlling microbial growth and microbial enzymatic activity in food is dependent upon many factors.[1-5] These factors can be arranged into three groups: those unique to low temperature, those related to the food environment, and those inherent in microorganisms. These factors not only aid in preventing or reducing the growth of microorganisms, but can also greatly influence the extent of sublethal and lethal injury that microorganisms incur in food preserved at low temperatures. An understanding of the influence of these factors and the interaction among them is important in designing an efficient and predictable method for the preservation of a specific food at low temperature.

A. Factors Unique to Low-Temperature Storage

At temperatures above freezing of free water (≤–2°C), different types of bacteria, molds, and yeasts can grow in a food. But the lag and exponential phases become longer as the temperature is reduced. In the low range, even a difference of <1°C can be highly important. A *Pseudomonas fluorescens* strain was reported to have a generation time of about 6.7 h at 0.5°C, but 32.2 h at 0°C. Thus a reduction in 0.5°C increased the generation time by about 4.5-fold. This is much more than the theoretical estimate, which suggests that the generation time doubles for every 10°C reduction. Spores of some spoilage *Bacillus* and *Clostridium* spp. can germinate even at refrigeration temperature (4.5°C or 40°F). Cells of some mesophiles and thermophiles can be sublethally injured as well as die as the temperature drops below 4.5°C.

As the temperature is reduced enough to cause a large portion of the water to freeze, the growth of most microorganisms stops except for some psychrophilic bacteria, yeasts, and molds. Although there are conflicting reports, slow growth probably can occur up to –10°C, especially for some molds. As the temperature drops further, to about –20°C, and water in a food freezes, more cells will have sublethal and lethal injury. The rate of freezing and the lowest temperature of freezing dictate the extent of microbial damage from ice crystals. Damage and death are more extensive at slower rates of freezing and at –20°C than at a rapid rate of freezing and at –78°C or –196°C. Death and sublethal injury are very high during initial storage (about 7 d), and subsequently slow down.

Fluctuation of temperature in a food during low-temperature storage has great impact on growth, sublethal injury, and death of microorganisms. This quite readily happens to foods during storage, transport, retail display, and at home. A fluctuation of temperature in food from ≤4.4 to 10 to 12°C not only stimulates rapid growth of psychrotrophic pathogenic and spoilage bacteria, but many mesophilic spoilage and pathogenic bacteria are also able to grow and their spores germinate in this range.

Just from the spoilage aspect, a 6- to 8-h temperature abuse (12°C) of a vacuum-packaged, refrigerated, low-heat-processed meat product can reduce its expected shelf life of 8 weeks by about 7 to 10 d. A fluctuation in temperature of a frozen food increases microbial death and injury due to repeated damaging solution effects and mechanical damage from larger ice crystals that form during repeated freezing and thawing. Dead microbial cells can also lyse, releasing intracellular enzymes, many of which (e.g., proteinases and lipases) can act on food components and reduce the acceptance quality of food.

The rate of cooling a food is also very important for effective control of the growth of pathogenic and spoilage microorganisms. A slow rate of cooling of foods has been implicated as a major cause of foodborne diseases. This can occur by trying to cool a large volume of hot or warm food in a big (deep) container in a refrigerator or overstuffing refrigerators with hot or warm foods. During thawing of a frozen food (such as an uncooked chicken), rapid thawing is desirable in order to control microbial the growth, especially the growth of pathogens. If the food is thawed slowly, the temperature on the food surface will soon increase, thereby allowing microbial growth, even when the inside is still frozen.

Refrigerated foods will have limited shelf life and, with time, microorganisms will grow and spoil the products. In frozen foods, microorganisms (only cells, not spores) will slowly die. However, even after long storage, there will be survivors in frozen foods.

B. Food Environment

Composition, pH, A_w, and the presence of microbial inhibitors (preservatives) in a food can greatly influence growth, sublethal injury, and viability of microorganisms during storage at low temperature. A food with a higher solid content (especially high proteins, carbohydrates, and lipids, but low ions), pH closer to 7.0, higher A_w, and the absence of microbial inhibitors will facilitate growth and survival of microorganisms at refrigeration temperatures and inflict less injury and cause less death at frozen temperatures. Thus the shelf life of refrigerated foods can be enhanced by using one or more of these factors, such as low pH, low A_w, incorporation of a suitable microbial inhibitor(s), and, when possible, vacuum or modified air packaging.

In packaged frozen foods, ice may form in the packages (package ice), especially if the storage temperature fluctuates. During thawing, the ice melts and is absorbed by the food, resulting in an increase in the A_w in a localized area (e.g., in a bread) and making it susceptible to microbial growth.

C. Inherent Characteristics of Microorganisms

While some microorganisms are capable of growing as low as –10°C, many mesophilic and thermophilic bacterial cells can be sublethally injured and may die with time at low temperatures above freezing. At temperatures below –10°C, vegetative cells of microorganisms can sustain sublethal injury and die. In general, Gram-negative or rod-shaped bacteria are more susceptible to the damaging effect of freezing than the Gram-positive or spherical-shaped bacteria. Also, cells from the early exponential phase of growth are more susceptible to freezing than those from the early stationary phase. Species and strains of microorganisms also differ greatly in sensitivity-resistance to freezing damage. Germination and outgrowth of spores of some *Clostridium* spp. can occur as low as 2°C and maybe at a little higher temperature for some *Bacillus* spp. spores. Spores will not lose viability in frozen foods. Some microbial enzymes, either heat stable or released by the dead and lysed cells, can catalyze reaction at temperatures above –20°C, but at a slow rate, and can reduce the acceptance quality of a food.[1-5]

V. METHODS USED

Foods are stored at low temperature in different ways in order to extend their shelf life. Many fresh fruits and vegetables are kept at temperatures between 10 and 20°C or lower, mainly to reduce their metabolic rate. Microorganisms to which these products are susceptible, namely yeasts and molds (and some bacteria), can grow at this temperature. Maintaining a low relative humidity to prevent moisture build-up on the food surface is very important in order to reduce their growth. Highly perishable products are generally stored at a low temperature below 7°C, often in combination with other preservation methods. The importance of rapid cooling of a food for microbiological safety has been mentioned before. The methods used for low-temperature preservation of foods and in food safety are briefly discussed.[1,3]

A. Ice Chilling

This is used in retail stores where the foods are kept over ice; the surface in contact with the ice can reach between 0 and 1°C. Fresh fish, seafoods, meats, cut fruits, vegetable salads (in bags), different types of ready-to-eat salads (prepared at the retail store), salad dressing (high pH, low caloric), sous vide, and some ethnic foods (e.g., tofu) are stored by this method. The trend is increasing.

Temperature fluctuation (due to the size of the container or melting of ice), duration of storage (fresh or several days), and cross-contamination (raw fish, shrimp, oysters, and ready-to-eat fish salads in an open container in the same display case at the retail store) can cause microbiological problems, especially from foodborne pathogens.

B. Refrigeration

The temperature specification for refrigeration of foods has changed from time to time. Previously, 7°C (≃45°F) was considered a desirable temperature. However, technological improvements have made it economical to have domestic refrigeration units at 4 to 5°C (40 to 41°F). For perishable products, ≤4.4°C (40°F) is considered a desirable refrigeration temperature. Commercial food processors may use as low as ≃1°C for refrigeration of perishable foods (such as fresh meat and fish). For optimum refrigeration in commercial facilities along with low temperature, the relative humidity and proper spacing of the products are also controlled.

Raw and processed foods of plant and animal origin, as well as a great variety of prepared and ready-to-eat foods, are now preserved by refrigeration. Their volume is increasing rapidly due to consumer preference for such foods. Some of these foods are expected to have a storage life of 60 days or more.

For refrigerated products with a long expected shelf life, additional preservation methods are combined with the lowest possible temperature that can be used (close to –1°C). However, as the products are nonsterile, even a very low initial microorganism population (i.e., ≤1 cells or spores per gram) capable of growing under the storage conditions can multiply to reach hazard (for pathogens) or spoilage levels, thereby reducng the safety and stability of the product. Any fluctuation in temperature or other abuse (e.g., a leak in a vacuum or modified atmosphere package, or oxygen permeation through the packaging materials) can greatly accelerate their growth. The processing and storage conditions may provide environments in which different types of spoilage and pathogenic microorganisms grow advantageously. This may increase spoilage and wastage of foods unless appropriate control measures are installed quickly.

C. Freezing

The minimum temperature used in home freezers (in the refrigerator) is –20°C, a temperature at which most of the free water in a food remains in a frozen state. Dry ice (–78°C) and liquid nitrogen (–196°C) can also be used for freezing: they are used for rapid freezing (instant freezing) and not for freezing a food to that low temperature. Following freezing, the temperature of the food is maintained around –20 to –30°C. Depending upon the type, foods can be stored at refrigerated temperature for months or even over a year. Raw produce (vegetables, fruits), meat, fish, processed products, and cooked products (ready-to-eat after thawing and warming) are preserved by freezing.

Microorganisms will not grow at –20°C in frozen foods. Instead, microbial cells will die during frozen storage. However, the survivors are capable of multiplying in the unfrozen foods. Accidental thawing or slow thawing can facilitate growth of survivors (spoilage and pathogenic microorganisms). Spores can also germinate and outgrow, depending upon the temperature and time following thawing. Enzymes, released by the dead microbial cells, can reduce the acceptance quality of the food.[1,3]

REFERENCES

1. Kraft, A. A., Refrigeration and freezing, *Psychrotrophic Bacteria in Foods,* CRC Press, Boca Raton, FL, 1992, 241.

1a. Ray, B., Kalchayanand, N., and Field, R. A., Meat spoilage bacteria: are we prepared to control them?, *The National Provisioner*, 206(2), 22, 1992.

1b. Ray, B., Kalchayanand, N., Means, W., and Field, R. A., The spoiler: *Clostridium laramie, Meat and Poultry*, 41(7), 12, 1995.

2. Ray, B., Enumeration of injured indicator bacteria from foods, *Injured Index and Pathogenic Bacteria,* Ray, B., Ed., CRC Press, Boca Raton, FL, 1989, 10.
3. Olson, J. C. and Nottingham, P. M., Temperature, in *Microbial Ecology,* Vol. I., Silliker, J. H., Ed., Academic Press, New York, 1980, 1.
4. Speck, M. L. and Ray, B., Effects of freezing and storage of microorganisms from frozen foods: a review, *J. Food Prot.,* 40, 333, 1977.
5. Kalchayanand, N., Ray, B., and Field, R. A., Characteristics of psychrotrophic *Clostridium laramie* causing spoilage of vacuum-packaged refrigerated fresh and roasted beef, *J. Food Prot.,* 56, 13, 1993.

CHAPTER 32 QUESTIONS

1. List the microbiological objectives of low-temperature preservation of food. How do these objectives differ from food preserved by heat?

2. Briefly discuss the mechanisms of microbial control by reducing the temperature of a food to 10°C, to –1°C, to –10°C, to –20°C.

3. List the major factors that need to be considered for effective control of microorganisms in a food at low temperature, and briefly discuss the importance of each.

4. Discuss the microbial implications of the following in low-temperature preservation of foods: (a) fluctuation of storage (refrigerated and frozen) temperature, (b) slow cooling of a warm food, and (c) slow thawing of a frozen food.

5. Briefly discuss the microbiological problem of foods stored by: chilling, refrigeration, and freezing.

CHAPTER 33

Control by Reduced A_w

CONTENTS

I. HISTORY

The ability of dried seeds, grains, tubers, and fruits to resist spoilage was probably recognized by humans even before their discovery of agriculture. Subsequently, this simple method (drying) was practiced in order to preserve the large volume of foods produced during the growing seasons to make them available during the nongrowing seasons. Later, reduced A_w was also extended to preserve other foods (e.g., meat, fish, and milk), not only by removing water but also by adding solutes (e.g., salt, honey, and starch) to bind water.

Since the beginning, reduced A_w has been used throughout human civilization in many ways, not only to preserve foods and stabilize the food supply, but also to develop different types of shelf-stable foods. Some of these include salted fish and meats; semidry and dry fermented sausages; dried fish, meat, vegetables, and fruits; evaporated and sweetened condensed milk; dry milk, cheeses, bread and bakery products, flour, cereals, molasses, jams and jellies, chocolate, noodles, crackers, dried potatoes, dried eggs, and confectioneries. In more recent years, new technologies have helped to produce foods with low A_w by freeze-drying, puffed drying, freeze-concentration, and osmotic-concentration methods. In addition, a better understanding of the relationship and influence between moisture and A_w on microbial growth has been instrumental in producing many types of intermediate-moisture ready-to-eat foods. Efforts are also in progress to use low A_w along with other microbial control parameters, such as low pH, vacuum-packaging, and low heat, to produce ready-to-eat meat, fish, and dairy products that can be stored at ambient temperature for relatively long periods of time. Because of the convenience, these foods are popular and the trend shows that their popularity will continue.

II. OBJECTIVES

The main objectives of reducing A_w in food are to prevent or reduce the growth of vegetative cells and germination and outgrowth of spores of microorganisms.[1-3] Prevention of toxin production by toxigenic molds and bacteria is also an important considerations. Microbial cells (not spores) also suffer reversible injury and death in foods with low A_w, although not in a predictable manner such as in heat treatment. Finally, reduced A_w is also used to retain the viability of starter culture bacteria for use in food bioprocessing; this aspect is discussed separately (see Chapter 12). In this chapter, preservation of food by controlling microbial growth at low A_w is described.

III. MECHANISM

Microorganisms need water for the transport of nutrients, nutrient metabolism, and removal of cellular wastes. In a food, the total water (moisture) is present as free water and bound water; the latter remains bound to hydrophilic colloids and solutes (it can also remain as capillary water or in a frozen state as ice crystals) and is not available for biological functions. Thus only the free water (which is related to A_w) is important for microbial growth. Microorganisms also retain a slightly lower A_w inside the cells than the external environment to maintain turgor pressure, and this is important for cell growth. If the free water in the environment is reduced either by removing water or by adding solutes and hydrophilic colloids, which cannot readily enter the cells, the free water from the cells flows outside in an effort to establish equilibrium. The loss of water will cause osmotic shock and plasmolysis, during which the cells do not grow. The water loss can be quite considerable even with a slight change in A_w. A 0.005 reduction in A_w from 0.955 to 0.950 in the environment reduced the intracellular water content by 50% in *Staphylococcus aureus,* and reduced the cell volume by 44% in *Salmonella typhimurium*. This is why even a slight reduction in A_w necessary for minimal growth of a microbial species or strain prevents its growth. Unless a microbial cell regains its intracellular turgor by reducing internal A_w, it will either remain dormant or die. This is often the case with microorganisms sensitive to slight A_w reductions. However, some other microorganisms have developed very effective mechanisms to overcome the plasmolysis and regain turgor. They achieve this by either transporting solutes inside or metabolizing solutes. The microorganisms that are relatively resistant to a great reduction in A_w and grow at relatively lower A_w have this capability.[1-3]

IV. INFLUENCING FACTORS

A. Factors Unique to A_w

Water activity and the total amount of water (% moisture) a food contains are different. The A_w of a food indicates the amount or fraction of the total amount of water that is available for some chemical or biochemical reactions. In pure water, both values are the same; but in food, A_w is always less than the total amount of water. Under a set of conditions, the relationship between the moisture content and the A_w of a food can be determined from the sorption isotherm. However, instead of a single line, the sorption isotherm forms a loop (hysteresis loop), depending on whether it is determined the removal of water from (desorption) or during the addition of water to (adsorption) a food (see Figure 6.1). At the same moisture level, the A_w value obtained by desorption is lower than that obtained by adsorption. In controlling microbial growth by reducing A_w, this is quite important.

Solutes differ in their ability to reduce A_w. The amounts (%w/w) of NaCl, sucrose, glucose, and inverted sugar necessary to reduce the A_w at 25°C of pure water to 0.99 are 1.74, 15.45, 8.9, and 4.11 g; and to 0.92, they are 11.9, 54.34, 43.72, and 32.87 g, respectively. These solutes do not freely enter the microbial cells and thus have a greater inhibitory effect on microbial cells as compared to solutes that enter freely into cells, (e.g., glycerol), which are required in higher amounts for similar inhibition.

Although microorganisms can die at reduced A_w, a lower A_w value is less detrimental. For a 90% reduction of salmonellae population in a product at 15°C, it took 27 d at A_w = 0.71, but 67 d at A_w = 0.34.

Studies on minimal A_w values to support the growth of specific microorganisms have generated conflicting data. This could be due to the inherent problems with different techniques used to measure A_w. However, with modern electronic hygrometers, this problem is expected to be minimized.[1-3]

B. Factors Related to Foods

Minimal A_w values for growth, as well as the influence of A_w on viability loss, of microorganisms vary with the food characteristics and the food environment. In a homogeneous food, A_w will remain unchanged provided other factors do not change. However, a heterogenous food with ingredients or items of different A_w (e.g., a sandwich or a meal with different items in the same package) will generate a gradient. This can lead to microbial growth in an item preserved by reduced A_w alone and stored with an item of high A_w containing a preservative. Also, condensation of water during storage with temperature fluctuation, followed by dripping of moisture in food, can alter safe A_w levels to an unsafe state.

The minimum A_w for growth of microorganisms in a food can be higher than that in a broth. Thus *Staphylococcus aureus* has a minimal A_w of growth of 0.86 in broth, but did not grow in shrimp at A_w = 0.89. As the A_w is reduced, anaerobic bacteria will require more of an anaerobic environment for growth. *Clostridium perfringens* grew in a broth with A_w = 0.995 at O–R = +194 mV; when the A_w was reduced to 0.975, it needed an O–R = +66 mV for growth. As the incubation temperature is moved in either direction from the optimum without changing the A_w, the microorganisms require a longer time to grow. In a broth of A_w = 0.975, a *Clo. botulinum* E strain grew in 6 d at 30°C, in 19 d at 15°C, and in 42 d at 7.2°C. The minimum A_w for growth of *Sta. aureus* was 0.865 at 30°C, but changed to 0.878 at 25°C. Reduced A_w and low pH interact favorably in inhibiting microbial growth. A *Clo. botulinum* B strain grew at A_w = 0.99 up to pH 5.3 and at A_w = 0.97 up to pH 6.0; but at A_w = 0.95, it failed to grow even at pH 7.0. Similarly, a spoilage strain of *Clo. butyricum* grew at A_w = 0.98 up to pH 3.8, but at A_w = 0.97 failed to grow at pH 4.5 even after 30 d at 30°C. Many chemical preservatives enhance the inhibiting effect of lower A_w on microbial growth. In the presence of low concentrations of sorbate, citrate, and phosphate, different microorganisms were found not to grow at the lowest A_w in which they grew in the absence of these chemicals.

Food composition can influence the microbial death rate even at the same A_w. At A_w = 0.33, *Escherichia coli* counts reduced by $\log_{10}$ 2.8 in ice cream powder, $\log_{10}$ 4.8 in dried potatoes, but over $\log_{10}$ 6 in coffee. Under the same conditions, the death rate of *Enterococcus faecalis* was much less. Although the survivors remain dormant in a low A_w food, as soon as it is rehydrated, the microorganisms regain the ability to metabolize and multiply. Thus a rehydrated food should be treated as a perishable food that, unless effective control methods are used, can be unsafe and spoiled.[1-3]

C. Factors Related to Microorganisms

Microorganisms differ greatly in their minimal A_w requirements for growth, sporulation, and germination (Table 33.1). In general, mold and yeasts can grow at lower A_w values than bacteria; among pathogenic and spoilage bacteria, Gram negatives require a slightly higher A_w than Gram positives for growth. *Staphylococcus aureus*, however, can grow at A_w = 0.86. Sporulation by sporeforming bacteria occurs at A_w values in which the species-strains will grow, while germination may occur at slightly lower A_w values.Toxin production may occur at the A_w of growth (by *Clostridium botulinum*) or at a slightly higher than minimum A_w of growth (*Sta. aureus* at 0.867 at 30°C and 0.887 at 25°C). The minimum A_w for microbial growth can vary with the type of solutes in a food. *Clo. botulinum* type E failed to grow below $A_w = 0.97$ when NaCl was used as a solute, but grew up to $A_w = 0.94$ when glycerol was the solute. *Pse. fluorescens* similarly showed minimum $A_w = 0.957$ for growth with NaCl, but 0.94 with glycerol. This is because glycerol enters freely inside the cell and thus does not cause as much osmotic stress as nonpermeable NaCl, sucrose, and similar solutes. Germination of spores in glycerol occurs at lower A_w since *Clo. botulinum* E spores, with glycerol, germinate at $A_w = 0.89$, but do not grow below $A_w = 0.94$. The growth rate of microorganisms also decreases as the A_w value is lowered. The growth rate of *Sta. aureus* reduces to about the 10% level at $A_w = 0.90$ of its optimum growth rate at $A_w = 0.99$.

Table 33.1 Minimal A_w for Microbial Growth at Optimum Growth Temperature

Microorganism Bacteria	A_w	Yeasts	A_w
Bacillus cereus	0.95	*Saccharomyces cerevisiae*	9.90
Bac. stearothermophilus	0.93	*Sac. rouxii*	0.62
Clostridium botulinum type A	0.95	*Debaryomyces hansenii*	0.83
Clo. botulinum type B	0.94		
Clo. botulinum type E	0.97	**Molds**	
Clo. perfringens	0.95		
Escherichia coli	0.95	*Rhizopus nigricans*	0.93
Salmonella spp.	0.95	*Penicillium chrysogenum*	0.79
Vibrio parahaemolyticus	0.94	*Pen. patulin*	0.81
Staphylococcus aureus	0.86	*Aspergillus flavus*	0.78
Pseudomonas fluorescens	0.97	*Asp. niger*	0.77
Lactobacillus viridescens	0.94	*Alternaria citri*	0.84

Among microbial groups, halophiles, osmophiles, and xerophiles grow better or grow preferentially at lower A_w values. Halophiles, such as some vibrios, need NaCl in varying amounts for growth. Osmophilie and xerophilie, yeasts, and molds grow at $A_w < 0.85$, as they do not have competition from bacteria.

Optimum growth of most microorganisms in foods occurs at $A_w \geq 0.98$. At $A_w \geq 0.98$, Gram-negatives, due to faster growth rate, predominate if other needs for optimum growth are available. As the A_w drops to 0.97, Gram-positive bacteria, such as bacilli, lactobacilli, micrococci, and clostridia, will be predominant. Below $A_w = 0.93$, Gram-positive bacteria, like micrococci, staphylococci, enterococci, and pediococci, as well as yeasts and molds, grow preferentially. As the A_w drops below 0.86, osmiophilic yeasts and xerophilic molds predominate. They will be able to grow in foods with A_w up to 0.6; as A_w drops below 0.6, microbial growth stops.[1-3]

V. METHODS USED

The water activity of foods can be reduced by using one or more of three basic principles: (1) removing water by dehydration, (2) removing water by crystallization, or (3) by adding solutes. Some of these methods and their effect on microorganisms are briefly described.[1-5]

A. Natural Dehydration

Natural dehydration is a low-cost method in which water is removed by the heat of the sun. It is used to dry grains as well as to dry some fruits (raisins), vegetables, fish, meat, milk, and curd, especially in warmer countries. The process is slow and, depending upon the conditions used, spoilage and pathogenic bacteria as well as yeasts and molds (including toxigenic types) can grow during drying.

B. Mechanical Drying

Mechanical drying is a controlled process, and drying is achieved in a few seconds to a few hours. Some of the methods used are tunnel drying (in which a food travels through a tunnel against the flow of hot air and the water is removed), roller drying (in which a liquid is dried by applying in a thin layer on the surface of a roller drum heated from inside), and spray drying (in which a liquid is sprayed in small droplets, which then come in contact with hot air that dries the droplets instantly). Vegetables, fruits, fruit juices, milk, coffee, tea, and meat jerky are some examples of foods that are dried by mechanical means. Liquids may be partially concentrated prior to drying by evaporation, reverse osmosis, freeze-concentration, and addition of solutes.

Depending upon the temperature and time of exposure, some microbial cells can die during drying, while some other cells can be sublethally injured. Also during storage, depending upon the storage conditions, microbial cells can die rapidly at the initial stage and then at a slow rate. Spores generally survive and remain viable during storage in a dried food.

C. Freeze-Drying

The acceptance quality of food is least affected by freeze-drying, as compared to both natural and mechanical drying. However, freeze-drying is a relatively costly process. It can be used for both solid and liquid foods. The process initially involves freezing the food, preferably rapidly at a low temperature, and then exposing the frozen food to a relatively high vacuum environment. The water molecules are removed from the food by sublimation (from solid state to vapor state) without affecting its shape or size. The method has been used to produce freeze-dried vegetables, fruits, fruit juices, coffee, tea, and meat and fish products, some as specialty products. Microbial cells are exposed to two stresses — freezing and drying — that reduce some viability as well as induce some sublethal injury. During storage, especially at high storage temperature and in the presence of oxygen, cells die rapidly initially and then more slowly. Spores are not affected by the process.

D. Foam Drying

The foam drying method consists of whipping a product to produce a stable foam and increase to the surface. The foam is then dried by means of warm air. Liquid products, such as egg white, fruit purees, and tomato paste, are dried in this manner. The method itself has very little lethal effect on microbial cells and spores. However, a concentration method prior to foaming, the pH of the products, and low A_w will cause both lethal and reversible damages to microbial cells.

E. Smoking

Many meat and fish products are exposed to low heat and smoke for cooking and depositing smoke on the surface at the same time. The heating process removes water from the products, thereby lowering their A_w. Many low-heat-processed meat products (dry and semidry sausages) and smoked fish are produced this way. Heat kills many microorganisms. The growth of the survivors is controlled by low A_w as well as the many types of antimicrobial substances present in the smoke.

F. Intermediate Moisture Foods (IMF)

These are foods that have an A_w value of 0.70 to 0.90 (moisture content, ~10 to 40%). They can be eaten without rehydration, but are shelf-stable for a relatively long period of time without refrigeration and are considered microbiologically safe. Some of the traditional IMF include salami, liverwurst, semidry and dry sausages, dried fruits, jams and jellies, and honey. However, in recent years, many other products have been developed, such as pop tarts, slim jims, ready-to-spread frosting, breakfast squares, soft candies, fruit rolls, food sticks, soft granola bars, and others. The low A_w value and relatively high moisture is obtained by adding water-binding solutes and hydrophilic colloids. Microorganisms can survive in the products, but due to low A_w, bacteria cannot grow. However, yeasts and molds can grow in some. To inhibit their growth, specific preservatives, such as sorbate and propionate, are added.[4,5]

REFERENCES

1. Christian, J. H. B., Reduced water activity, in *Microbial Ecology of Foods,* Vol. I, Silliker, J. H., Ed., Academic Press, New York, 1980, 70.
2. Gould, G. W., Drying, raised osmotic pressure and low water activity, in *Mechanisms of Action of Food Preservation Procedures,* Gould, G. W., Ed., Elsevier Applied Science, New York, 1989, 97.
3. Sperber, W. H., Influence of water activity on foodborne bacteria: a review, *J. Food Prot.,* 46, 142, 1983.
4. Leistner, L. and Russell, N. J., Solutes and low water activity, in *Food Preservatives,* Russell, N. J. and Gould, G. W., Eds., Van Nostrand Reinhold, New York, 1990, 111.
5. Erickson, L. E., Recent developments in intermediate moisture foods, *J. Food Prot.,* 45, 484, 1982.

CHAPTER 33 QUESTIONS

1. List the microbiological objectives of reducing A_w in foods.

2. Discuss the mechanisms by which a low A_w value produces an antimicrobial effect.

3. Define sorption isotherm. Describe the hysteresis loop, and discuss how one can have problems in selecting the correct A_w in a food.

4. Discuss the influences of different solutes in lowering the A_w value of a food and controlling microbial growth (use NaCl, sucrose, and glycerol as examples).

5. Briefly describe the interaction of A_w with other intrinsic and extrinsic factors in controlling microorganisms in a food.

6. List how different microorganisms are controlled in a food as the A_w value drops from 0.99 to 0.60. Discuss the minimal A_w value for growth, toxin production, sporulation, and germination by a bacterial species (*Clo. botulinum*).

7. Briefly explain the microbiological concerns of: naturally dried foods, freeze-dried foods, smoked foods, rehydrated foods, and IMF.

CHAPTER 34

Control by Low pH and Organic Acids

CONTENTS

I. HISTORY

During the early stages of human history when food was scarce, our ancestors probably recognized that some foods from plant sources, especially fruits, resisted spoilage. Later, they observed that the fermented foods and beverages prepared from fruits, vegetables, milk, fish, and meat were much more shelf-stable than the raw materials from which they were produced. That probably was an incentive and instrumental in developing large varieties of fermented foods, especially in tropical areas where, unless preserved, foods spoil rapidly.[1]

When the microbial involvement in food spoilage and foodborne diseases was recognized, methods to control their growth as well as to kill them in food were studied. It was observed that over a rather restricted pH range, many microorganisms present in food can grow; but at lower pH ranges many of them die. Once this was recognized, many organic acids were used as food additives. In addition to their effectiveness as food preservatives, they are also used to improve the acceptance qualities of foods. The amounts and types of organic acids that can be added to foods are governed by the regulatory agencies.

The organic acids can be present in foods in three ways. They can be present naturally, such as citric acid in citrus fruits, benzoic acid in cranberries, and sorbic acid in rowan berries. Some, like acetic, lactic, and propionic acids, are produced in different fermented foods by the desirable food-grade starter culture bacteria. Many acids are also added to foods and beverages to reduce the pH. Among the organic acids used in food as preservatives are acetic, propionic, lactic, citric, sorbic, and benzoic acids, their salts, and some derivatives of benzoic acid (e.g., paraben).The influence of these acids in reducing food pH and producing antimicrobial effects are briefly discussed in this chapter.

II. OBJECTIVES

The major antimicrobial objective in using weak organic acids is to reduce the pH of food in order to control microbial growth.[1] As the pH drops below 5.0, some bacteria die. However, the death rate in low pH is not predictable as in the case of heat. Thus it could not be used with the objective of destroying a predictable percentage of a microbial population in the normal pH range of foods.[1-6].

III. MECHANISMS

The antimicrobial action of the weak organic acid is produced by the combined actions of the undissociated molecules and the dissociated ions.[1-6] Microorganisms that are important in food tend to maintain an internal cytoplasmic pH around 6.5 to 7.0 in acidophiles, and 7.5 to 8.0 in neutrophiles. The internal pH (pHi) is tightly regulated and drops by about 0.1 unit for each 1.0 unit change in the environmental pH (pHo). For nutrient transport and energy synthesis, the microorganisms also maintain a transmembrane pH gradient (about 0.5 to 1.0 unit with alkaline pHi) and a proton gradient (about 200 mV); together, they form the proton motive force (PMF).[3,4]

When a weak organic acid is added to the environment (in a food), depending upon the pH of the food, the pK of the acid, and the temperature, some of the molecules dissociate while others remain undissociated (Table 34.1). At the pH of most foods (pH 5 to 8), except for paraben, the organic acid molecules remain generally dissociated, as a result, $[H^+]$ in the environment increases, which interferes with the transmembrane proton gradient of the microbial cells. To overcome this, the cells transport proton, through the proton pump that causes depletion in energy

Table 34.1 Influence of pH on the Amount (%) of Dissociated Ions of Weak Organic Acids

Acid	pK	% Dissociated at pH		
		4	5	6
Acetic	4.8	15.5	65.1	94.9
Propionic	4.9	12.4	58.3	93.3
Lactic	3.8	60.8	93.9	99.3
Citric	3.1	81.1	99.6	>99.1
Sorbic	4.8	18.0	70.0	95.9
Benzoic	4.2	40.7	87.2	98.6
Paraben[a]	8.5	<0.1	0.1	0.3

[a] Paraben: esters of benzoic acids.

and a decrease in pHi. The structures on the cell surface, outer membrane or cell wall, inner membrane or cytoplasmic membrane, and periplasmic space are also exposed to $[H^+]$. This can adversely affect the ionic bonds of the macromolecules and thus can interfere with their three-dimensional structures and some functions. At pH < 5.0, the undissociated molecules of some acids can be considerably high. They, being lipophilic (except citric), enter freely through the membrane as a function of the concentration gradient. The pHi being much higher than the pK of the acid results in the dissociation of the molecules and the release of proton and the anions. Some anions (e.g., acetate and lactate) are utilized by several microorganisms as a C-source. If they are not metabolized, the anions are removed from the cell interior. However, the $[H^+]$ will reduce the internal pH and adversely affect the proton gradient. To overcome this problem, the cells pump out the protons by expending energy. At lower pHo (pH 4.5 or below), this represents a large amount of energy that cells may not be able to generate. As a result, the internal pH will drop, adversely affecting the pH gradient. The low pH can also act on the cellular components and adversely affect their structural (by interfering with the ionic bonds) and functional integrity.[1,3,4,7]

These changes can interfere with nutrient transport and energy generation, and in turn interfere with microbial growth. In addition, low pH can reversibly and irreversibly damage cellular macromolecules that subsequently can inflict sublethal injury as well as lethal injury to cells.

Low pH can alter the ionic environment of the spore coat by replacing its ions with H^+ and make the spores unstable toward other environmental stresses, such as heat and low A_w.[1-6]

IV. INFLUENCING FACTORS

A. Factors Associated with Acids

(See also Chapter 15.) The weak organic acids used in food vary in antimicrobial effectiveness due to their differences in pK. An acid with higher pK, with proportionately higher amounts of undissociated molecules at a food pH, will be more antimicrobial. Limited studies have revealed, that in general, under similar conditions the antimicrobial effectiveness of four acids follows the order: acetic > propionic > lactic > citric. Similarly, at lower pH and higher concentrations, an acid is more antimicrobial. The solubility of the acids is also important for the desirable effect. While acetate, propionate, lactate, and citrate are very soluble in water, benzoate (50 g%), sorbate (0.16 g%), and paraben (0.02 to 0.16 g%) and poorly soluble in water and thus, at the same concentration, will have different effectiveness. In many studies, the effectiveness of these acids against microorganisms is studied on a percent basis (g in 100 ml). However, they vary in molecular weight and thus, at the same concentration, they have different numbers of molecules and will produce different concentrations of undissociated molecules as well as dissociated ions. For comparison, it is better to use the acids on a molar concentration basis.

The organic acids also differ in their lipophilic properties, which in turn regulate their ease in entering inside the cells. Acetic and propionic acids are more lipophilic than lactic acid and have more antimicrobial effectiveness than lactic acid. Citrate is transported through the membrane by a specific transport mechanism (citrate permease) and is less effective than the lipophilic acids. Many microorganisms can metabolize the anions of weak acids, such as acetate, lactate, and citrate. Use of salts of these acids may have less antimicrobial effects at higher pH. Some acids show synergistic effects when used in suitable combinations (e.g., acetic and lactic acids, propionic and sorbic acids) or with another preservative (e.g., benzoic acid with nisin; propionic, acetic, or lactic acid with nisin or pediocin AcH; propionate, or benzoate with CO_2).[1,4,5,7]

B. Factors Associated with Foods

The normal pH of foods varies greatly from the very acid range (3.0; citrus juice) to the alkaline range (pH 9.0; egg albumen). The initial pH can strongly influence the antimicrobial effect of an acid. In a food at a lower pH, an acid will be more inhibitory than in a food at a higher pH. The buffering action of the food components also reduces the effectiveness of low pH. The nutrients can also facilitate repair of sublethal acid injury of microorganisms.

C. Factors Associated with Microorganisms

Microorganisms important in food vary greatly in the lower limit of pH that allows growth (Table 34.2). In general, Gram-negative bacteria are more sensitive to low pH than Gram-positive bacteria, and yeast and molds are the least sensitive. Fermentative bacteria are more resistant to lower pH than the respiring bacteria probably because they are able to resist changes in external pH as well as withstand slightly lower internal pH. The ability of yeasts and molds to withstand low pH is also due to these factors.[1,3,4,6]

The antimicrobial property of an organic acid is enhanced by heat, low A_w, the presence of some other preservatives, and low storage temperatures.

The inhibitory effect of weak acids is reduced at higher microbial load. Also, in a mixed microbial population, the metabolism of an acid (such as lactate) by one

Table 34.2 Minimum pH at which Growth Will Occur

Microorganism	Minimum growth pH	Microorganisms	Minimum growth pH
Gram-negative bacteria		**Yeasts**	
Escherichia coli	4.4	*Candida* spp.	1.5–2.3
Pseudomonas spp.	5.6	*Saccharomyces* spp.	2.1–2.4
Salmonella spp.	4.5	*Hansenula* spp.	2.1
Vibrio spp.	4.8	*Rhodotorula* spp.	1.5
Serratia spp.	4.4		
Gram-positive bacteria		**Molds**	
Bacillus cereus	4.9	*Aspergillus* spp.	1.6
Bac. stearothermophilus	5.2	*Penicillium* spp.	1.6–1.9
Clostridium botulinum	4.6	*Fusarium* spp.	1.8
Clo. perfringens	5.0		
Enterococcus faecalis	4.4		
Lactobacillus spp.	3.8		
Staphylococcus aureus	4.0		
Listeria monocytogenes	4.6		

resistant species can reduce its effective concentration against another sensitive species, allowing the latter to grow. Some microorganisms important in food, such as some *Salmonella* strains, seem to have genetic determinants that enable them to grow at higher acid concentrations (or lower pH) than other strains of the same species. The acid tolerance seems to be related to the overproduction of a group of proteins (stress proteins) by these strains.

Finally, microorganisms differ in their sensitivity to different organic acids. Yeasts and molds are particularly sensitive to propionic and sorbic acids, while bacteria are more sensitive to acetic acid. Bacterial spores at lower pH become susceptible to heat treatment, and do not germinate or grow at minimal A_w of growth. Also the inhibitory effect of NO_2 against spores is more pronounced at the lower pH range of growth.

V. ACIDS USED

A. Acetic Acid

Acetic acid is usually used as vinegar (5 to 10% acetic acid) or as salts of sodium and calcium at 25% or higher levels in pickles, salad dressings, and sauces. It is more effective against bacteria than against yeasts and molds. Those bacteria that grow better above pH 6.0 are more inhibited. The inhibitory concentrations of undissociated acid are 0.02% against *Salmonella* spp., 0.01% against *Sta. aureus*, 0.02% against *Bac. cereus*, 0.1% against *Aspergillus* spp., and 0.5% against *Saccharomyces* spp. The inhibitory action of acetic acid is produced through neutralizing the electrochemical gradient of the cell membrane as well as denaturing proteins inside the cells. Besides its use in food, acetic acid has been recommended for use (1 to 2% level) in carcass wash to reduce bacterial levels.[1,3,7]

B. Propionic Acid

Propionic acid is used as salts of calcium and sodium at a level of 1000 to 2000 ppm (0.1 to 0.2%) in bread, bakery products, cheeses, jam and jellies, and tomato puree. It is effective against molds and bacteria, and almost ineffective against yeasts at concentrations used in foods. The inhibitory concentration of undissociated acid is 0.05% against molds and bacteria. The antimicrobial action is produced through the acidification of cytoplasm as well as destabilization of membrane proton gradients.[1,3,7]

C. Lactic Acid

Lactic acid is used as acid or the sodium salt up to 2% in carbonated drinks, salad dressings, pickled vegetables, low heat-processed meat products, and sauces. It is less effective than acetic, propionic, benzoic, or sorbic acids, but more effective than citric acid. It is more effective against bacteria yet quite ineffective against yeasts and molds. It produces an inhibitory effect mainly by neutralizing the membrane proton gradient. The sodium salt of lactic acid may also reduce A_w. L(+)-lactic acid is preferred over D(–)-lactic acid as a food preservative. It has also been recommended at 1 to 2% levels to wash carcasses of food animals to reduce microbial load.[1,3,7]

D. Citric Acid

Citric acid is used at 1% level (or more) in nonalcoholic drinks, jams and jellies, baking products, cheeses, canned vegetables, and sauces. It is less effective than lactic acid against bacteria, as well as yeasts and molds. It produces an antibacterial effect probably by different mechanisms than the lipophilic acids. The antibacterial effect is partially due to its ability to chelate divalent cations. However, many foods have sufficient divalent cations to neutralize this effect.[3,7]

E. Sorbic Acid

Sorbic acid is an unsaturated acid and used either as acid or as salts of sodium, potassium, or calcium. Some of the foods in which it is used are nonalcoholic drinks, some alcoholic drinks, processed fruits and vegetables, dairy desserts, confectioneries, mayonnaise, salad dressings, spreads, and mustards. The concentrations used vary between 500 and 2000 ppm (0.05 to 0.2%). It is more effective against molds and yeasts than against bacteria. Among bacteria, catalase-negatives (e.g., lactic acid bacteria) are more resistant than catalase-positive species (e.g., aerobes, *Sta. aureus,* and *Bacillus* spp.). Also, aerobic bacteria are more sensitive than anaerobic bacteria. The inhibitory concentrations of dissociated acid are: $\leq$0.01% (100 ppm) for *Pseudomonas* spp., *Sta. aureus, Esc. coli,* and *Serratia* spp.; $\simeq$ 0.1% for *Lactoba-*

cillus spp. and *Salmonella* spp.; and ≤0.02% for most yeasts and molds, but for *Clostridium* spp., ≃ 1.0%.

The antimicrobial effect of sorbate is produced through its inhibitory action on some enzyme functions, some being from the citric acid cycle. It also interferes with synthesis of cell wall, protein, RNA, and DNA. In addition, like other organic acids, it also interferes with the membrane potential and inhibits spore germination.[3,5,8]

F. Benzoic Acid

Benzoic acid is used as an acid or a sodium salt at a concentration of 500 to 2000 ppm (0.05 to 0.2%) in many low pH products, such as nonalcoholic and alcoholic beverages, pickles, confectioneries, mayonnaise and salad dressings, mustards, and cottage cheese. It is more effective against yeasts and molds than against bacteria. The inhibitory effect is produced by both the undissociated and dissociated acids. The inhibitory concentrations of undissociated acid are 0.01 to 0.02% against bacteria, and 0.05 to 0.1% against yeasts and molds. The inhibitory action is produced in several ways. It is inhibitory to the functions of many enzymes necessary for oxidative phosphorylation. It also, like other acids, destroys the membrane potential. In addition, it inhibits functions of membrane proteins.[3,5,7]

G. Parabens (Esters of *p*-Hydroxybenzoic Acid)

Parabens are used as methyl, ethyl, butyl, or propyl parabens. They are broad-spectrum antimicrobial agents. Due to high pK values, they are effective at high pH, and against bacteria, yeasts, and molds. They are used at 100 to 1000 ppm (0.01 to 0.1%) in nonalcoholic and alcoholic beverages, fruit fillings, jams and jellies, pickles, confectioneries, salad dressings, spreads, and mustards. The undissociated inhibitory concentrations are 0.05 to 0.1% against yeasts, molds, and bacteria. Propyl- and butyl-parabens are more inhibitory.

Parabens produce antimicrobial action by acting on several targets in microbial cells. They may inhibit functions of several enzymes. They dissolve in membrane lipids and interfere with membrane functions, including transport of nutrients. They also interfere with the synthesis of proteins, RNA, and DNA. In addition, they destroy the membrane potential, similar to other weak organic acids.[3,5,7]

REFERENCES

1. Ray, B. and Sandine W. E., Acetic, propionic and lactic acid of starter culture bacteria as biopreservatives, in *Food Biopreservatives of Microbial Origin,* Ray, B. and Daeschel, M. A., Eds., CRC Press, Boca Raton, FL, 1992, 103.
2. Corlett, D. A. and Brown, M. H., pH and acidity, in *Microbial Ecology,* Vol. I., Silliker, J. H., Ed., Academic Press, New York, 1980, 92.
3. Baird-Parker, A. C., Organic acids, in *Microbial Ecology*, Vol. I., Silliker, J. H., Ed., Academic Press, New York, 1980, 126.

4. Booth, I. R. and Kroll, R. G., The preservation of foods by low pH, in *Mechanisms of Action of Food Preservation Procedures,* Gould, G. W., Ed., Elsevier Applied Science, New York, 1989, 119.
5. Eklund, T., Organic acids and esters, in *Mechanisms of Action of Food Preservation Procedures,* Gould, G. W., Ed., Elsevier Applied Science, New York, 1989, 181.
6. Brown, M. H. and Booth, I. R., Acidulants and low pH, in *Food Preservatives,* Russell, N. J. and Gould G. W., Eds., Von Nostrand Reinhold, New York, 1990, 22.
7. Doors, S. Organic acid, in *Antimicrobials in Foods,* Marcel Dekker, New York, 1993, 95.
8. Sofos, J. N. and Busta, F. F., Sorbic acid and sorbates, in *Antimicrobials in Foods,* Marcel Dekker, New York, 1993, 49.

CHAPTER 34 QUESTIONS

1. Discuss the mechanisms of antimicrobial action of weak organic acids and low pH. Explain the differences in sensitivity to low pH among Gram-negative and Gram-positive bacteria, bacterial spores, yeasts, and molds.

2. Define the pK of an organic acid and, using acetic and lactic acids as examples, describe their differences in antibacterial effect at pH 5 and 6.

3. Briefly discuss the influence of the following factors on antimicrobial effectiveness of low pH: acids, foods, and microorganisms.

4. List the specific acids to be used to inhibit growth of: Gram-negative bacteria, yeasts, and molds.

5. Discuss the antimicrobial properties of: acetic acid, propionic acid, lactic acid, benzoic acid, and parabens.

CHAPTER 35

Control by Modified Atmosphere (or Reducing O–R Potential)

CONTENTS

I. HISTORY

That many foods, when stored in air, were susceptible to quality loss was probably recognized by humans in the early ages of agriculture and animal husbandry. This probably led to the preservation of foods around 6000 B.C. by initially excluding air in pits and, later, with the invention of pottery and baskets, in large vessels. Storage of grains, semidry products, some fermented products, concentrated syrups and molasses, and similar products are still stored in large, closed, air-tight vessels in many parts of the world to prevent insect infestation and the growth of molds and yeasts. However, this method of changing the gaseous environment to preserve more perishable products, such as fresh meat, fish, fruits, and vegetables and other processed products, originated about 70 years ago. In the 1920s and 1930s, studies revealed that by using CO_2 in higher concentrations (4 to 100%), the growth of molds on fresh meats could be greatly reduced and the ripening of fruits and vegetables could be prolonged. In the 1960s refrigerated beef carcasses were transported by ship from New Zealand and Australia to other countries in controlled CO_2-rich environments.

In the 1960s, the availability of suitable plastic materials for packaging and the necessary technology helped to develop methods to preserve foods in different sized packages in altered atmospheres. This was achieved either by vacuum-packaging or by gas flushing the package with one gas or a mixture of gases. During storage, the gaseous environment could change, due to the metabolic activities of the food products and microorganisms and also to the permeability of gases from the air through the packaging materials.[1,2]

In recent years, modification of the storage environment to preserve foods has become one of the most used methods. The desire of consumers for fresh foods, foods that have been given less processing treatments and do not have undesirable preservatives, as well as the availability of the necessary technology, have helped in the economical production of many convenient and ready-to-eat foods that are preserved by the alteration of the environment. As the demand for such food is expected to grow in the future, the use of an altered atmosphere to preserve these foods will increase.[1,2]

Three terminologies are used in relation to the methods used to alter the atmosphere in foods in order to increase their acceptance quality. To avoid confusion, these terms are defined here.

1. Controlled atmosphere packaging (CAP): In this method, the atmosphere in a storage facility is altered and the levels of the gases are continually monitored and adjusted as required. This is expensive to operate and is used for long-term storage of fruits and vegetables to maintain their freshness.
2. Modified atmosphere packaging (MAP): This method, as in CAP, does not require a high degree of control of the gaseous environment during the entire storage period. In this method, a food is enclosed in a high gas-barrier packaging material, the air is removed from the package, which is then flushed with a particular gas or combination of gases, and the package is hermetically sealed.
3. Vacuum packaging involves removing of the air from the package and then sealing the package hermetically.

MAP and vacuum-packaging have been described as two separate methods by some, but as one method (i.e., MAP) by others. In this chapter, MAP has been used both for vacuum-packaging and gas-flush packaging.

II. OBJECTIVES

The objectives of MAP are to control or reduce the growth of undesirable microorganisms in food. The technique also helps to retard enzymatic and respiratory activities of fresh foods. The growth of aerobes (molds, yeasts, and aerobic bacteria) are prevented in products that are either vacuum-packaged or flushed with 100%

CO_2, 100% N_2, or a mixture of CO_2 and N_2. However, under these conditions, anaerobic and facultative anaerobic bacteria can grow unless other techniques are used to control their growth.

III. MECHANISM OF ANTIMICROBIAL ACTION

The antimicrobial action in MAP foods can be produced by the changes in redox potential (Eh) and CO_2 concentrations based on the methods used. Aerobes and anaerobes, depending upon the microbial species and reducing or oxidizing state of food, have different Eh requirements for growth, while facultative anaerobes grow over a wide Eh range. Vacuum-packaging and gas flushing, especially with CO_2 or N_2 or their mixture and no O_2, discourage growth of aerobes but encourage growth of facultative anaerobes and anaerobes. However, even under these conditions of packaging, tissue oxygen and dissolved and trapped oxygen can allow initial growth of aerobes to produce CO_2 even by the proteolytic microbes. In addition, natural reducing components in foods, such as –SH group in protein-rich foods, and ascorbic acid and reducing sugars in fruit and vegetable products, can alter the Eh of a food and encourage growth of anaerobes and facultative anaerobes. Thus just by changing the Eh it is not possible to control microbial growth. Other methods, in addition to modification of environment, are necessary for effective preservation of foods. However, by controlling the growth of aerobic bacteria, many of which have short generation times, the shelf life of the product is greatly extended.

When CO_2 is used in high concentrations (20 to 100%) alone or in combination with N_2 and/or O_2, the shelf life of MAP foods is also extended. Several mechanisms by which CO_2 increases the lag and exponential phases of microorganisms have been proposed. Some of these include rapid cellular penetration of CO_2 and alteration in cell permeability, solubilization of CO_2 to carbonic acid (H_2CO_3) in the cell with the reduction of the pH inside the cells and interference of CO_2 with several enzymatic and biochemical pathways, which in turn slow the microbial growth rate. The inhibitory effect of CO_2 on microbial growth occurs at the 10% level and increases as the concentrations increase. Too high a concentration can inhibit growth of facultative spoilage bacteria and stimulate growth of *Clostridium botulinum*.[2-6]

IV. INFLUENCING FACTORS

A. Factors Associated with the Process

These factors include efficiency of vacuum, permeability of packaging materials to O_2, and the composition of gas (in the gas flushing method) used.[2-6] High vacuum can effectively control the growth of aerobes by removing O_2 from the products, except probably trapped and dissolved O_2. Similarly, packaging films that prevent or considerably reduce permeation of O_2 during storage effectively control the growth of aerobes. Minute leaks in the film are found to adversely affect the protective effect of vacuum-packaging or gas-flush packaging. The development of polymeric film has helped to greatly reduce the O_2 permeability. In gas flushing, CO_2 and N_2, and in the case of fresh meat along with some O_2, are used in a mixture. N_2 is used as an inert filler and O_2 to give the red oxymyoglobin color in meat. CO_2 is used for its antimicrobial effect. As low as 20% CO_2 has been found to control the growth of aerobes such as *Pseudomonas, Acenatobacter,* and *Moraxella;* however, in general, CO_2 at 40 to 60% gives better results. In some cases, 100% CO_2 is used.

B. Factors Associated with Foods

The amounts of oxygen (dissolved and trapped), metabolizable carbohydrates, and other nutrients, and the reducing components present will influence the growth of microbial types in an MAP food. In the presence of O_2, aerobes such as *Pseudomonas* will utilize glucose and lactate. Growth of facultative anaerobes, however, will discourage the growth of aerobes but can stimulate the growth of anaerobes. Reducing components will also encourage the growth of anaerobes. The presence of specific inhibitors, either present in a food or produced by associated bacteria (such as by some lactic acid bacteria), low A_w, and low pH will also influence the microbial ability to grow or not grow under specific packaging conditions.

C. Factors Associated with Microorganisms

Aerobes have limited initial growth depending upon the oxygen present in the vacuum-packaged or gas-flushed packaged product.[2-6] However, their growth will stop as soon as the facultative anaerobes start growing. These include lactic acid bacteria, *Brochothrix thermosphacta*, some *Enterobacteriaceae,* and *Corynebacteriaceae*. In low pH products, growth of lactic acid bacteria can reduce or prevent growth of other bacteria due to their ability to produce acids as well as other antibacterial substances (e.g., bacteriocins). Also, the composition of the gas mixture influences the predominant types; with 100% CO_2, lactic acid bacteria, especially *Leuconostoc* and *Lactobacillus* spp., predominate. CO_2 at about 20% concentration controls growth of *Pseudomonas* spp., but $CO_2 > 60\%$ is required to control the growth of *Enterobacteriaceae*. As the Eh of a product starts reducing, the anaerobes, especially *Clostridium* spp., start growing. The presence of reducing agents in the foods also favors growth of anaerobes. The possible growth of *Clo. botulinum* in MAP refrigerated foods, especially type E and nonproteolytic type B, is a major concern. Nonsporeforming psychrotrophic pathogens, especially the facultative anaerobes (*Listeria monocytogenes* and *Yersinia enterocolitica*) can also multiply in the MAP foods. In addition, some mesophilic facultative anaerobic pathogens with growth capabilities at 10 to 12°C (some *Salmonella* strains and *Staphylococcus aureus*) can also grow if the MAP foods are temperature abused during storage. Suitable preservatives, low pH, and/or low A_w should be used as additional hurdles for control.[2-6]

V. METHODS USED IN FOODS

A. Vacuum Packaging

Vacuum-packaging is predominantly used in retail packs in many fresh and ready-to-eat meat products that include beef, pork, lamb, chicken, and turkey. Red meat, due to change in color to purple (reduced myoglobin), is not very popular with consumers . The refrigerated storage life of these products varies greatly: in fresh meat, about 3 to 4 weeks; while in processed meats, as long as 8 weeks or more. If the products have low A_w and/or low pH, are produced under sanitary conditions, and are kept at temperatures around 1.5°C, the storage life can be much longer. In low pH products, *Leuconostoc* spp. and *Lactobacillus* spp. usually predominate; while in high pH products, *Brochothrix thermosphacta, Serratia liquifaciens,* and *Hafnia* spp. are isolated, along with lactic acid bacteria. In fresh beef and pork and roasted beef, psychrotrophic *Clostridium* spp. have also been found to grow, even at temperatures below –1°C. Vacuum packaging is also used in different types of cheeses, sausages, and low pH condiments to control the growth of yeasts and molds.[2-6]

B. Gas Flushing

Gas flushing method has been used in both bulk and retail packs to increase the shelf life of many refrigerated foods in European countries. In the United States, it is being used in products such as fresh pasta, bakery products, cooked poultry products, cooked egg products, fresh and cooked fish and seafoods, sandwiches, raw meats, and some vegetables. The gases usually used are a CO_2 and N_2 mixture, with some O_2 for packaging red meats. Generally, the composition of gas mixtures must be tailored for each product. In raw meats, a composition of 75% CO_2, 15% N_2, and 10% O_2 was found to effectively prevent growth of *Pseudomonas fragi*. Products flushed with CO_2 alone were found to be dominated by lactic acid bacteria, principally *Leuconostoc* spp. and *Lactobacillus* spp. When N_2 and CO_2 are used together, along with lactic acid bacteria, *Brochothrix thermosphacta*, some coryneforms, and *Enterobacteriaceae* were also found. In the presence of some O_2, lactic acid bacteria, *Bro. thermosphacta, Enterobacteriaceae,* and *Pseudomonas* spp. were isolated. The product storage life, depending upon a product, can be 4 weeks or more for fresh products and 8 weeks for processed products.[2-6]

REFERENCES

1. Ooraikul, B. and Stiles, M. E., *Modified Atmosphere Packaging of Food,* Ellis Harwood, New York, 1991.
2. Daniels, J. A., Krishnamurthi, R., and Rizvi, S. S. H., A review of the effects of carbon dioxide on microbial growth and food quality, *J. Food Prot.,* 46, 532, 1985.

3. Anonymous, Is current modified/controlled atmosphere packaging technology applicable to the U.S. food market?, IFT Symposium, *Food Technol.,* 42(9), 54, 1988.
4. Farber, J. M., Microbiological aspects of modified atmosphere packaging technology — a review, *J. Food Prot.,* 54, 58, 1991.
5. Brown, M. H. and Emberger, O., Oxidation-reduction potential, in *Microbial Ecology of Foods*. Vol. I., Silliker, J. H., Ed., Academic Press, New York, 1980, 112.
6. Gill, C. D. and Molin, G., Modified atmosphere and vacuum-packaging, in *Food Preservatives,* Russell, N. J. and Gould, G. W., Eds., Van Nostrand Reinhold, New York, 1990, 172.

CHAPTER 35 QUESTIONS

1. Describe the methods used for the alteration of atmosphere to preserve foods.
2. Discuss the mechanisms of antimicrobial action in the modified atmosphere packaging (MAP) method used in foods.
3. Explain the different factors that can inhibit or influence microbial growth in an MAP food.
4. List the predominant microorganisms that can grow during the storage of vacuum-packaged and gas-flushed packaged foods.
5. What is the major microbiological concern of MAP foods? How can this be solved?

CHAPTER 36

Control by Irradiation

CONTENTS

I. HISTORY

A. Irradiation (Radiation) and Radioactivity

In the electromagnetic spectrum, energy exists as waves and the intensity of the energy increases as the waves get shorter. On either side of visible rays (~400 to 800 nm) are invisible long waves (>800 nm; IR and radio waves for radio, TV, microwave, and radar) and invisible short waves (<300 nm; UV rays, X-rays, β-rays, γ-rays, and cosmic rays). The long waves, visible light waves, and UV rays do not cause any change in atomic structures. In contrast, X-rays, β-rays, and γ-rays can remove electrons from the outer shell of an atom and thus form an ion pair (negatively charged and positively charged). Ion formation or ionization does not make an atom radioactive. For radioactivity, the nucleus of an atom has to be disrupted by much higher energies, such as by neutrons; X-rays, β-rays, γ-rays do not have that much energy.

For application in food preservation, X-rays, β-rays, and γ-rays were studied for their ability to penetrate inside foods and kill microorganisms, their efficiency, and their effect on food quality. β-rays (actually electrons, similar to cathode rays) have very little penetration power; they cannot penetrate inside metal cans and are thought to be ineffective in food preservation. X-rays, although they have good penetration power, cannot be effectively focused on foods and, due to this low efficiency, are not considered for application in food. In contrast, γ-rays (photons) have high penetration power (~40 cm thick) and may be considered effective and economical for use in foods. Cobalt-60 (^{60}Co) and cesium-137 (^{137}Cs) are considered to be good sources of γ-rays. ^{60}Co, an artificially induced radioactive isotope, is produced for use in nuclear medicine; and when its energy level becomes so low that it cannot be further used in medicine, it is used for irradiation of foods. Due to the easy availability of ^{60}Co, over ^{137}Cs, it is preferred for food irradiation. Thus foods irradiated with ^{60}Co do not become radioactive. Instead, the atoms (and molecules) in a food and in microbial cells form ion pairs and other components that inhibit the multiplication of microorganisms (and food cells, such as the inhibition of sprouting in potatoes).[1]

B. Use of Irradiation in Food

The ability of X-rays, γ-rays, and β-rays to kill microorganisms was recognized soon after their discovery in the late 19th and early 20th centuries. Some were even tested to determine their effectiveness in killing microorganisms in food in those early years. However, until the late 1940s and early 1950s their use in food preservation was not actively studied. One of the reasons for this was the unavailability of a large economical supply of radioisotopes, the technology of which was developed during World War II. Early studies showed that different foods can be irradiated to extend shelf life without making them radioactive, but the process affected the flavor qualities of the foods. Later these problems were overcome by reducing exposure time and lowering product temperature. At the time, it was thought that irradiation of food would give storage stability for indefinite periods of time. The U.S. Army started a research program to determine if storage-stable, acceptable, safe army rations could be produced by irradiation. Although foods do not become radioactive at low doses of irradiation, it was found that food accumulated radiolytic products. Extensive studies showed that they are present in very low concentrations (~3 ppm). Most of them are not unique as they are present naturally in different foods and also are produced in heated foods (thermolytic products). Well-designed feeding studies, with animals as well as humans for a fairly long period, showed that irradiated foods did not cause any toxic effects or genetic defects. An expert committee on the wholesomeness of irradiated foods of the World Health Organization, after extensive review of more than 200 well-designed studies conducted throughout the world, has recommended irradiation of food up to a certain dose level (10 kilogray). This level is currently approved in over 30 countries for use in meat, fish, vegetables, fruits, and grains. Many countries are regularly selling irradiated foods, especially some vegetables, fruits, and grains. However, consumer resistance has been quite high. This is mainly due to the lack of information and understanding that irradiated foods are not radioactive and that radiolytic products are not unique to only irradiated foods. This is probably the only food preservation method that has been studied for safety for a long time (over 40 years) prior to its use in many countries and has been found to be safe by the world body of expert scientists. The other reason is just political. A small, but vocal group of lobbyists are waging a very successful campaign against the acceptance of irradiated food by large numbers of consumers. As a result, food industries are in a wait-and-watch situation.

There is no controversy among the experts that food irradiation is an economical and effective food preservation method. As the world population gets larger and food production gets lower, effective preservation methods must be used to feed the hungry mouths; otherwise, social unrest may start. Politics and lobbying will not stop hunger, but food preservation methods (like irradiation) will.

II. OBJECTIVES

A food is irradiated because of the destructive power of the microorganisms it harbors. Depending on the method used, it can either completely or partially destroy molds, yeasts, bacterial cells and spores, and viruses. In addition, irradiation is capable of destroying worms, insects, and larvae in food. It also prevents sprouting of some foods such as potatoes and onions. However, irradiation cannot destroy toxins or undesirable enzymes in a food, and in that respect it differs from heat treatment (heat also does not destroy heat-stable toxins and enzymes). Irradiation is a cold sterilization process inasmuch as the temperature of a food does not increase during treatment and thus the irradiated foods do not show some of the damaging effects of heat on food quality. However, irradiation can cause oxidation of lipids and denature food proteins, especially when used at higher doses.[1-3]

III. MECHANISMS

When an object (food or microorganism) is exposed to high-energy γ-rays (10^{-1} to 10^{-2} nm), the energy is absorbed by thousands of atoms and molecules in a fraction of a second, which strips electrons from them. This produces negative and positive ion pairs. The released electrons can be highly energized and thus can remove electrons from other atoms and convert them into ions. This energization and ionization can adversely affect the normal characteristics of biological systems.

The ionizing radiation produces both direct and indirect effects on microorganisms. The direct effect is produced from the removal of electrons from the DNA, thereby inducing damage to these molecules. The direct effect is produced from the ionization of water molecules present in the cell. The hydrogen and hydroxyl radicals formed in this process are highly reactive and cause oxidation, reduction, as well as the breakdown of carbon-to-carbon bonds of other molecules, including DNA. Studies have shown that both single- and double-strand breaks in DNA at the sugar-phosphate backbone can be produced by the hydroxyl radical. In addition, the radicals can change the bases, such as thymine to dihydroxydihydrothymine. The consequence of these damages is the inability of microorganisms to replicate DNA and reproduce, resulting in death.

In addition to DNA damage, ionizing radiation has also been shown to cause damage in the membrane and other structures, causing sublethal injury. Some microorganisms are capable of repairing damage to the DNA strands, especially the single-strand breaks, and in the bases, and are designated as radiation-resistant microorganisms.

Microbial death by ionizing radiation, as in heat treatment, occurs at a predictable rate, which, like heat, is dependent on dose (strength and time), microbial species, and environmental factors. Because of this, the D value (minutes to reduce 1 log of species population at a given exposure) can be derived. This in turn can be used to determine the time necessary to reduce the population to a desirable level under a specific condition of treatment.

When microorganisms are exposed to UV radiation ($\simeq$260 nm), the energy is absorbed by the nucleotide bases in the DNA. The bases can react with each other to form dimers (e.g., thymine dimers) and cause DNA strand breaks. Microbial death and injury are mainly associated with DNA damage.[1-3]

IV. INFLUENCING FACTORS

A. The Process Itself

As mentioned before, among the several methods available (X-rays, β-rays, γ-rays), γ-rays have the potential for effective and economical use in food preservation. Cobalt-60 is predominantly used in food irradiation because it is more readily and economically available. It has a half-life of ~5.3 years. It continuously emits γ-rays, and thus energy can be lost even when it is not used. Cesium-137 can also be used in foods, but it is relatively difficult to obtain and is also required in larger amounts. The antimicrobial efficiency of ionizing radiation increases as the dose is increased. The antimicrobial efficiency, however, decreases in the absence of oxygen (due to reduced oxidizing reactions) and at low A_w (due to reduced free radical formation with less water). Freezing also reduces the efficiency due to the reduced availability of reactive water molecules.[1-3]

B. Factors Related to Foods

γ–Rays have a penetration capability of about 40 cm and can penetrate through paper, plastic, and cans. Thus foods can be exposed to γ-radiation in packages, cans, baskets, and bags. As indicated above, frozen, dry, or anaerobically packaged foods will need higher doses of treatment to obtain the desirable antimicrobial effect. In contrast, treatments such as curing, high hydrostatic pressure, higher temperature, and lower pH enhance the antimicrobial effect of radiation in food. Food compositions (thickness and particle size) also determine the efficiency of irradiation in reducing microbial numbers.

C. Microbial Factors

Microorganisms vary greatly in their sensitivity to ionizing (and UV) radiation. Due to size differences, molds are more sensitive than yeasts, which are more sensitive than bacterial cells; bacterial cells are more sensitive than viruses (including phages). Among bacteria, Gram-negative cells are more sensitive than Gram-positive bacteria, and rods are more sensitive than cocci. Species and strains of bacterial cells vary greatly in their sensitivity to irradiation. Some strains, designated as radiation resistant, have effective metabolic systems to repair the cellular damages (especially single- and double-strand breaks of DNA and base damage). These include some bacterial strains that are important in foods, such as *Salmonella typhimurium, Escherichia coli, Enterococcus faecalis, Staphylococcus aureus,* and others. Spores are quite resistant to irradiation, probably due to the content of very little water. Among the sporeformers, spores of *Clostridium botulinum* type A and *Bacillus pumilus* are probably the most resistant to irradiation. Generally, *Bacillus* spores (aerobes) are less resistant than *Clostridium* spores (anaerobes).

The rate of death of microorganisms by irradiation follows similar straight line patterns as the thermal destruction curve and thus can be expressed as a D value (time necessary to destroy 90% of viable microorganisms). Accordingly, it can be influenced by a higher initial population, relative numbers of resistant cells in the population, number of spores present, and age and condition of growth of a strain. Toxins of microorganisms are not destroyed by ionizing radiation at the dose levels recommended in foods. Although irradiation can cause mutations in some microbial cells in a population, either a possible increase in pathogenicity or an induction of a gene that transcribes for a toxin is not expected to occur, as was observed in the many studies conducted in these areas.

V. METHODS USED

A. Doses

Radiation dose was originally designated as "rad," and 1 rad was defined as the quantity of ionizing radiation that results in the absorption of 100 ergs energy per gram irradiated material. The current unit is Gray (Gy), and 1 Gy is equivalent to 100 rad. When 1 kg food absorbs the energy of 1 joule (1 joule = 10^7 ergs), it has received a dose of 1 Gy. According to international health and safety authorities, foods irradiated up to 10,000 Gy (10 kGy) are considered safe.

The relative sensitivity of microorganisms to irradiation dose is a function of their size and water content. Approximate lethal dose levels for insects and different microorganisms have been suggested as follows: insects, ≤1 kGy; molds, yeasts, bacterial cells, 0.5 to 10 kGy; bacterial spores, 10 to 50 kGy; viruses, 10 to 200 kGy. Thus, at the recommended level of 10 kGy, *Clostridium botulinum* spores are not destroyed in foods (need about 30 to 60 kGy), although cells of pathogenic (and spoilage) bacteria are destroyed. The products thus treated should have other barriers (low pH, low A_w, temperature ≤4°C) to control germination and growth of sporeformers. But some sporeformers (such as *Clo. laramie*) can germinate and multiply at <4°C. At present a low dose level (<1 kGy) is used to control insects in fruits and grains, parasites in meat and fish, and sprouting in vegetables. Medium doses (1 to 10 kGy) are used to control foodborne pathogens and spoilage microorganisms to extend the safety and stability of refrigerated foods. Higher doses to destroy spores (above 10 kGy) are not used in foods except in spices and vegetable seasonings used in very small quantities in foods.

Irradiated foods, like postheat contamination in heated foods, can be contaminated with pathogenic and spoilage microorganisms from various sources unless proper precautions (such as packaging) are used.[1-4]

B. Specific Terms

1. Radurization

Radiation pasteurization is intended to destroy spoilage bacteria in high pH-high A_w foods, especially Gram-negative psychrotrophs in meat and fish, and yeasts and molds in low pH-low A_w foods. The treatment is generally milder ($\simeq$1 kGy). The products should be packaged and chilled to prevent growth of pathogens, which were previously thought to be mesophiles. However, with the recognition of psychrotrophic pathogens and the importance of psychrotrophic Gram-positive spoilage bacteria, this treatment may not be effective.

2. Radicidation

Radiation of foods to destroy vegetative foodborne pathogens. The dose level used is about 2.5 to 5.0 kGy. Although it is effective against pathogenic vegetative bacterial cells and molds, spores of the pathogens will not be destroyed. Also, some radiation-resistant strains of pathogens can survive, such as some *Salmonella typhimurium* strains. The irradiated products thus need to be stored at ≤4°C, especially to prevent germination and outgrowth of spores of *Clo. botulinum*.

3. *Radappertization*

This method involves the radiation of food at high doses (≈30 kGy) to destroy *Clo. botulinum* spores in order to get safety similar to 12D heat treatment. However, this is not recommended for use in food.

C. Current Recommendations

Irradiated fresh fruits, vegetables, meat, and fish have been approved in 37 countries and are used in many countries, such as France, Belgium, and the Netherlands. In the United States, irradiated foods have been used by the Army and in the space program for some time. Prior to 1985, irradiation was permitted by the regulatory agencies for spices, wheat, wheat flour (to destroy insects), and potatoes (to prevent sprouting). Subsequently, permission for irradiation was approved in 1985 for pork (against trichina), in 1986 for fresh foods (fruits and vegetables to destroy insects and larvae), and in 1992 for poultry and poultry parts (to destroy pathogens, specifically *Salmonella*). Currently, radiation of seafood (to destroy pathogens and spoilage bacteria) is being considered. In addition, in the United States radiation sterilization of beef steaks in the space program and refrigerated shelf-stable food for the military is being studied.

Currently, irradiated fruits (e.g., strawberry, mango, and papaya) and poultry are being marketed in limited amounts, and consumer response has been favorable. Irradiated foods will have a special logo (Figure 36.1) along with the words "Treated with Radiation" or "Treated by Irradiation" for consumer information.

Figure 36.1 Logo for irradiated foods. The Codex Alimentarius, an international committee on food safety, has developed this logo to put in green color on the package of irradiated foods. Irradiated foods from the United States will be labeled with this logo, along with the words "Treated with Radiation" or "Treated by Irradiation."

D. UV Radiation

Microorganisms are especially susceptible to UV light between 200 and 280 nm. Due to low penetration power, it has been used to inactivate microorganisms on the surface of foods (meat, fish, and bread), as well as in air and on walls, shelves, and equipment in the food handling and processing area. In addition, liquids, such as water and syrups, in thin layers have been treated with UV irradiation.

REFERENCES

1. Urbain, W. M., *Food Irradiation,* Academic Press, New York, 1986.
2. Ingram, M. and Roberts, T. A., Ionizing irradiation, in *Microbial Ecology,* Vol. I., Silliker, J. H., Ed., Academic Press, New York, 1980, 46.
3. Moseley, B. E. B., Ionizing radiation: action and repair, in *Mechanisms of Action of Food Preservation Process,* Gould, G. W., Ed., Elsevier Applied Science, New York, 1989, 43.
4. Anonymous, *Facts About Food Irradiation,* International Atomic Energy Agency, Vienna, Austria, 1991.

CHAPTER 36 QUESTIONS

1. Briefly discuss the reasons why irradiated foods are not commercially successful in many countries.

2. Explain the mechanisms of antimicrobial actions of irradiation by γ-ray and UV light.

3. List the factors that can reduce or increase the antimicrobial effectiveness of irradiation in food.

4. What recommended levels of γ-irradiation are approved in different foods and what are their objectives? What is the status of handling spore problems in irradiated foods?

5. Briefly explain the importance of the following in relation to food irradiation:

(a) Irradiated products
(b) Radioactive products
(c) Radiolytic products
(d) Thermolytic products
(e) Free radicals
(f) Radurization
(g) Radicidation
(h) Radappartization

CHAPTER 37

Control by Antimicrobial Preservatives

CONTENTS

I. INTRODUCTION

Many chemical compounds, either present naturally or formed during processing or legally added as ingredients, are capable of killing microorganisms or controlling their growth in foods. They are, as a group, designated as antimicrobial inhibitors or preservatives.[1,2] Some of the naturally occurring preservatives can be present in sufficient amounts in foods to produce antimicrobial action, such as lysozyme in egg white, and organic acids in citrus fruits. Similarly, some of the antimicrobials can be formed in enough quantities during food processing to control undesirable microbial growth, such as lactic acid in yogurt fermentation. However, some others can be naturally present or formed in small quantities and they essentially do not produce antimicrobial action in foods. Examples of these are lysozyme in milk (13 μg/100 ml) and diacetyl in some fermented dairy products. Among the many food additives, some are specifically used to preserve foods against microorganisms (such as NO_2 in cured meat to control *Clostridium botulinum*), while others are added principally to improve the functional properties of a food (such as butylhydroxyanisol, BHT, used as an antioxidant although it has antimicrobial properties). Use of some of the preservatives started as far back as 6000 B.C. Salting (which had nitrate as contaminant) of fish and meat, burning sulfur to generate SO_2 to sanitize the environment and equipment used for baking and brewing, smoking of fish and meats, adding spices, herbs, acids, and alcohol to foods (from fermentation) are some of the examples.

Following the discovery of the microbial role in the spoilage of food and in foodborne diseases, many chemicals were introduced since the late 1900s to control microorganisms in foods. The safety of the chemicals to humans was not tested prior to addition to foods, and some of the chemicals used were later found to be harmful to humans. To protect consumers, in 1958 in the United States, the Food Additive Amendment was passed (see Appendix E). According to this amendment, a food processor who wants to use a food preservative as well as other additives must prove its safety, by the procedures recommended, before its incorporation. This law, however, granted an exemption to substances used for a long time, found to be safe, and considered as GRAS (generally regarded as safe) substances.

In recent years, the possible effect of different preservatives and other additives on human health from long-term use through different foods has been questioned. Many of them are of nonfood origin or added at a level not normally present in foods, and some are synthetic. The effect of interactions of the different preservatives in the body consumed through different foods, especially in children, debilitated and elderly people and the possible cumulative effect from the consumption for many years have not been determined. Many health-conscious consumers are interested in foods that do not contain any preservative, especially those that are not normally found in foods. This has resulted in the search for preservatives that are either naturally present in the food of plant and animal origin or produced by safe food-grade microorganisms used to produce fermented foods. They are also designated as biopreservatives. Some of them have been used, not always as preservatives, for a long time in foods and have been found to be safe (such as lactic, acetic, and propionic acids), while others have not been used directly as preservatives. Although many of them have been unknowingly consumed through foods without any adverse effect (such as bacteriocins of many lactic acid bacteria), they probably have to be tested for safety and regulatory approval prior to use in foods.

II. OBJECTIVES

Antimicrobial chemicals are used in food in relatively small doses, either to kill undesirable microorganisms or to prevent or retard their growth in food. They differ greatly in their ability to act against different microorganisms. While some are effective against many microorganisms, others are specifically effective against either molds and yeasts or only bacteria. Similarly, there are compounds that are effective against either Gram-positive or Gram-negative bacteria, or bacterial spores, or viruses. Those that are capable of killing microorganisms are designated as germicides (kill all types), fungicides, bactericides, sporicides, and viricides, depending upon their specificity of killing actions against specific groups. Those that inhibit or retard microbial growth are classified as fungistatic or bacteriostatic. However, under the conditions in which antimicrobials are used in foods, they cannot completely either kill all the microorganisms or prevent their growth for a long time during storage.

III. FACTORS TO CONSIDER

There are several factors that need to be considered in evaluating the suitability of an antimicrobial agent as a food preservative.[1-4] These are based on their antimicrobial properties, suitability for application in a food, and ability to meet the regulatory requirements. For the antimicrobial properties, a compound that kills (-cidal) instead of controlling growth (-static) is preferable. Similarly, a compound with a broader antimicrobial spectrum, so that it is effective against many types of microorganisms important in foods (namely, molds, yeasts, bacteria, and viruses) as compared to one that has a narrow spectrum, will be more suitable for application in foods. Also, a compound that is effective not only against vegetative cells, but also against spores as well is preferred. Finally, it should not allow development of resistant strains. Most compounds do not meet all these requirements. Many times, more than one compound is used in combination to increase the inhibitory spectrum. In addition, food environments may restrict the growth of many types of microorganisms. Under such circumstances, a preservative that can effectively control the growth of the microorganism(s) of concern can be used alone.

To be suitable for application in a food, a compound should not only have the desired antimicrobial property, but it also should not affect the normal quality of a food (texture, flavor, or color). It should not interact with food constituents and become inactive. It should have a high antimicrobial property at the pH, A_w, Eh, and storage temperature of the food. It should be stable during the storage life of the food. Finally, it should be economical and readily available.

The regulatory requirements include the expected effectiveness of an antimicrobial agent in a food system. It should be effective in small concentrations and should not hide any fault of a food (i.e., conceals poor quality and spoilage). Most importantly, it should be safe for human consumption. Finally, when required, it should be listed on the label and indicate its purpose in the food.

IV. EXAMPLES OF ANTIMICROBIAL PRESERVATIVES

Foods can contain antimicrobial compounds in three ways: present naturally, formed during processing, or added as ingredients. Those that are added have to be GRAS-listed and approved by the regulatory agencies. Some of these are added specifically as antimicrobial preservatives, while others, although they have antimicrobial properties, are added for different reasons. Those in the latter group are also called indirect antimicrobials. In this chapter, the antimicrobial effectiveness, mode of action, and uses in foods of some inorganic and organic antimicrobials added to foods are briefly discussed.[1-6] A list of antimicrobial preservatives (direct or indirect) used in foods is presented in Table 37.1.[1-6]

Table 37.1 Some Antimicrobial Chemical Preservatives Used in Foods

Acetaldehyde	Dehydroacetate	Lactic acid	Propylene glycol
Acetic acid	Diacetate	Lauric acid	Propylene oxide
Ascorbic acid	Diacetyl	Lysozyme	Propyl gallate
Bacteriocins	Diethyl dicarbonate	Malic acid	Smoke
Benomyl	Diphenyl	Methyl bromide	Sodium chloride
Benzoic acid	Ethyl alcohol	Monolaurin	Sorbic acid
Betapropiolactum	Ethyl formate	Natamycin	Spices
BHA, BHT, and TBHQ	EDTA	Nitrite and nitrate	Succinic acid
Boric acid	Ethylene oxide	Parabens	Sucrose
Caprylic acid	H_2O_2	Peracetate	Sulfites and SO_2
Chitosan		Polyphosphates	Tetracyclines
Citric acid		Propionic acid	Thiabendazole
CO_2 and CO			Tylosin

Note: Not all are permitted in the United States.

A. Nitrite ($NaNO_2$ and KNO_2)

Curing agents that contain nitrite, and together with NaCl, sugar, spices, ascorbate, and erythorbate, are permitted for use in heat-processed meat, poultry, and fish products to control growth and toxin production by *Clostridium botulinum*. Nitrate and nitrite are also used in several European countries in some cheeses to prevent gas blowing by *Clo. butyricum* and *Clo. tyrobutyricum*. The mechanisms of antibacterial action of nitrite is not properly understood, but there is indication that the inhibitory effect is produced in several ways. These include reactions with some enzymes in the vegetative cells and germinating spores, restriction of the bacterial use of iron, and interference with membrane permeability limiting transport. In addition to clostridial species, nitrite is inhibitory to some extent to *Staphylococcus aureus, Escherichia, Pseudomonas,* and *Enterobacter* spp. at 200 ppm; *Lactobacillus* and *Salmonella* spp. seem to be resistant to this concentration of NO_2.

The antibacterial effect of NO_2 is enhanced at lower pH (5.0 to 6.0), in the presence of reducing agents (e.g., ascorbate, erythorbate, and cysteine) and with sorbate. The current regulatory limit in the United States is 156 ppm of NO_2, but varies widely in other countries. This amount can also be reduced by supplementing NO_2 with other reducing agents, as well as sorbates. The NO_2 effect is also enhanced by reducing A_w and at low Eh. In cured meat products, NO_2 reacts with myoglobin to form a stable pink color of nitrosyl hemochrome during heating. In bacon, nitrite can lead to the formation of carcinogenic compounds, nitrosoamines. Because of this, there is a trend to reduce NO_2 or to use other preservatives to control *Clo. botulinum* in low-heat-processed meat products.

B. Sulfur Dioxide (SO_2) and Sulfites (SO_3)

Sulfur dioxide, sodium sulfite (Na_2SO_3), sodium bisulfide ($NaHSO_3$), and sodium metabisulfite ($Na_2S_2O_5$) are used to control microorganisms (and insects) in soft fruits, fruit juices, lemon juices, beverages, wines, sausages, pickles, and fresh shrimp.

Currently, these additives are not permitted in the United States in meat, as they destroy vitamin B1. They are more effective against molds and yeasts than bacteria; and among bacteria, the aerobic, Gram-negative rods are the most susceptible. The antimicrobial action is produced by the undissociated sulfurous acid that rapidly enters inside the cell and reacts with thiol groups in structural proteins, enzymes and cofactors, as well as with other cellular components. At low pH (≤4.5) and low A_w, the fungicidal effect is more pronounced. In bacteria, they are effective at high pH (≥5.0) but probably are bacteriostatic at lower concentrations and bactericidal at higher concentrations. The concentrations used in foods vary greatly in different countries. In the United States, 200 to 300 ppm is generally permitted for antimicrobial use.

Sulfur dioxide and sulfites are also used as antioxidants in fresh and dried fruits and vegetables (salads) to prevent browning. However, people with respiratory problems can be mildly to severely allergic. The products need to be labeled to show the presence of sulfites.

C. H_2O_2

(See also Chapters 15 and 29) Solutions of H_2O_2 (0.05 to 0.1%) are recommended as antimicrobial agents in raw milk to be used in cheese processing, liquid egg to facilitate destruction of *Salmonella* by low heat pasteurization, packaging material used in aseptic packaging of foods, and food processing equipment. In raw milk and liquid egg, catalase is used prior to pasteurization to hydrolyze H_2O_2 to water and oxygen. It is a strong oxidizing agent and the germicidal action is associated with this property.

Recently, H_2O_2 has been used to produce modified plant fiber flour from straws for use in low-calorie foods and for bleaching and color improvement of grains, chocolate, instant tea, fish, sausage casings, and many others, and to reduce sulfite in wines. In the future, the use of H_2O_2 in foods is expected to increase.

D. Epoxides (Ethylene Oxide and Propylene Oxide)

Ethylene oxide and propyylene oxide are used as fumigants to destroy microorganisms (and insects) in grains, cocoa powder, gums, nuts, dried fruits, spices, and packaging materials. They are germicidal and effective against cells, spores, and viruses. Ethylene oxide is more effective. They are alkylating agents and react with various groups (e.g., –SH, $-NH_2$, and –OH) in cellular macromolecules, particularly structural proteins and enzymes, and adversely affect their functions. They can react with some food components, such as chlorides, and form toxic compounds that can remain as residue in treated foods. They can be toxic at high concentrations (as residue), particularly to people who are sensitive to them.

E. Acids

Acetic, propionic, lactic, citric, benzoic, and sorbic acids are discussed in Chapters 15 and 34.

F. Parabens

Methyl and propyl esters of *p*-hydroxybenzoic acid are discussed in Chapter 34.

G. Bacteriocins of Lactic Acid Bacteria

Nisin and pediocins are discussed in Chapter 15.

H. Diacetyl

Diacetyl is discussed in Chapter 15.

I. CO_2

CO_2 is discussed in Chapter 35.

J. Butylated Hydroxyanisol (BHA), Butylated Hydroxytoluene (BHT), and t-Butyl Hydroquinone (TBHQ)

BHA, BHT, and TBHQ are primarily used at 200 ppm or less as antioxidants to delay oxidation of unsaturated lipids. Additionally, they have antimicrobial properties and thus can be regarded as indirect antimicrobials. In concentrations ranging from 50 to 400 ppm, BHA inhibits growth of many Gram-positive and Gram-negative bacteria; however, some species may be resistant to it. They also effectively prevent growth and toxin production by molds and growth of yeasts, but BHA seems to be more effective. The antimicrobial action is most likely produced by their adverse effect on the cell membrane and enzymes. Their antimicrobial effectiveness increases in the presence of sorbate, but decreases in foods with high lipids and at low temperature.

K. Chitosan

Chitosan, a polycationic polymer, is obtained by alkaline hydrolysis of chitin from the shells of *Crustaceae*. It has many applications in foods, including food preservation due to its antimicrobial capability. It causes destabilization of the cell wall and cell membrane functions and is effective against bacteria, yeasts, and molds.

L. Ethylenediaminetetraacetate (EDTA)

The sodium and calcium salts of EDTA at 100 ppm are approved for use in foods to chelate trace metals in order to prevent their adverse effect on food quality. At low doses (5000 ppm), EDTA appears to have no toxic effect and mostly passes through the GI tract unabsorbed. By itself, EDTA may not have much antimicrobial effect, but due to its ability to chelate divalent cations, it can destabilize the barrier functions of the outer membrane of the Gram-negative bacteria and, to some extent, the cell wall of Gram-positive bacteria. In this way, it enhances the antibacterial action of other chemicals, especially those that are membrane acting, such as surface-active compounds, antioxidants, lysozymes, and bacteriocins. EDTA is also inhibitory for germination and outgrowth of spores of *Clo. botulinum*. In the presence of divalent cations in the environment (e.g., dairy products), the effectiveness of EDTA is greatly reduced.

M. Lysozyme

The enzyme lysozyme (a muramidase) is present in large quantities in some foods such as egg white and shellfish (oysters and clams), and in small amounts in milk and some plant tissues. It hydrolyzes the mucopeptide layer present in the cell wall of Gram-positive bacteria and in the middle membrane of Gram-negative bacteria. However, Gram-negative bacteria become sensitive to the lysozyme effect only after the barrier function of the outer membrane is destabilized by chemical (i.e., EDTA) and physical (e.g., freezing or heating) stresses. The antimicrobial effect is manifested by the lysis of cells. Lysozyme is most effective at pH 6.0 to 7.0 and at concentrations of about 0.01 to 0.1%. It can be used directly to control Gram-positive bacteria and, with EDTA and other similar compounds, to control Gram-negative bacteria. It has been used in wine (sake) to prevent the growth of undesirable lactic acid bacteria.

N. Monolaurin (Glycerol Monolaurate)

Monolaurin, the ester of lauric acid and glycerol, is one of the more effective bactericidal agents among the different derivatives of lauric acid tested in foods. Its effectiveness in deboned meat, chicken, sausages, minced fish, and other foods has been observed against undesirable bacteria, particularly the anaerobes. It also enhances the thermal inactivation of spores of *Bacillus* spp. The antimicrobial property of this lipophilic compound is enhanced with lactate, sorbate, ascorbate, and nisin, but may be reduced by starchy and proteinaceous compounds. In combination with monolaurin, the fungistatic activity of several antifungal compounds is enhanced. The antimicrobial activity of monolaurin is produced through its ability to destabilize the functions of the membrane. At lower concentrations, it acts as a bacteriostatic by interfering with the uptake of nutrients. It can be utilized up to 500 ppm without affecting the taste of the food.

O. Antibiotics (Tetracyclines, Natamycin, and Tylosin)

Several classical antibiotics that do not include bacteriocins of Gram-positive bacteria (nisin, pediocin, sakacin, and subtalin) were studied as antimicrobial food preservatives. Tetracyclines (about 10 ppm) were approved by the Food and Drug Administration (FDA) to extend the refrigerated shelf life of seafoods and poultry in the 1950s. However, due to the possible increase in antibiotic resistant bacteria, the use of these antibiotics in food was later banned. Natamycin, a microlid produced by *Streptomyces natalensis,* is an antifungal agent. Its use as a dip or spray to prevent growth of molds and formation of mycotoxins on the surface of some cheeses, sausages, and in raw peanuts has been approved by the Expert Committee of the World Health Organization. It is customarily used at 500 ppm, which leaves detectable but safe levels of the antibiotic on the product surface. Tylosin, a microlid that inhibits protein synthesis, is a bactericidal antibiotic that is more effective against Gram-positive than against Gram-negative bacteria and is also an inhibitor of outgrowth of germinated endospores. Due to its high heat resistance, it has been studied in low concentrations (1 ppm) to determine its effectiveness in controlling the growth of sporeformers in low-acid canned products.

P. Wood Smoke

Many processed meat products and fishes are processed with smoke generated by burning hardwood such as hickory, oak, maple, walnut, and mahogany. As an alternative, liquid smoke, obtained as a distillate of hardwood smoke, is also used with the ingredients of the products. A main reason for smoking meat, fish, and cheese is to impart desirable flavor, texture, and color to the products. The other benefit is the long shelf life of smoked products, especially those exposed to smoke during heating. The smoke contains several different types of chemicals that deposit on the food surface, many of which have antibacterial properties. The most important antibacterial agents are formaldehyde, phenols, and cresols. Depending upon the temperature and time of heating, degree of surface drying (A_w), and the concentrations, smoking can be both bacteriostatic and bactericidal against bacterial cells. Although smoke has a slight antifungal action, it does not have any adverse effects on the survival or germination of bacterial spores. Liquid smoke, under similar conditions, is less antimicrobial than wood smoke. Smoke also contains some chemicals that are carcinogenic, such as benzopyrene and dibenzanthracene. One of the recommendations to reduce colon cancer is to minimize the consumption of foods treated with smoke.

Q. Spices

Many spices, condiments, and plant extracts are known to contain antimicrobial compounds. Some of these include cinnamic aldehyde in cinnamon; eugenol (2-methoxy-4-allyl phenol) in cloves, allspice, and cinnamon; and paramene and thymol in oregano and thyme. Depending upon an active agent, they have bacteriostatic and fungistatic properties. Due to the small amounts used as spice in foods, they probably do not produce any antimicrobial effects. However, the antimicrobial components can be used in higher concentrations as oleoresins or essential oils.

The antimicrobial properties of garlic, onion, and ginger, as well as cabbage, brussels sprouts, carrots, and others have generated interest for their possible use as natural preservatives. It is expected that in the future the antimicrobial compounds from plants, especially food plants, will be studied more effectively and thoroughly.

REFERENCES

1. Davidson, P. M. and Branen, A. L., Eds., *Antimicrobials in Foods,* 2nd ed., Marcel Dekker, New York, 1993.
2. Dillon, V. M. and Board, R. G., Eds., *Natural Antimicrobial Systems and Food Preservation,* CAB International, Wallingford, UK, 1994.
3. Russell, N. J. and Gould, G. W., Eds., *Food Preservatives,* Van Nostrand Reinhold, New York, 1990, 4–6, 9, and 11.
4. Silliker, J. H., Ed., *Microbial Ecology of Foods,* Vol. I., Academic Press, Inc., New York, 1980, 8–10.
5. Giese, J., Antimicrobials: assuring food safety, *Food Technol.,* 49(6), 102, 1994.
6. Benedict, R. C., Biochemical basis for nitrite-inhibition of *Clostridium botulinum* in cured meat, *J. Food Prot.,* 43, 877, 1980.

CHAPTER 37 QUESTIONS

1. Discuss the uses of antimicrobial chemicals before and after 1865 A.D., after 1958, and since the 1980s.

2. Read the labels of ten different foods from at least five food categories and identify the chemicals that have antimicrobial action (direct or indirect).

3. Define the following terms: bacteriostatic, germicidal, bactericidal, fungicidal, sporicidal, viricidal, and sporostatic. Give one example of preservative for each.

4. List five characteristics one should consider in selecting a suitable chemical preservative for a specific food.

5. Discuss the antimicrobial properties and any health concerns of the following chemical preservatives, and list two foods (for each) in which they are used: NO_2, sulfites, bacteriocins, parabens, benzoic acid, lysozyme, natamycin, and lactic acid.

CHAPTER 38

Control by New Nonthermal Methods

CONTENTS

I. INTRODUCTION

Many of the conventional food preservation methods used at present and discussed here have several disadvantages. High heat treatment given to foods for safety and long shelf life results in loss of heat-sensitive nutrients (e.g., thiamine, riboflavin, folic acid, and vitamin C), denatures proteins and causes changes in texture, color, and flavor, and induces formation of new compounds through covalent bondings (e.g., lysinoalanine). Low heat processing, such as pasteurization, minimizes the disadvantages of high heat processing of foods, but the foods have limited shelf life even at refrigerated storage. Drying and freezing also reduce the nutritional and acceptance qualities of food, especially when stored for a long time. Irradiated foods have not been well accepted by consumers. Many of the chemical preservatives used are of nonfood origin and have limited efficiency.

Since the 1980s health-conscious consumers, especially in developed countries, have been concerned about the possible adverse effects that "harshly produced" and "harshly preserved" foods might have on their health and on the health of future generations. There is a concern that the cumulative effects of different types of food preservatives on the human body during one's life-time is not properly understood. The revelation about the harmful effect of some of the additives that were once allowed to be incorporated in foods has shattered consumer confidence. The philosophy of consumers has changed from, "How long will I live?" to "How well will I live?", which in turn has shifted the desire of these consumers to nutritious, natural, and minimally processed foods that have not been subjected to "harsh processing" or "harsh preservation" techniques. Due to changes in socioeconomic patterns and lifestyles, many consumers are also interested in foods that have a long shelf life and take very little time to prepare.[1]

The suitability of several nonthermal processing and preservation methods are being studied to produce such foods. Some of these include high electric field pulses, oscillating magnetic field pulses, intense light pulses, and ultrahigh hydrostatic pressure. The principal advantages and disadvantages and the current status of these four methods are briefly described here. Major emphasis is given to their antimicrobial properties.

II. FOOD PRESERVATION BY HIGH ELECTRIC FIELD PULSES

The antimicrobial effect of high electric field pulses is not due to the electric heat or electrolytic products, but rather to the ability to cause damage to the cell membrane.[2,3] When microbial cells, in a suspension, are exposed to pulses of high-voltage electric fields, a potential difference occurs between the outside and inside of the membrane. When the external electric field strength is moderately higher, so that transmembrane potential does not exceed the critical value by 1 V, pore formation occurs in the membrane, but the process is reversible (this principle is used in electroporation of cells to introduce foreign DNA). However, if a much higher external electric field strength is applied so that membrane potential exceeds the critical value, the pore formation becomes irreversible, causing the destruction of membrane functions and cell death. For destruction of microbial cells, an electric field strength of about 15 to 25 kV/cm for 2 to 20 μs is necessary. Destruction of bacterial and fungal spores requires a higher voltage and a longer period of time.

The lethal effect of pulsed electric field against microorganisms has generated interest to use this for nonthermal pasteurization and commercial sterilization of foods. During the process, the temperature of the suspension increases very little. However, to obtain greater microbial destruction, the temperature of the suspension can be increased to 60°C or higher. Also, by increasing the number of pulses, greater microbial destruction can be achieved.

A process designated as "Elsterile" was developed in Germany for the microbial destruction by pulsed electric fields in liquid food. The liquid food, in a treatment chamber that has two carbon electrodes, is subjected to high-voltage electric pulses. A four log reduction was obtained for *Lactobacillus brevis* in milk by treatment with 20 pulses at 20 kV/cm for 20 μs. A similar reduction was also observed by treating *Saccharomyces cerevisiae* in orange juice with five pulses of 20-μs duration at 4.7 kV/cm. The increased reduction of the yeast cells, as compared to bacterial cells, was thought to be due to the low pH of orange juice and the larger cell size of yeasts.

In the United States a method was developed to preserve fluid foods such as dairy products, fruit juices, and liquid egg products. In this patented method, a food is subjected to high-voltage electric field pulses in a chamber fitted with two electrodes, at 12 to 25 kV/cm for 1 to 100 μs. In addition, the products are subjected to pasteurization followed by rapid cooling. The shelf life of these products is much longer than conventionally pasteurized products.

These limited studies have shown that microorganisms in a liquid can be killed nonthermally by exposing them to high electric field pulses for a very short time. However, the method does not inactivate bacterial and fungal spores. Due to their nonthermal nature, it does not affect the texture, flavor, or color of the products.[2,3]

III. OSCILLATING MAGNETIC FIELD PULSES

Exposure of microorganisms to high-intensity oscillating magnetic fields tends to energize cell macromolecules (e.g., DNA and proteins) to the extent of breaking covalent bonds and rendering them metabolically inactive.[1,3] For antimicrobial effect, a magnetic field intensity should be between 5 to 50 tesla (unit of magnetic field intensity) at a frequency of 5 to 500 kHz exposed for a total time of 25 μs to a few milliseconds. Limited studies have shown that such a treatment can reduce microbial population by 2 logs. This is a nonthermal process inasmuch as the temperature increases very little during the treatment. The antimicrobial efficiency at a certain magnetic field is dependent upon the total time of exposure and not on the number of pulses. Orange juice, milk, and yogurt were tested for antimicrobial effectiveness and acceptance quality following treatment with oscillating magnetic fields. In one study, the foods were sealed in plastic bags and subjected to 1 to 100 pulses in an oscillating magnetic field with a frequency of 5 to 500 kHz at 0 to 50°C for a total exposure time of 25 μs to 10 ms. The temperature of the foods increased by 2 to 5°C and there were no changes in visual or organoleptic properties. Microorganisms were reduced by 3 to 4 logs. This method is safe and can be used for long storage of food along with other suitable methods (such as low heat).[2,4]

IV. INTENSE LIGHT PULSES

Energy released in short high-intensity pulses of light is known to kill microorganisms and deactivate enzymes.[2] The light pulses are used for a fraction of a second and have a wavelength spectrum between 170 and 2600 nm. The wavelength spectrum used depends upon the type of food materials. For treating packaging materials, the wavelengths in the UV range are used; but for UV-sensitive food materials, wavelengths beyond UV range are used. The treatment does not increase the temperature of a food. The method can be used in addition to sterilizing packaging materials; for the surface sterilization of beef, pork, and poultry carcasses, fresh fish, vegetables, fruits, bakery goods, and solid dairy products in transparent packages; and bulk sterilization of transparent liquids devoid of solid particles. Test results revealed that bread slices treated through packaging materials maintained a fresh appearance for more than 2 weeks, while untreated controls became moldy. Fresh uncut tomatoes exposed to pulsed light remained acceptable at refrigerated temperature for over 30 d.

The pulsed light process units contain a series of hooded lamps that can be installed on-line for continuous on-site operation. Like other nonthermal processes, light pulses do not increase the temperature or adversely affect the texture, flavor, color, or nutrient content of a food.[2]

V. ULTRAHIGH HYDROSTATIC PRESSURE (UHP)

Microbial cells, when exposed to high hydrostatic pressure inside a pressure vessel containing water, die rapidly, especially at 14,500 psi and above (14.5 psi = 1 Bar = 1 Atm = 1 kg/cm^2 = 750 torr = 100 kilo Pascal or 0.1 mPa).[2,5,7] Microbial death has been attributed to the damage and loss of activity of the cytoplasmic membrane. In addition, damage to the cell wall (or outer membrane), deactivation of intracellular enzymes, and the inability of amino-acyl t-RNA to bind to ribosomes were observed. The death curve of a microbial species shows characteristics of thermal destruction curve; i.e., initially there is a rapid rate of reduction of viability, followed by a slow rate and then tailing. In general, the death rate is higher: for young cells as compared to stationary phase cells; for rods as compared to cocci; in water suspension as compared to in a broth; at a higher pressure as compared to a lower pressure; and with longer time as compared to a shorter time. Surviving cells show repairable injury in the wall and membrane. Bacterial spores are killed at very high pressures (generally >100,000 psi). However, spores of some *Bacillus* spp. show more death in the 15,000- to 45,000-psi range than at higher pressure. This is thought to be due to induction of spore germination at lower pressure, followed by outgrowth of cells which, like vegetative cells, are then killed by the pressure. Depending upon the different factors mentioned above, as much as 5 to 6 logs or more reduction of bacterial cells can be achieved at pressures ≤100,000 psi.

Application of ultrahigh hydrostatic pressure (between 15,000 and ≤150,000 psi) has been advocated to reduce microbial counts for the preservation of foods. In the process, the food in a bag or a container is suspended in a liquid (generally water mixed with a little oil) in a pressure chamber (Figure 38.1). After closing the chamber, the pressure is raised by pumping more liquid into the closed chamber. The pressure is immediately and uniformly transmitted into the food (and microorganisms present in it). At high pressure, water shrinks relatively little (as compared to gas); specifically, about 4% at 15,000 psi, 11% at 60,000 psi, and 15% at 87,000 psi. Also, the temperature remains essentially unchanged. Since the pressure is uniform around a food exposed to hydrostatic pressure, it does not undergo substantial changes in acceptance qualities, especially in a low pressure range (≤60,000 psi).

Among food components, proteins are denatured by high hydrostatic pressure due to the destruction (and reformation) of hydrogen bonds, ionic bonds, and hydrophobic bonds, which only affect the tertiary structures; covalent bonds are not affected. The changes in the original tertiary structure from breaking and reformation can change coagulation or gelation characteristics of some foods, giving them a

Figure 38.1 Photograph of an ultrahigh hydrostatic pressure unit. (A) The unit showing different sections: (a) high pressure pump, (b) control panel, (c) pressure vessel with heating system, (d) yoke, (e) time-pressure control unit. (B) Pressure vessel showing the basket containing the packaged food is being immersed in chamber (4 × 12") containing the liquid. The unit has the capacity to pressurize food up to 100,000 psi at 20 to 90°C for the desired time.

unique and novel texture. However, high-pressure treatment does not reduce the flavor or nutrient content of a food.

Applications of UHP for the processing and preservation of different types of foods are being studied in many countries. Studies as early as 1914 revealed that the microbial level in milk was reduced by about 6 logs when exposed to 100,000 psi for 10 min at room temperature, and meat exposed to 78,000 psi for 1 h at 52°C in a bag did not show any deterioration during the 3 months of the study periods. Also, fruit juices and fruits subjected to hydrostatic pressure at ≤100,000 psi had a long shelf life. In recent years, the UHP process has been effectively applied to many products to reduce microbial load and increase shelf life. These include fresh fruit juices, fresh jellies and jams, whipped frozen strawberries, apple pieces in syrup, crushed fresh tomatoes, fresh vegetable juices, raw fish, meat, ground meat, meat products, fruit cocktail, spaghetti in sauce, rice in sauce, and vegetables. These products have more natural color,more flavor, and an extended shelf life. It has been successfully applied to the commercial production of fresh fruit juices, jams and jellies, and coffee and tea. The antimicrobial effect of high hydrostatic pressure has also been studied as it relates to the inhibition of undesirable growth of lactic acid bacteria in yogurt and possible accelerated ripening of cheeses. Pressure treatment is also being tested to quickly kill insects and parasites in foods.

As indicated before, ultrahigh hydrostatic pressure, in addition to antimicrobial action, induces some textural and color changes in some foods, especially proteinaceous foods. Some of these changes are being used to develop innovative and better products. When eggs are subjected to high pressure (≈90,000 psi), the white and yolk coagulate to form a gel. The gel is more lustrous than heat-induced gels, has a natural taste (without cooked flavor), and a soft, elastic (as adhesive) texture that does not break easily. The nutrient content remains unchanged. This gelling property can be used to produce egg-containing products with better texture, flavor, and nutrients. Fish paste, meat paste, and soy protein paste also form gels following high hydrostatic treatment that have desirable rheological and organoleptic properties than heat-induced gels. The gels are also smooth and glossy. This method is being used to make different types of surimi.

Pressure treatment of proteins and starches can make them susceptible to the hydrolyzing action of proteinases and α-amylase. In addition, starches also undergo gelatinization. Thus pressure treatment can also be used to enhance enzyme digestion of proteins and starches where required.

Beef muscle is tenderized at low hydrostatic pressure (15,000 psi for 4 min at 35°C prerigor or 22,000 psi for 1 h at 60°C postrigor). The pH also drops by 0.3 at 44,000 psi. Tenderization is probably brought about by the proteolytic action of cathepsins released from the lysosomes in the cytoplasm. Tenderization in beef by pressure seems to occur by a different mechanism than that which occurs during conditioning. The increase in flavor in the pressurized meat may be due to the increase in flavor compounds (amino acids or peptides). Tenderization of lower grade beef can be achieved by high-pressure treatment. However, myoglobin in fresh meat may lose the bright red color above 44,000 psi. Beef treated at 58,000 psi for 10 min may look like ham without any change in beef flavor; when lightly baked, it tastes like rare steak without loss of liquid.

In addition, high hydrostatic pressure is being studied for other innovative food processing technologies. At high pressure, the freezing point of water decreases: namely –10°C at 20,000 psi and –20°C at 29,000 psi. For a long shelf life, foods can be stored at low temperature (–10°C or below) by moderately low hydrostatic pressure without freezing. Moderately low pressure was observed to cause rapid thawing of frozen foods without causing any loss of liquid or other changes.

At present, high hydrostatic pressure in food can be used for batch processing. Technology for this process is rapidly progressing and equipment that will achieve pressures >150,000 psi is being produced. Pressure vessels with a capacity of several thousand liters are now in operation. In addition, methods for semi-continuous operation are being developed. As the process is instantaneous and may not take a long time, there is suggestion that a unit can be recycled rapidly. However, that might reduce the service life of a unit due to metal fatigue.

The influence of other parameters are being studied along with the pressure at a lower range (≤60,000 psi). They include several chemical antimicrobial agents. Units are now available in which the temperature of the liquid in the chamber can be raised to 90°C or higher. In addition, biopreservatives, such as bacteriocins of lactic acid bacteria, lysozyme, and chitosan (a polycationic derivative of chitosan), can be added to enhance the antimicrobial effect of hydrostatic pressure treatment.[5-7]

Ultrahigh hydrostatic pressure technology has great potential in the preservation and processing of foods. From the progress made since the mid-1980s and the interest it has generated in many countries among food processors, government agencies, and researchers, ultrahigh hydrostatic pressure treatment may be one of the most important nonthermal processing and preservation methods of food in the next century.[2,5-7]

REFERENCES

1. Miller, S. A., Science, law and society: the pursuit of food safety, *J. Nutrition*, 123, 279, 1993.
2. Mertens, B. and Knorr, D., Development of nonthermal processes for food preservation, *Food Technol.,* 46(5), 124, 1992.
3. Castro, A. J., Barbosa-Canovas, G. V., and Swanson, B. G., Microbial inactivation of foods by pulsed electric fields, *J. Food Process. Preservation,* 17, 47, 1993.
4. Pothakamary, U. R., Barbosa-Canovas, G. V., and Swanson, B. G., Magnetic-field inactivation of microorganisms and generation of biological changes, *Food Technol.,* 47(12), 85, 1993.
5. Anonymous, Use of hydrostatic pressure in food processing: an overview, *Food Technol.,* 47(6), 150, 1993.
6. Farr, D., High pressure technology in the food industry, *Trends Food Sci. Technol.,* 1, 16, 1990.
7. Kalchayanand, N., Sikes, T., Dunne, C. P., and Ray, B., Hydrostatic pressure and electroporation have increased bactericidal efficiency in combination with bacteriocins, *Appl. Environ. Microbiol.,* 60, 4174, 1994.

CHAPTER 38 QUESTIONS

1. List four nonthermal processes currently being studied for application in food processing and preservation. What advantages do they have over the conventional methods now being used?

2. Briefly describe the principles of antimicrobial action of the four processes listed in this chapter.

3. Discuss why UHP treatment does not adversely affect food quality. List six foods being tested for preservation by the UHP method.

4. Describe three other applications of UHP in food processing besides preservation and discuss the advantages.

CHAPTER 39

Control by Combination of Methods (Hurdle Concept)

CONTENTS

I. INTRODUCTION

Many of the factors used to kill microorganisms or control microbial growth in foods are also able to facilitate microbial survival and growth. These factors, which include temperature, A_w, pH, Eh, and some preservatives, have a range in which microbial growth varies from optimal to minimal. Beyond the range, depending upon the factor, the microorganisms either do not grow or die. When just a single method is used to preserve food, the condition beyond the growth range of microorganisms is generally used. Examples of such methods are high heat treatment to produce a sterile food (preferably in a container to prevent postheat contamination) or drying a food below A_w 0.6. Although microbiologically these products can have a very long shelf life and be safe, their acceptance and nutrient qualities are generally lower and may not be preferred by many consumers. This is particularly true for those consumers interested in foods that are "fresh, natural, healthy and convenient" and not "harshly processed or harshly preserved." There is general interest in producing foods low in salt, sugar, and fat, and that can be stored for a desirable period of time by refrigeration or chilling or even at room temperature and be eaten readily or by microwave heating. There is also interest in producing such fresh-like foods for the U.S. Army.

II. MECHANISMS OF ANTIMICROBIAL ACTION

Many foods invariably contain different types of spoilage bacteria and there is a good chance that they can have some pathogens (e.g., *Staphylococcus aureus* cells and pathogenic clostridial spores). For safety and stability, the growth of pathogens and spoilage microorganisms must be controlled during the storage life expected for these products. To achieve this goal, two or more antimicrobial agents are employed together at a level in which individual agents may facilitate some growth of the concerned microorganism, but not when used in combination.[1,2] As each of these agents or methods is used "gently" (at a lower level) as opposed to "harshly," they do not adversely affect the acceptance and nutritional qualities of a food but do retain their desirable safety and stability. The mechanism(s)[1,2] by which the combination of factors, or hurdle concept, works can be explained with an example (Figure 39.1).

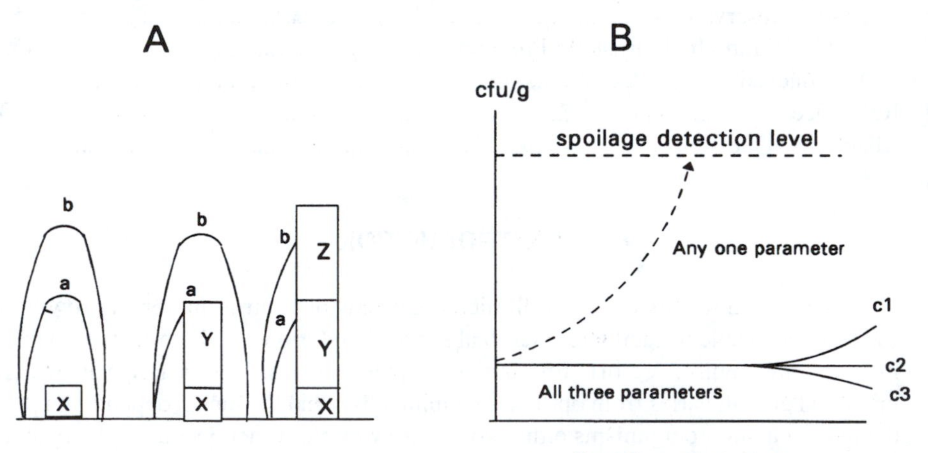

Figure 39.1 (A): Schematic presentation showing how the combination of several preservation techniques (x, y, z) is more effective in controlling different microbial groups (a, b). (B): An effective combination of several parameters can either reduce the growth rate, prevent growth, or even kill microorganisms that may not be achieved by any one parameter alone.

Two target microorganisms, "a" and "b", can grow when preservation methods X, Y, or Z are used as individual hurdles. However, if X and Y are combined, the growth of "a" is arrested; and when X, Y, and Z are used in combination, both microorganisms fail to grow.

It can also be explained with a separate example (Figure 39.1B). Suppose a food has several types of spoilage bacteria that can grow rapidly, either at 30°C or at pH 6.0 or at A_w 0.99, and spoil the food rather quickly. If any one of them is reduced (e.g., if the temperature is reduced to 10°C, keeping the other two unaltered) the growth rate can be slightly reduced, but the product can still be spoiled before the desired shelf life time (say, 60 d). However, if all three parameters are reduced slightly, which will allow the target microorganisms to grow if used individually, such as 10°C, pH 5.0, and A_w 0.94, their growth will either stop (C_2), fall to a very slow rate (C_1), or even show some death during storage (C_3). With this combination, the product will remain stable during its expected shelf life. Each of the three parameters used in this example can be ineffective individually; but when used together, their effect becomes additive, which the target organism(s) fail to accommodate for growth. If more and more such parameters are added, the growth of microorganisms in foods can be reduced drastically for a considerable period of time.

The different parameters or factors used in combination include the intrinsic factors (i.e., A_w, pH, Eh, and natural inhibitors), processing factors (i.e., heating, drying, fermentation, and preservatives), and extrinsic factors (i.e., temperature and aerobic/anaerobic). Along with these, the competitive flora and the nonthermal processing methods (e.g., irradiation, ultrahigh hydrostatic pressure, and electric, magnetic, and light pulses) could be added. Food preservation by combining several of these factors has been used for a long time. An example is the preservation of jams and jellies where high heat, low pH (of fruits), low A_w (sugar in fruits are added), and anaerobic packaging are used to reduce microbial numbers as well as the growth of the survivors. However, the greatest challenge will be the long-term preservation of meat, dairy, and fish products at refrigerated temperature or even at ambient temperature, where refrigeration is not always available, such as during a military operation, or in some developing countries. To achieve this goal, it will be necessary to study the effect of combining different factors to control target microorganisms in a food. In conducting such studies, one should also recognize that the factors could have additive, synergistic, or even adverse (neutralizing) effects when used in combination.

III. CURRENT STATUS

These aspects have been studied in a rather limited way. Results of some of these studies are briefly described here.[3,4]

A. Low Heat Processing

Low heat (≤100°C) will not kill many pathogenic and spoilage bacterial spores. They can be heat activated, which leads to germination and outgrowth. However, if the pH of the food is reduced to 4.5 or NO_2 (and NaCl) is added, the heat shock spores will not germinate.

B. Low Storage Temperature

Clostridium botulinum will grow at 35°C and A_w 0.95. However, if the storage temperature is reduced to 20°C, it will not grow unless the A_w is increased to 0.97. Similarly, *Listeria monocytogenes* will grow at 25°C in a broth containing 6.5% NaCl in 3 d, but fails to grow under similar conditions at 14°C.

C. Low pH

Clo. botulinum grows at pH 7.0, 37°C, and A_w 0.94; however, when the pH dropped to 5.3, no growth was observed, even at A_w 0.99. A *Salmonella* species grew at pH 5.8 in A_w 0.97, but when the pH was reduced to 5.0, an A_w 0.99 was required for growth. *Clo. botulinum* produced toxin during incubation at 16°C in 28 d at pH 5.5; but at pH 5.2 under the same conditions, no toxin was produced.

In using acid to reduce pH, it is important to recognize that organic acids, such as acetic, propionic, and lactic acids, are more effective than HCl and phosphoric acid. Also, acetic and propionic acids are more effective than lactic acid. Some acids, like citric and phosphoric acids, act by both lowering pH and chelating divalent cations that are important for microbial functions.

D. Low A_w

Some of the examples used with other parameters show how reduced A_w interacts with other factors in adversely affecting microbial growth. *Staphylococcus aureus* can grow at A_w 0.86 and high pH. However, it fails to grow even at A_w 0.93 (in the presence of NaCl) at pH 4.6, a pH that favors growth of this bacterium at higher A_w. Similarly, *Sta. aureus* will grow at 12°C, pH 7.0, and A_w 0.93. However, if A_w is reduced to 0.90, it will not grow under the same conditions.

It is also important to recognize that at low A_w, some microorganisms may develop resistance to the killing effect of heat.

E. Modified Atmosphere

In vacuum-, N_2-,or CO_2-packaged foods, growth of aerobes is prevented and that of facultative anaerobes can be reduced. But there is concern that it can selectively facilitate growth of anaerobic spoilage and pathogenic bacteria. Thus suitable combinations must be used with modified atmosphere to control their growth.

F. Preservatives

Some preservatives such as NaCl and BHA act synergistically to increase antimicrobial action of sorbates. The organic acids are also effective at lower pH due to higher concentrations of undissociated molecules. Some preservatives may not be effective at higher pH and some may lose potency during storage. Bacteriocins may be destroyed by proteolytic enzymes present in the food (raw food). The bactericidal effect of bacteriocins can be enhanced when used with acids and EDTA.

G. Ultrahigh Hydrostatic Pressure

The antibacterial effect of ultrahigh hydrostatic pressure can be increased by incorporating bacteriocins, lysozyme, and chitosan. Spores are difficult to destroy unless very high pressure is used (≃200,000 psi). However, after low-pressure treatment (≃15,000 psi), the spores may be induced to germinate, which can then be killed by higher pressure, heat, bacteriocins, or other antibacterial agents.

IV. CONCLUSION

Many of the methods listed here, such as low heat, low pH, low A_w, low temperature, and low hydrostatic pressure, induce sublethal stress among the surviving cells and spores. Cells and spores injured by one parameter become sensitive to other parameters and are killed in their presence. Sublethal injury to microbial cells and spores and the increased susceptibility of injured cells and spores to one or more preservation methods could play an important role in controlling microorganisms by a combination of factors.[5]

It will be important to study the beneficial effect of combining different parameters at different levels to determine their antimicrobial efficiency against target microorganisms in food systems. However, due to the large numbers of possibilities, actual product challenge studies may not be possible for all. Instead, computer simulation using multifactorial parameters can be designed that take into consideration the characteristics of a food. These methods are now being developed.

REFERENCES

1. Leistner, L. and Rodel, W., The stability of intermediate moisture foods with respect to microorganisms, *Intermediate Moisture Food,* Applied Science Publishers, Englewood, NJ, 1976, 120.
2. Mossel, D. A. A., Essentials and perspectives of the microbial ecology of foods, *Food Microbiology: Advances and Prospects,* Academic Press, London, 1983, 1.
3. Scott, V. N., Interaction of factors to control microbial spoilage of refrigerated food, *J. Food Prot.,* 52, 431, 1989.
4. Gould, G. W. and Jones, M. V., Combination and synergistic effects, *Mechanisms of Action of Food Preservation Procedures,* Gould, G. W., Ed., Elsevier, New York, 1989, 401.
5. Ray, B., Sublethal injury, bacteriocins and food microbiology, *ASM News*, 59, 285, 1993.

CHAPTER 39 QUESTIONS

1. Describe the principle of using a combination of factors, or hurdle concept, in the preservation of "fresh-like" foods.
2. In combining two or more factors to preserve foods, what criteria need to be considered?
3. Give three examples to show how the combination of factors has been found to control growth of foodborne pathogens.
4. Spores of *Clostridium laramie*, associated with spoilage of roast beef at refrigerated temperature, were not destroyed by UHP treatment at 80,000 psi for 20 min. However, when the spores were subjected to 10 min UHP treatment at 20,000 psi and then treated with a bacteriocin containing biopreservatives, the growth medium was free of this bacterium (spores or cells). Explain the possible causes of these differences.
5. How can microbial sublethal injury be advantageously used to control microbial growth in food preserved by the hurdle concept?

SECTION VII

APPENDICES

This section includes brief discussions of topics that are considered either relatively new in food microbiology, or are taught in special and advanced food microbiology courses or in laboratory courses. Thus these topics can be used as additional reading materials.

APPENDIX A

Microbial Attachment to Food and Equipment Surfaces

I. IMPORTANCE IN FOOD

The normal tendency of a microbial cell, when it comes in contact with a solid surface, is to attach itself to the surface in an effort to compete efficiently with other microbial cells for space and nutrient supply and to resist any unfavorable environmental conditions. Under suitable conditions, almost all microbial cells are capable of attaching to solid surfaces, which is achieved through their ability to produce extracellular polysaccharides. As the cells multiply, they form microcolonies, giving rise to a biofilm on the surface containing microbial cells, extracellular polysaccharide glycocalyx, and entrapped debris. In some situations, instead of forming a biofilm, the cells may attach to contact surfaces and other cells by thin, thread-like exopolysaccharide materials, also called fimbriae.

Attachment of microorganisms on solid surfaces has several implications on the overall microbiological quality of food. Microbial attachment and biofilm formation to solid surfaces provide some protection of the cells against physical removal of the cells by washing and cleaning. These cells seem to have greater resistance against sanitizers and heat. Thus spoilage and pathogenic microorganisms attached to food surfaces, such as carcasses, fish, and meat, cannot be easily removed by washing, and later they can multiply and reduce the safety and stability of the foods. Similarly, microbial cells attached to equipment surfaces, especially those that come in contact with the food, may not easily be killed by chemical sanitizers or heat designed to be effective against unattached microbial cells and thus can contaminate food. Finally, microbial attachment and biofilm formation in the food processing environment, such as floors, walls, and drains, enable the cells to establish in the environment and become difficult to control effectively by the methods designed against the unattached microorganisms. These places in turn can be the constant source of undesirable microorganisms to foods handled in the environment.

The concept and importance of microbial attachment and biofilm formation in solid food, equipment, and food environments are now being recognized.[1-3] The limited studies have shown that under suitable conditions many of the microorganisms important in food can form biofilm. Several species-strains of *Pseudomonas* were found to attach to stainless steel surfaces, some within 30 min at 25°C to 2 h at 4°C. *Listeria monocytogenes* was found to attach to stainless steel, glass, and rubber surfaces within 20 min of contact. Attachment of several pathogenic and spoilage bacteria has also been demonstrated on meat and carcasses of poultry, beef, pork, and lamb. The microorganisms found to attach to meat surfaces include *Lis. monocytogenes, Micrococcus* spp., *Staphylococcus* spp., *Clostridium* spp., *Bacillus* spp., *Lactobacillus* spp., *Brochothrix thermosphacta, Salmonella* spp., *Escherichia coli, Serratia* spp., *Pseudomonas* spp., and others. It is apparent from the limited data that microbial attachment to solid food and food contact surfaces is quite wide and needs to be taken into consideration in controlling microbiological quality of food.

II. MECHANISMS OF ATTACHMENT

Several possible mechanisms by which microbial cells attach and form biofilm on solid surfaces have been suggested.[1,3] One suggestion is that the attachment occurs in two stages. In the first stage, which is reversible, a cell is weakly held to the surface by weak forces (electrostatic and Van der Waals forces). In the second stage, a cell produces complex polysaccharide molecules to attach its outer surface to the surface of a food or equipment and the process is irreversible. A three-step process that includes adsorption, consolidation, and colonization has been suggested by others. In the reversible adsorption stage, which can occur in 20 min, the cells attach loosely to the surface. During the consolidation stage, the microorganisms produce threadlike exopolysaccharides fimbriae and firmly attach the cells to the surface. At this stage, the cells cannot be removed by rinsing. In the colonization stage, which is also irreversible, the complex polysaccharides may bind to metal ions on equipment surfaces and the cells may metabolize products that can damage the surfaces.

III. INFLUENCING FACTORS

The level of attachment of microorganisms to food processing equipment surfaces is found to be directly related to contact time. As the contact time is prolonged, more cells attach to the surface, the size of the microcolony increases and attachment between cells increases. Fimbriae formation by the cells occurs faster at optimum temperature and pH of growth. Limited studies also showed that when microorganisms, such as *Pse. fragi* and *Lis. monocytogenes*, are grown together, they form a more complex biofilm than when either is grown separately.[1-3]

The factors associated with the attachment of spoilage and pathogenic bacteria to meat and carcass surfaces can be divided into three groups: associated with bacteria, associated with meat, and extrinsic factors.[1-3] Many Gram-positive bacteria (*Clostridium, Micrococcus, Staphylococcus,* and *Lactobacillus* spp., and *Bro. thermosphacta*) and Gram-negative bacteria (*Esc. coli, Proteus, Pseudomonas, Ser-*

ratia, Salmonella, Enterobacter, Shewanella, and *Acenatobacter* spp.) are capable of attaching to skin and meat surfaces of chicken, pork, beef, and lamb. Some studies showed that several Gram-negatives, such as *Proteus* and *Pseudomonas* spp., attach more rapidly and in higher numbers than some Gram-positive bacteria, such as *Lactobacillus, Staphylococcus,* and *Micrococcus* spp. However, there are differences in observations and some researchers think the more negative charge the cells of species-strains have, the faster and higher ability they have to attach to muscle surfaces. Similarly, results differ among researchers with respect to influences of species or types of food animals, birds, and microbial species to attachment. Bacterial attachment to lean tissues is generally higher than to adipose tissues; however, there is probably no difference in the attachment of both Gram-positive and Gram-negative bacterial species in the skin and muscle of different types of food animals and birds. Among the extrinsic factors, the number of attached cells on meat surfaces is directly related to contact time and the cell concentrations used. At optimum growth temperature a bacterial species shows a higher attachment rate. Electrical stimulation of the carcasses may favor attachment, but there are differences in opinion on this.[1-3]

IV. CONTROL MEASURES

Microbial biofilm formation on equipment surfaces and attachment to solid food surfaces can resist their removal and killing by some of the methods used but designed to remove or kill unattached microbial cells. Some modifications in these methods may be necessary to overcome the problems, especially in food contact and processing equipment surfaces. Biofilms are difficult to remove if they are left to grow. They should be removed when the biofilms are young. In a food processing operation, cleaning and sanitation needs to be done every few hours. This needs to be determined for a specific food and the nature of the operation. To break down the glycolalyx minerals, a treatment with some suitable hydrolyzing enzymes may be initially administered. EDTA, along with quaternary ammonium compound treatment following enzyme treatment, can be helpful. In addition, processing equipment needs to be designed to prevent or control biofilm formation. To reduce attachment of microorganisms to solid food surfaces, suitable control measures need to be adopted. Use of suitable preservatives, low A_w, low pH, and low storage temperature can be used judiciously in combinations to reduce the rate of biofilm formation in food.

REFERENCES

1. Zottola, E. A., Microbial attachment and biofilm formation: a new problem for the food industry. Scientific status summary, *Food Technol.,* 48(7), 107, 1994.
2. Mattila-Sandholm, T. and Wirtanen, G., Biofilm formation in the industry: a review, *Food Rev. Intl.,* 8, 573, 1992.
3. Notermans, S., Dormans, J. A. M. A., and Mead, G. C., Contribution of surface attachment to the establishment of microorganisms in food processing plants: a review, *Biofouling*, 5, 1, 1991.

APPENDIX B

Predictive Modeling of Microbial Growth in Food

I. IMPORTANCE

In 1990, over 12,000 new food products were introduced to U.S. consumers over and above the existing hundred thousand food items. Some of these products obviously have limited shelf life due to the possibility of microbial growth under the conditions in which they are stored and handled following production. The initial microbial population in many of these products may include both spoilage and pathogenic types. To ensure stability and safety, it is important that proper control measures are used to prevent the growth of microorganisms from the time of production to the time of consumption of these products.[1,2]

Among the thousands of new products marketed every year, some are definitely produced to satisfy consumers who are "looking for foods that are healthy, natural, low fat, low salt, tasty, convenient, do not contain harsh preservatives, are not given harsh treatment, have long storage life, and are safe." To produce these new types of foods (such as the new generation of refrigerated foods) with desirable safety, stability, and acceptance quality, the influence of many parameters, such as different ingredients at different concentrations under different processing and storage conditions and at different A_w, pH and other intrinsic and extrinsic factors, on microbial growth and destruction in foods must be determined.

It is almost impossible to conduct such studies for each product by the methods traditionally used to ensure safety and stability of food products. However, with the aid of computers, mathematical models can be developed to determine the influence of combinations of several parameters on microbial growth. Although they may not be accurate, they can be effective in obtaining first-hand information very rapidly, helpful in eliminating many of the possibilities, and in selecting a few that appear more promising. This information can then be used to conduct a modest traditional study that is feasible, both experimentally and economically.

II. TRADITIONAL STUDIES

Several methods are used to determine influence of control parameters on growth and/or toxin production by target microorganisms in foods. Three of these methods are briefly described here.[3,4]

A. Challenge Studies

In this method, a food is first inoculated with the microorganism(s) that is expected to be the major cause of loss of stability and/or safety under retailing conditions. A realistic number of microbial cells/spores are used as inoculum. The product is then stored under conditions in which it is expected to be stored normally and then examined for the duration of storage for microbial growth or toxin production.

To determine effective control measures against the target microorganisms a large number of variables need to be studied. These include: ingredients used, intrinsic and extrinsic factors, possible abuse, and others. As a result, the study can be quite complicated, time consuming, and costly.

B. Storage Studies

A normal product is stored under conditions that it is supposed to be stored and, at selected intervals, samples are examined for the level of spoilage and pathogenic microorganisms that are expected to be normally present in the product. The results can be used to predict expected shelf life and safety under the suggested storage conditions. However, this study will not provide information if the product is temperature abused. In addition, if some of the samples being tested by chance do not contain the spoilage and pathogenic microorganisms normally expected to be present, then the results are of no practical value.

C. Accelerated Shelf Life Studies

For products expected to have relatively long shelf life, accelerated tests provide rapid information about their storage stability. The products are generally held at a higher temperature in order to increase the rate of microbial growth and accelerate the spoilage. However, foods normally contain mixed microflora that differ in optimum growth rate at different temperature, and spoilage of a food is generally caused by a few species, among those present, that grow optimally at a given condition. To make the results valid, in designing this study, the temperature should be increased only to a level that does not affect the growth of the microorganisms of interest.

III. PREDICTIVE MICROBIOLOGY

Several mathematical models have been developed to predict growth of pathogenic and spoilage microorganisms in foods from the data generated by studying microbial growth rate at different pH, A_w, temperature, and preservative concentrations in laboratory media. The availability of suitable computers has helped in the rapid analysis of multifactorial data. Two kinetic-based models are discussed here. They indicate the effect of culture parameters on the growth rate of a microorganism, especially at the lag and exponential growth phases.[1-4]

A. Square Root Model

It is based on the linear relationship between the square root of the growth rate and temperature, and is expressed as: $\sqrt{r} = b\,(T - T_{min})$, where r is the rate of growth, b is the slope of the regression line, T is the temperature of the food, and T_{min} is the theoretical minimum temperature of growth. The model has been further expanded to include effects of temperature, pH, and A_w on growth rate:

$$\sqrt{r} = b\sqrt{\left[\left(A_w - A_{w\,min}\right)\left(pH - pH_{min}\right)\right]}\left(T - T_{min}\right)$$

where A_w and pH are water activity and pH of a food respectively, and A_{wmin}, pH_{min}, and T_{min} are minimum or lower limits of A_w, pH, and temperature, respectively, for growth.

In general, this model is quite effective when just one or two parameters are used. However, its effectiveness is reduced when a variety of parameters, such as pH, organic acid, modified atmosphere, A_w, preservatives, salts, and others, are used in combination to control microbial growth.

B. Sigmoidel Model (Gompertz: USDA Model)

This model has been developed by the United States Department of Agriculture to predict microbial growth in a food environment containing many controlling parameters. It has been tested in laboratory media to determine the growth rate of *Listeria monocytogenes, Staphylococcus aureus, Salmonella* spp., *Shigella flexneri, Escherichia coli* 0157:H7, and *Aeromonas hydrophila* under variable temperatures, pH, A_w, NaCl and $NaNO_2$ concentrations, and aerobic and anaerobic atmospheres. The growth curves were statistically fitted using nonlinear regression analysis in conjunction with Gompertz functions. The results were then analyzed to develop this model. It gives the lag time, the maximum growth rate constant, and the maximum microbial load directly from nonlinear regression of the numbers vs. time data. For these four parameters, the asymmetric, sigmoidal equation is:

$$N = A + C\exp\left(-\exp\left[-B(t - M)\right]\right)$$

where N is $\log_{10}$ colony-forming units (cfu) per milliliter at time t, A is initial $\log_{10}$ cfu/ml, C is $\log_{10}$ cfu difference between time t and initially, "exp" is exponential, B is the relative growth rate at M, and M is time at which the growth rate is maximum. This model has been extensively used due to its simplicity and overall effectiveness.

IV. CONCLUSIONS

Predictive microbiology has generated great interest because of its many advantages, most important of which is the quick availability of computer-generated information for both simple and complex studies. Interactions between different intrinsic and extrinsic factors, and the importance of changing a parameter on growth and survival of microorganisms can be determined very quickly. However, current models have been designed using laboratory media. Thus their effectiveness in different food systems is not exactly known. Limited studies have shown that in a food system (e.g., ham salad), while growth responses of *Listeria* spp. and *Salmonella* spp. obtained by predictive modeling and by actual experiment were favorable, for *Sta. aureus* the growth rate predicted by the model was lower than that obtained in actual experiments. Thus, at present, the predictive data needs to be adopted with caution. Model data can be used in initial studies using large numbers of variables to eliminate a great majority of them. Experimental data can then be obtained from actual product studies using limited numbers of variables. This will help reduce the time and cost, and make the study more manageable and effective.

REFERENCES

1. Whiting, R. C. and Buchanan, R. L., Microbial modeling: scientific status summary, *Food Technol.,* 48(6), 113, 1994.
2. Buchanan, R. L., Predictive food microbiology, *Trends in Food Sci. Technol.,* 4, 6, 1993.
3. Smittle, R. B. and Flowers, R. S., *Scope*, January issue. Silliker Laboratories Technical Bulletin, Chicago Heights, IL, 1, 1994, 1.
4. Labuza, T. P., Fu, B., and Taoukis, P. S., Prediction of shelf life and safety of minimally processed CAP/MAP chilled foods — a review, *J. Food Prot.,* 55, 741, 1992.

APPENDIX C

Hazard Analysis Critical Control Points (HACCP)

I. INTRODUCTION

The traditional method of examining microbiological safety, storage stability, and sanitary quality of a food is to test representative portions (or samples) of the final product for the presence of some pathogens (e.g., *Salmonella*) or the number or level of certain pathogens (e.g., *Staphylococcus aureus* and *Vibrio parahaemolyticus*), different microbial groups (e.g., aerobic plate counts, psychrotrophic counts, thermoduric counts, and yeasts and molds), and indicator bacteria (e.g., coliforms and enterococci) per gram or milliliter of product. The major disadvantage with these types of end-product testings is that they do not provide close to 100% assurance about the safety, stability, and sanitary quality of the products. This aspect is particularly important with respect to pathogenic bacteria. By analyzing samples from a batch of a product, according to sampling plan and testing methods recommended by the regulatory agencies, it is not possible to give a high degree of assurance (close to 100%) that the untested portion of the end-product is free of pathogens. Under certain circumstances, such as foods consumed by the crew in a space mission, or by the military, especially engaged in an important mission, as close to 100% assurance of safety of a food is required.

HACCP was originally developed by the Pillsbury Company, the National Aeronautics and Space Administration (NASA), and the U.S. Army Natick Laboratories to produce foods with high assurance of safety for use in the space program. It has two components: first, to identify or analyze the hazards (HA) associated with the production and processing of a food and then to identify critical control points (CCP) (i.e., the places during processing of a food where proper control measures need to be implemented in order to prevent any risk to the consumers). It is regarded as a systemic approach to assure safety and better than end-product testing.

Since the introduction of the concept in 1971, the HACCP has undergone several changes according to specific needs. In 1980, the National Academy of Science suggested that food industries in the United States could use the HACCP principles to produce safer foods, and the regulatory agencies could help develop programs suitable for a particular processing operation. Currently, the Food Safety and Inspection Service (FSIS) of the USDA advocates the use of HACCP in meat and poultry inspection programs during both slaughtering and processing of carcasses for products. The Food and Drug Administration (FDA) is also developing HACCP programs that cover the various products it regulates. Internationally, HACCP is also being used in Canada, Australia, New Zealand, and several European countries.

It seems that the HACCP concept can be used universally to develop food safety programs. This will help reduce problems associated with foodborne pathogens, as well as the global trade of foods.[1-3]

II. HACCP PRINCIPLE OF THE NACMCF

As indicated above, there have been some changes in the original HACCP system. The National Advisory Committee on Microbiological Criteria for Foods (NACMCF) has developed a document for HACCP system.[1-3] It is expected that by implementing this system in the production and processing of food, the safety of the products can be assured. This system is briefly presented here.

A. Seven Principles of HACCP

The principles of HACCP include:

1. Conduct a hazard analysis to determine risks associated at all stages, from growing raw materials and ingredients to final product ready for consumption.
2. Identify critical control points to control these hazards.
3. Implement conditions to control hazards at each critical control point.
4. Implement effective procedures to monitor control for each point.
5. Implement corrective measures to be taken if a deviation occurs at a point.
6. Implement effective record-keeping systems for HACCP plan activities.
7. Implement procedures to verify the plan is working effectively.

B. Brief Description of Each Principle

1. Principle 1

Determine the possible hazards that exist for a specific food or ingredient from pathogenic microorganisms and their toxins. This is performed in two segments: first by ranking the specific food into six categories and then by assigning each a risk category on the basis of ranking for hazard categories.

Ranking for the six hazard categories is based on the following assessments for a food: (1) does it contain microbiologically sensitive ingredients; (2) does the processing method contain a step that is effective in destroying the pathogen; (3) does the processing method have a step for post-processing contamination of the product with pathogens or their toxins; (4) is there a possibility of abusing the product during subsequent handling (transportation, display at retail stores, handling and preparation by the consumers) that can render the product harmful for consumption; and (5) is the product, after packaging, given an effective heat treatment prior to consumption?

The hazard characteristics of a food is based on the following:

Hazard A: A nonsterile food intended for consumption by high-risk consumers (infants, elderly, sick, and immunocompromised individuals).
Hazard B: The product contains "sensitive ingredients" in terms of microbial hazard.
Hazard C: The process does not contain a step to destroy harmful microorganisms.
Hazard D: The product is subject to recontamination after processing and before packaging.
Hazard E: The possibility exists for abuse prior to consumption that can render the product harmful if consumed.
Hazard F: The product is not given a terminal heat treatment following packaging and prior to consumption.

A food is ranked according to hazards A through F by using a plus (+) for each potential hazard. The risk category is determined from the number of pluses.

Category VI: A special category that applies to nonsterile food to be consumed by high-risk consumers.
Category V: A food that has five hazard characteristics: B, C, D, E and F.
Category IV: A food that has a total of four hazard characteristics in B through F.
Category III: A food that has a total of three hazard characteristics in B through F.
Category II: A food that has a total of two hazard characteristics in B through F.
Category I: A food that has a total of one hazard characteristic in B through F.
Category O: No hazard.

2. *Principle 2*

To control microbiological hazard(s) in a food production system, it is necessary to determine critical control point (CCP). A CCP is any point or procedure in a specific food system where effective control must be implemented in order to prevent hazards. All hazards in a food production sequence, from growing and harvesting to final consumption, must be controlled at some point(s) to ensure safety. The CCP includes heat treatment, chilling, sanitation, formulation control (pH, A_w, preservatives), prevention of recontamination and cross-contamination, employee hygiene, and environmental hygiene. Application of one or more CCP will destroy hazardous microorganisms or prevent their growth and toxin formation in the product.

3. *Principle 3*

To control microbiological hazard at each identified CCP, it is necessary to set up critical limits (such as temperature, time, A_w, pH, preservatives, and in some cases, aroma, texture, and appearance). One or more critical limit may be necessary at each CCP, and all of them should be present to ensure that the hazard(s) is under control.

4. Principle 4

The CCP used for a specific food production must be monitored to determine if the system is effective or not effective in controlling the hazards. The monitoring can be continuous, such as in case of heat treatment (temperature and time) of a product. If continuous monitoring cannot be adopted for a processing condition, it can be done at reliable intervals. The interval for a food production has to be developed on a statistical basis to ensure that the potential hazards are under control. Most on-line monitoring is done using rapid chemical and physical methods. Generally, microbiological methods are not effective due to the long time required for testing. However, some current methods, such as the bioluminescence method, could be effective under some conditions. All results of monitoring must be documented and signed by persons doing the monitoring, as well as by an official of the company.

5. Principle 5

In case a deviation from the present HACCP plan is identified from the monitoring system in a food production operation, effective corrective actions must be taken to assure product safety. The corrective action(s) must demonstrate that it is effective in controlling the potential hazard(s) resulting from the deviation of CCP. The deviation(s) as well as the specific corrective measure(s) for each must be documented in the HACCP plan and agreed upon by the regulatory agency prior to application of the plan. Also, the product produced by the new plan should be placed on hold until its safety has been ensured by proper testing.

6. Principle 6

Records of the HACCP plan developed for a specific food production should be kept on file in the plant. In addition, documents of CCP monitoring and any deviation and corrective procedures taken should be kept at the establishment. If necessary, the records must be made available to the regulatory agencies.

7. Principle 7

Verification systems should be established to ensure that the HACCP system developed for a specific food production system is working effectively to ensure safety to the consumers. Both the food producer and the regulatory agency have to be involved in the verification of the effectiveness of the HACCP in place. Verification methods include testing samples for physical, chemical, sensory, and microbiological criteria as established in the HACCP plan.

III. CONCLUSION

In a biological hazard category, the main objective of the HACCP system is to provide a high degree (close to 100%) assurance that a food ready to be consumed would be free of pathogenic microorganisms or their toxins.[4,5] However, the same principle can be applied to design control of spoilage microorganisms in foods, especially for those that are expected to have extended shelf life. Currently, there is an increase in the production of vacuum and modified-air packaged unprocessed and low heat-processed food products, many with low fat content and high pH, some with an expected shelf life over 50 d at refrigerated temperature. There is evidence that many of these products are showing unusual types of microbial spoilage and some are by species that were not known before. Although limited, there is evidence that the incidence of such spoilage is on the rise. It would be commercially advantageous for the food industry to adopt HACCP, not only with pathogens in mind, but also to control spoilage microorganisms. It is also logical to think about the controlling microbiological problems as a whole which would include both pathogens and spoilage microorganisms (total microbial quality, TMQ). One cannot devise methods to control only pathogens or only spoilage microorganisms. The ISO (International Organization of Standardization) 9000 series standards can be used to supplement HACCP systems and develop effective process control and process assurance systems.[5] Finally, the HACCP system can be combined with the hurdle concept of food preservation for effective control of pathogenic and spoilage microorganisms for foods that otherwise could be produced commercially (see Chapter 39). Although food safety is of primary importance in countries with abundant food supplies, food spoilage is equally important in many countries where large quantities of food are lost due to spoilage. In these countries, the HACCP concept to reduce food spoilage will be particularly important.

REFERENCES

1. Pierson, M. D. and Corlett, D. A., Jr., Eds., *HACCP Principles and Applications,* Van Nostrand Reinhold, New York, 1992.
2. Sperber, W. H., The modern HACCP System, *Food Technol.,* 45(6), 116, 1991.
3. Sofos, J. N., HACCP, *Meat Focus Int.,* 5, 217, 1993.
4. Golomski, W. A., ISO 9000 — the global perspective, *Food Technol.,* 48(2), 57, 1994.
5. Leistner, L. E. E., Linkage of hurdle technology with HACCP. *Proc. 45th Annu. Reciprocal Meat Conf.,* June 14–17, Colorado State University, Ft. Collins, Am. Meat Science Assoc. and National Live Stock and Meat Board, Chicago, 1992, 1.

APPENDIX D

Detection of Microorganisms in Food and Food Environment

I. IMPORTANCE

Microbial population in foods, food ingredients, and the food contact environment normally constitute many different species coming from different sources. The total microbial population in a food varies greatly depending upon the level of sanitation used at all phases, the degree of abuse that leads to microbial growth, and the processing and preservation methods used to kill and prevent growth of microorganisms (Chapters 29 to 39). Similarly, contamination of a food by specific types or species of microorganisms is dependent upon the presence of source of these microorganisms and their entrance into the food, mostly due to poor sanitation, during handling and processing. Microbiological examination of foods and food ingredients helps one assess their safety to the consumers, their stability or shelf life under normal storage conditions, and the level of sanitation used during handling. In addition, the microbiological load and type can be important in determining if a food and food ingredient meet acceptable standards, specifications, and guidelines. Microbiological evaluation of raw materials also provides important information about the heat-processing parameters that would be necessary to meet the microbiological standard, guideline, or specification of a product. Finally, microbiological evaluation of a food, food ingredient, and environment helps determine possible sources of a specific microbial type in a food and, in the case of a heated food, the source and nature of post-heat treatment contamination.

II. METHODS USED

The methods used for the microbiological evaluation or detection of foods, food ingredients, and environment are broadly grouped as quantitative and qualitative methods. The quantitative methods are designed to enumerate or estimate directly or indirectly the microbial load in a test material. It should be recognized that none of the quantitative methods that are used now do enumerate or estimate "total microbes," "total bacteria," or "total viable population" present in a food. Rather, each method enumerates or estimates a specific group among the total microbial population present in a food and that grows or multiplies preferentially under the conditions or methods of testing. These include composition of a medium, temperature and time of incubation and oxygen availability during incubation, pH, and treatments given to a sample prior to enumeration and estimation. Examples of some of the quantitative methods used are: aerobic plate counts (or standard plate counts for dairy products), anaerobic counts, psychrotrophic counts, thermoduric counts, coliform counts, *Staphylococcus aureus* counts, and yeast and mold counts.

In contrast, qualitative methods are designed to determine if a representative amount (a sample) of food or a certain number of samples in a batch of food contain a specific microbial species among the total microbial population or not. These methods are used to detect the possible presence of certain foodborne pathogens, especially those capable of inflicting high fatality rates among consumers. *Salmonella, Clostridium botulinum, Escherichia coli* 0157:H7, and probably *Listeria monocytogenes* are some that fall into this group.

III. STANDARD AND RECOMMENDED METHODS

Many methods and variations of different methods that can be used for quantitative and qualitative detection of microorganisms in foods are reported in the literature. However, it is desirable to use methods that have been approved by the regulatory agencies. They can be either standard or recommended methods. In the United States, for the microbiological examinations of milk and milk products, one needs to use methods that have been designated as standard methods by the regulatory agencies. For other foods, however, specific microbiological methods have been recommended or approved by the regulatory agencies. Although it is advisable to use such a method, there is no mandatory requirement. Some of the books in the United States are: "*Standard Methods for the Examination of Dairy Products,*" "*Standard Methods for the Examination of Water and Waste Water,*" "*Standard Methods for the Examination of Seawater and Shellfish,*" "*Compendium of Methods for the Microbiological Examination of Food*" (all four are published by the American Public Health Association, Washington, D.C.), and "*Bacteriological Analytical Manual of Food and Drug Administration*" (prepared by the FDA and published by Association of Official Analytical Chemists, Arlington, VA). The last two publications include recommended methods. All these books are revised when necessary to incorporate the new methods and to update the old methods by different groups of expertise. Many other countries have similar publications for the standard and recommended methods. Similar books are published by international organizations, such as specific branches of the World Health Organization and the Food and Agricultural Organization of the United Nations.

A food microbiology course invariably contains a laboratory component. The methods described in one or more of these books for the microbiological examination of food, food ingredients, and environment can be used in the laboratory. This will help the students or interested individuals become familiar with the standard and recommended methods approved by the regulatory agencies in the United States. Some of the methods are briefly discussed here. Details of these methods are available in the books listed above and references listed at the end of this appendix.

IV. QUANTITATIVE METHODS FOR MICROBIAL ENUMERATION IN FOODS

A. Direct Enumeration

1. Microscopic Counts

Either stained cells under a bright field or live cells under a phase contrast microscope can be counted and, using an appropriate microscopic factor, these counts can be expressed as microscopic counts per milliliter or gram food sample. However, viable and dead cells can not be differentiated by this method. In addition, a sample must have large numbers of microorganisms for effective use of this method.[1-3]

2. Colony-Forming Units (CFU) in Nonselective Agar Media

Aliquots from a serially diluted sample are either pour plated or surface plated using nonselective media such as plate count agar, tryptic soy agar, nutrient agar, and others. However, plate count agar (PCA) is recommended for cfu determination of several groups. The temperature and time of plate incubation required for the colonies to develop differ with the microbial groups being enumerated. For standard plate count (SPC), it is 32°C for 48 h; for aerobic plate counts (APC), it is 35°C for 48 h; and for psychrotrophic counts, it is 7°C for 10 d or 10°C for 7 d. The same procedure with specific modifications can be used to determine thermophilic, thermoduric, anaerobic, and facultative anaerobic groups present in a food sample. The specific group(s) to be tested depends upon its relative importance in a food. For a vacuum-packaged refrigerated food, the most important groups will be psychrotrophic, anaerobic, and facultative anaerobic groups[1-3].

3. CFU in Nonselective Differential Agar Media

A nonselective medium, is supplemented with an agent capable of differentiating the colonies produced by specific group(s) of microorganisms that differ in metabolic or physiological characteristics from other(s) in the population. pH indicators or oxidation-reduction indicators are often used in the medium. Thus, the colonies of cells capable of metabolizing lactose to lactic acid are differentiated from those that do not ferment lactose by growing them in an agar medium supplemented with lactose as carbon source and a pH indicator such as bromocresol purple. The lactose-fermentating colonies will be yellow and the others that can grow will be white against a purple background. Differential methods are also used for the enumeration of proteolytic, lipolytic, and pectinolytic microbial groups in a food.[1-3]

4. *CFU in Selective Agar Media*

A medium can be supplemented with one or more selective or inhibitory agents and used by pour or surface plating of serially diluted samples. In the presence of such an agent(s), only the microorganism(s) resistant to it can grow. Incubation conditions to stimulate colony formation differ with the organisms being studied. Enumeration of aciduric bacteria in a medium at pH 5.0, yeasts and molds in a medium at pH 3.5, and *Clo. perfringens* in the presence of cycloserine are examples of selective enumeration of specific groups present in a food. Halophilic and osmophilic microorganisms can also be enumerated by such selective procedures.[1-3]

5. *CFU in Selective-Differential Agar Media*

In this method, a medium is supplemented with one or more selective agents to allow selective growth of specific resistant microbial group(s) while inhibiting growth of other sensitive associative microorganisms. In addition to selective agent(s), a medium is also supplemented with agent(s) that enable each type among the selective microbial groups to produce colonies that differ in characteristics from one another. Violet-red bile agar for coliforms, KF-azide agar for *Enterococcus* spp., V-J agar or Baird-Parker agar for *Staphylococcus aureus*, and media recommended for the enumeration of some pathogens (e.g., *Yersinia enterocolitica, Campylobacter jejuni, Salmonella* spp., *Listeria monocytogenes, Clostridium perfringens, Aeromonas hydrophila*) are selective as well as differential agar media. Due to the presence of one or more selective compound(s), they allow selective growth and colony formation of several closely related species. The differential agent(s) then helps to differentiate these species or groups from one another due to their specific colony characteristics.[1-3]

B. Indirect Estimation

1. *Dilution to Extinction in Nonselective Broths*

The method consists of serial dilution of a sample and transfer of an aliquot (usually 1.0 or 0.1 ml) to 5 or 10 ml of a nonselective broth, such as tryptic soy broth. This is followed by incubation of the tubes at a specific temperature for a specific period of time, which depends upon the specific microbial group being investigated. The tubes are then examined for the presence and absence of growth (from the turbidity of the broth). Using highest sample dilution that gave growth and assuming that this tube had one to nine viable cells of the group of interest, microbial numbers per milliliter or gram sample are estimated. The estimated numbers, however, can vary widely from the actual numbers. This method is not used much for the microbiological estimation in food.[1-3]

2. *Most Probable Number (MPN) in Selective Broth*

In this method aliquots from a serially diluted sample are inoculated in a broth (in tubes) having one or more selective agents that facilitate growth of selected microbial groups present in a food. Generally, three or five broth tubes in each dilution and a minimum of three consecutive dilutions are used. After incubation at the recommended temperature and time, the broth tubes in each dilution are scored for the presence and absence of growth. From the number of tubes showing growth in each of the three successive dilutions, the number of viable cells of the specific microbial group can be estimated from the statistically calculated available tables. This method also gives wide variation. MPN methods are quite often used for the estimation of coliforms and fecal coliforms in foods using brilliant green lactose bile broth and E-C broth, respectively.[1-3]

3. *Dye Reduction Test*

The method is based on the principle that some dyes like methylene blue and resazurin, are colored in the oxidized state but colorless under reduced conditions. This change can occur due to microbial metabolism and growth. It is assumed that the rate of reduction during incubation of a specific concentration of Methylene Blue added to a food is directly proportional to the initial microbial load in the food. However, as microbial groups differ in the rate of metabolism and growth and ability to reduce the environment, this method is not considered very accurate or effective with different foods. This method is generally used to determine the microbiological quality of raw milk.[1-3]

C. Enumeration of Injured Microbial Groups by Selective Media

Sublethally injured coliforms and pathogenic bacterial cells, when enumerated by selective agar media, may die due to their developed sensitivity to selective agents in the media. They are first allowed to repair the injury in nonselective media (broth or agar) for a short period, which enables the cells to become resistant to the selective agents. Following repair, they can be exposed to selective media. For the enumeration of coliforms in food that may contain injured coliforms, the diluted sample can first be pour plated in nonselective plate count agar and incubated for 1 to 2 h at 25 to 35°C, thus enabling the repair of the cells. Double-strength violet red bile agar then can be overlaid on the plates and the plates incubated at 35°C for 24 h for selective growth of coliforms.[1-3]

V. QUALITATIVE METHODS TO ISOLATE MICROORGANISMS IN FOODS

A. Isolation of Pathogens

The main objective of this method is to determine if a sample does or does not contain viable cells or spores of a specific pathogen. Foods are tested for several pathogens, including *Salmonella* spp., *Escherichia coli* 0157:H7, *Clostridium botulinum, Listeria monocytogenes, Vibrio cholerae,* and *Shigella* spp., by the isolation procedure when necessary. For other pathogens, such as enteropathogenic *Esc. coli, Yersinia enterocolitica,* and *Campylobacter jejuni*, isolation procedures are not generally used.

An isolation procedure generally contains several steps, such as nonselective preenrichment followed by selective enrichment and then testing on an agar medium containing selective and differential agents. It is assumed that a food normally contains a low population of a pathogen as compared to the associative microorganisms and the pathogens could be in injured state. The food sample (e.g., 25 g) is first preenriched in a nonselective broth and incubated for the injured cells to first repair and then multiply in order to reach moderately high numbers (along with many other associated microorganisms). An aliquot is then transferred from the preenrichment broth to a selective enrichment broth and incubated. It is expected that during incubation, the specific pathogen and related microorganisms will selectively grow to a high number, while many of the associated microorganisms will not grow. A small amount (~0.01 ml) of enrichment broth is then usually streaked on the surface of a selective-differential agar medium plate, which is then incubated for the colonies to develop. From the colony characteristics, the presence of a pathogen can be tentatively established.

This is generally considered a presumptive test. For confirmation, the cells from the characteristic colonies are purified and examined for biochemical reaction profiles and serological reaction against a specific antibody. Isolation of a pathogen using the conventional methods can take 10 to 12 d, depending upon a particular species.[1-3]

VI. TEST FOR BACTERIAL TOXINS IN FOODS

Staphylococcus aureus and *Clostridium botulinum* strains, while growing in foods, are capable of producing toxins and causing intoxication or food poisoning among the consumers. Specific methods have been developed to test for the presence of toxins in the suspected foods.[1-3]

To detect *Sta. aureus* enterotoxin in foods, recommended methods are used, first to extract the toxin from 100 g food and then to concentrate the toxin in 0.2 ml sterile water or 0.2 *M* saline. The presence of toxin in the concentrated extract is then assayed against specific antibody by the microslide precipitation method. This method is sensitive at the level of 0.05 μg enterotoxin per milliliter (food poisoning probably can occur from the consumption of less than 1.0 μg enterotoxin A). Methods such as radioimmunoassay (RIA), enzyme-linked immunosorbent assay (ELISA), and reverse passive latex agglutination assay (RPLA) methods can also be used in place of the microslide precipitation method.

To detect botulin, the toxin of *Clo. botulinum* strains, the recommended procedure is used to extract toxin from a food. To activate toxin of nonproteolytic types (B and E), a portion of the extract is then treated with trypsin. Both trypsinized and nontrypsinized (for proteolytic types A and B), extracts in 0.5-ml portions are then injected into mice intraperitoneally (two mice for each sample). Heated (100°C for 10 min) extracts are injected similarly. The mice are then observed up to 48 h for botulism symptoms and death. Typical botulism symptoms in mice, in sequence, are ruffling of fur, rapid and gasping breathing, weakness of limbs, and death due to respiratory failure.

VII. RAPID METHODS AND AUTOMATION

The traditional or conventional methods used for the quantitative or qualitative detection of microorganisms in foods and toxins take a relatively long time and are labor intensive. To overcome these difficulties, different rapid methods, many of which are automated, have been developed to detect microbial loads, foodborne pathogens, and their toxins.[1,4,5] Some of these methods have been approved by the regulatory agencies. In addition to being rapid, they are quite specific, sensitive, accurate, and less labor intensive. The principles, procedures and applications in food of some of these methods are briefly presented.

A. Immunoassays for Rapid Detection of Pathogens

Several rapid and automated methods have been developed that rely on the specific antigen-antibody reaction and production of agglutination, color formation from chromogenic substrate, formation of an immunoband, or fluorescence.

1. Immunofluorescence Method

In this method, commercially available specific fluorescence-conjugated antibodies (against somatic or flagellar antigens of a pathogen) are mixed with an enriched medium suspected to contain the specific pathogen, such as *Salmonella* (antigen), on a glass slide. Following incubation and removal of reagents, the slide is examined under a fluorescence microscope for cells showing fluorescence on the wall and/or flagella.

2. Reverse Passive Latex Agglutination (RPLA) Method

This method has been developed to detect toxins of several foodborne pathogens, namely *Sta. aureus, Clo. perfringens, Bac. cereus, Vib. cholerae,* and enteropathogenic *Esc. coli*. The antibody of specific toxin is immobilized on latex particles that are then mixed with a sample preparation suspected to contain toxin (antigen) in wells of microtitration plates. If the specific toxin is present in the sample, a diffuse pattern will appear in the bottom; in the absence, a ring or button will appear.

3. Immunoimmobilization Method

This method was developed for motile pathogens such as *Salmonella* and *Esc. coli* 0157:H7. The motile cells and antibody against flagellar antigen are applied on soft agar to enable them to diffuse in the same direction from opposite sides. When the bacterial cells and specific antibody meet, they form a visible arc due to immobilization of cells.

4. Enzyme Immunoassay (EIA) or Enzyme-Linked Immunosorbent Assay (ELISA) Methods

These methods were commercially developed and used for the detection of several foodborne pathogens and their toxins, namely *Salmonella, Listeria monocytogenes,* and *Campylobacter jejuni*. The specific antibody is first allowed to bind on a solid surface (on a wall of a microtitration plate) using the direction of a commercial producer. Blocking agents (bovine serum albumin) are then added to block the additional protein binding sites on the surface. The sample suspected of containing the antigen (pathogens or their toxins) is prepared as per requirement and then added to the well and incubated for the antibody-antigen reaction. After removing the unbound antigen, another antibody labeled with a specific enzyme (such as peroxidase) is added and incubated for its binding to antigen to form a sandwich (antibody-antigen-antibody-enzyme). The unbound enzyme-linked antibody is then removed. The sandwich complex is finally detected by adding a chromogenic substrate specific for the enzyme (such as 4-chloro-1-naphthol for peroxidase), incubating for a specified time, and adding an enzyme inactivator to stop the reaction. The intensity of color development can then be measured to identify the presence of a specific pathogen or toxin. Instead of a chromogenic substrate, a specific fluorogenic substrate (such as 3-*p*-hydroxyphenyl propionic acid for peroxidase) can be used and the reaction can be measured with a fluorimeter. Automated systems for the ELISA test are available.

5. Magnetic Immunobeads Method

The objective of the method is to immunocapture the cells (antigen) of a pathogen present in a food by the help of magnetic immunobeads. The immunobeads are plastic-covered ferrous beads coated with specific antibody. When the beads are mixed with a sample, they capture the specific antigen (cells). The immunocaptured beads are then removed by a magnet and tested for the presence of a target pathogen. This method was developed for *Lis. monocytogenes*.

B. Nucleic Acid Probe for Detection of Pathogens

1. Hybridization Method

In this method, a DNA probe consisting of a 20- to 4000-nucleotide base sequence unique to a group of similar pathogens, such as *Salmonella* spp., is prepared. The unique sequence can be identified in the DNA or rRNA in the cells or from the amino acid sequence of a toxin produced by the bacterial group. The DNA probe can then be obtained from the cell DNA or synthesized from the nucleotides. The probes are radiolabeled with ^{32}P if, after hybridization, autoradiography is used for their detection. In the nonisotopic colorimetric method, an enzyme (e.g., peroxidase) is bound to a specific protein (e.g., Streptavidin) that in turn binds to specific ligands (e.g., biotin) on the DNA probe and can be used to produce color reaction to aid in measurement.

Both isotopic and nonisotopic DNA probes specific for *Salmonella* and *Lis. monocytogenes* are commercially available and approved for use by the regulatory agencies. DNA probes for toxigenic *Esc. coli* and *Yer. enterocolitica* are also available but not yet approved by the regulatory agencies. The commercial companies provide step-by-step procedures for each technique. In general, it involves an initial preenrichment step; followed by an enrichment step (which can be just 6 h or more) to facilitate growth of target bacteria; then lysis of cells, hybridization of the DNA probe with cell DNA or rRNA; capture of the hybridized DNA containing the probe; and measurement of either radioactivity or color production, depending upon the probe (i.e., if it is isotopic or nonisotopic). The methods are specific and relatively more rapid than traditional methods.

2. Polymerase Chain Reaction (PCR) Technique

The PCR technique helps to amplify a segment of DNA located between two DNA segments of known nucleotide sequence. It is thus possible to obtain large numbers of copies of a specific DNA segment from a small sample (50 ng), which in turn facilitates its detection by gel electrophoresis (nonisotopic) or by Southern hybridization (isotopic). The method is very sensitive and may not need preenrichment or enrichment steps. However, to eliminate confusion between dead and viable cells of a pathogen as the source of DNA, a short preenrichment step to increase viable cell number of target pathogen may be necessary. The method has been developed for the identification of *Lis. monocytogenes*, pathogenic *Esc. coli,* and *Vib. vulnificus*.

C. Measurement of Microbial Level

Several indirect methods to estimate microbial levels in a food have been developed. They are relatively more rapid than the traditional cfu enumeration method and some are also automated.

1. Electrical Impedance Measurement

This method is based on the theory that as microorganisms grow in a liquid medium, they metabolize uncharged or weakly charged substrate to produce highly charged small products (e.g., amino acids and lactic acids). This causes a change in the electrical impedance (resistance to flow of an alternating current) and conductance of the medium. By measuring these two parameters, bacterial growth in a medium as a function of time at a given temperature can be monitored. Impedance detection time (IDT) is the time required by microorganisms initially present in a food to reach $\geq 10^6$/ml. It is inversely proportional to the initial microbial load; a higher IDT means a lower initial load. From the IDT and the slope of the curve, the initial load as well as generation time of a microbial population can be calculated. Several automated systems are now commercially available that provide a description of the technique as well as methods for interpretation of the data. Using selective media, the population of a specific bacterial group can also be determined by this method.

2. Bioluminescence Method

The bioluminescence method measures the ATP content in a sample as an indirect measurement of microbial load.[1,6] As only the viable cells retain ATP, the amount of ATP is regarded as directly related to the microbial load in the sample. Using the luciferin-luciferase (from firefly) system in the presence of Mg^{2+}, the ATP concentration in a sample is measured. The method can detect as low as 10^{2-3} viable bacterial cells and about 10 yeast cells per gram or milliliter of food. It can also be used to determine the microbial population on equipment surfaces. The method is very rapid, and several automated systems are now commercially available.

In another method, the genes encoding bacterial bioluminescence (*lux*-gene) in luminous bacteria (e.g., *Vibrio* species) have been cloned in some pathogenic, indicator, and spoilage bacteria important in food. The bacterial luciferase catalyzes oxidation of reduced flavin mononucleotide ($FMNH_2$) and a long-chain aliphatic aldehyde by molecular O_2 and emits light. As only live cells can produce light, the bacterial strains containing the cloned *lux*-gene can be used to determine the effectiveness of methods employed to kill cells or remove in food and food environments very rapidly. It is also being used to detect injured bacteria incapable of producing light, but able to do so following repair of injury. Automated systems to measure light from the reaction are now commercially available.

Many other rapid methods, automated systems, and miniatures of traditional methods are commercially available. Some have been discussed in more detail in the references included and can be consulted.

REFERENCES

1. Vanderzant, C. and Splittstoesser, D. F., Eds., *Compendium of Methods for the Microbiological Examination of Foods,* 3rd ed., American Public Health Association, Washington, DC, 1992.
2. Richardson, G. H., Ed., *Standard Methods for the Examination of Dairy Products,* 15th ed., American Public Health. Association, Washington, DC, 1985.
3. Food and Drug Administration, *Bacteriological Analytical Manual,* 6th ed., Assoc. Offic. Anal. Chem., Arlington, VA, 1992.
4. Fung, D. Y. C., *Rapid Methods and Automation in Food Microbiology,* Marcel Dekker, 1994, 357.
5. Vasavada, P. C. and White, C. H., Rapid methods and automation in dairy microbiology, *J. Dairy Sci.,* 76, 3101, 1993.
6. Stewart, G. S. A. B. and Williams, P., Shedding new light on food microbiology, *ASM News,* 59, 241, 1993.

APPENDIX E

Regulatory Agencies Monitoring Microbiological Safety of Foods in the United States

I. FOOD SAFETY REGULATIONS

Annual sales of food in the United States are over $350 billion. Before foods reach the consumer they are processed and handled by over 25,000 processors; 35,000 wholesalers; 250,000 retailers; 275,000 eating establishments; and in many millions of homes. If one adds to them the various types of pathogens that can cause foodborne diseases and their natural presence in food animals, birds, food producing environments, and among food handlers, it becomes quite apparent that the chances of a foodborne pathogen being present in a food could be high. Yet the numbers of reported foodborne disease outbreaks and the cases per year are relatively low. A major reason for this could be credited to the stringent food safety laws in the United States.[1,2]

Prior to 1906, there were no food laws in this country and the marketing of unsafe food was quite common. In 1906, the federal government passed the Pure Food and Drug Act and the Meat Inspection Act and gave the United States Department of Agriculture (USDA) the power of supervision of the laws. According to these laws, selling of unwholesome and unsafe foods and meat in interstate commerce was made illegal. The Meat Inspection Act also required inspection of the slaughter and meat processing facilities. However, as the burden of proof was with the USDA that a substance had been added to a food at an unsafe level, the laws were not very effective. In 1938, the federal government passed the Food, Drug and Cosmetic Act and gave additional power to the Food and Drug Administration (FDA) to administer the law. This law prevents the manufacture and shipment of unsafe and spoiled food and food ingredients, in interstate commerce as well as for export and import. It permits the sale of foods that are pure, wholesome, safe, and produced in a sanitary environment. According to this law, contamination of certain foods with specific pathogens, such as *Clostridium botulinum* and *Salmonella* spp., is considered to be unwholesome and selling such food is illegal.

In 1959, the Federal Poultry Product Inspection Act was passed. This act requires federal inspection of poultry processing facilities that ship products in interstate commerce.

The Wholesome Meat Act (in 1965) and the Wholesome Poultry Act (in 1968) were passed, requiring inspection of the processing facilities, even those engaged in intrastate commerce. These facilities are to be inspected by state inspectors in the same manner as the processors engaged in interstate commerce and inspected by the federal inspectors. At the federal level, the inspection of meat, poultry, eggs, and milk remains with the Food Safety and Inspection Service (FSIS) branch in the USDA.[1,2]

II. THE AGENCIES

A. Federal Agencies

In the United States several federal agencies are delegated the responsibilities to monitor and regulate the origin, composition, quality, quantity, safety, labeling, packaging, and marketing of foods. Two agencies, the FDA and USDA, are directly involved with the microbiological safety of foods, and their responsibilities in this regard are discussed here.[2] Other agencies are involved in areas not directly related to microbiological safety of foods. Some of these are:

1. Bureau of Alcohol, Tobacco and Firearms (ATF): responsible for enforcing the laws that cover production and labeling of all alcoholic beverages except wine.
2. Department of Justice: In case of a violation of food law, seizures of a product and criminal proceedings are conducted by this department.
3. Environmental Protection Agency (EPA): Determines safety and tolerance levels of pesticide residues in foods and establishes water quality standards for drinking water (not bottled water).
4. Federal Trade Commission (FTC): Regulates the advertising of foods.
5. National Marine Fisheries Service (NMFS): Responsible for seafood quality, habitat conservation, and aquaculture production (not microbiological quality).
6. FDA: At present, the FDA is a part of the Federal Public Health Service in the Department of Health and Human Services. It is delegated the power to administer the federal Food, Drug and Cosmetic (FD&C) Act. In that capacity, the FDA ensures the safety and wholesomeness of all foods, except meat, poultry, and eggs sold in interstate commerce, as well as exported and imported foods. The safety and wholesomeness of foods served on planes, trains, and buses are also inspected by the FDA. It also conducts research to improve detection and prevention of contamination of food, develops regulatory testing procedures, specifications, and standards, and enforces regulations on processing and sanitation of food and processing facilities. The FDA sets up specific microbiological criteria that contain specifications (toleration levels) for aerobic plate counts (APC), coliforms, *Esc. coli*, coagulase-positive *Sta. aureus*, pathogens, and mycotoxins in various foods.

FDA inspectors inspect processing plants once every 2 to 3 years, unless a processor is in violation and a food has hazard potential. The FDA also helps processors develop their effective sanitation programs as well as gives contracts to state regulatory agencies to inspect plants. For foreign countries interested in exporting food to the United States, the FDA sends inspectors to help improve their food quality according to U.S. regulations.

In case of violations, the agency sends formal notice of violation to the offender and sets a date (usually 10 d) by which the problem needs to be corrected. In case of failure, depending upon the severity of the offense, the FDA is authorized to take legal action, through the Department of Justice, that includes seizure of products, destruction of products, and criminal prosecution of the offenders. This includes products produced in the United States as well as imported products.

In addition to microbiological quality and safety of foods, the FDA is also responsible for testing foods for composition, nutritional quality, food additives, food labeling, and pesticide residues (some with other agencies). FDA also conducts market basket survey to test certain number of foods (imported and produced in the United States) for safety and other qualities.

7. USDA: Under the Wholesome Meat Act and the Wholesome Poultry Product Act, the FSIS is responsible for inspection of slaughtering facilities of food animals and birds, and the production of wholesome and safe meat, poultry, eggs, and egg products that are sold in interstate commerce or imported. The FSIS also sets up microbiological specifications, particularly pathogen levels, for some of these products. It also conducts research on food safety and quality.

   Based on eating quality, the USDA also grades vegetables, fruits, milk and dairy products, and meat and egg products either produced in the United States or imported.
8. Centers for Disease Control (CDC): The CDC is a branch of the Department of Health and Human Services. In the event of a foodborne disease outbreak, this agency investigates the causative agent of the outbreak as well as the sequence of events that resulted in the outbreak.
9. U.S. Army: The Wholesomeness and Safety of Foods (ration) consumed by U.S. military personnel are the responsibility of the Department of Defense. The microbiology branch at Natick Army Research facility has established necessary microbiological specifications for APC, coliforms, *Esc. coli*, pathogens, and other microbial groups for different types of foods procured from outside suppliers.

B. State and Local Government Agencies

Some of the branches in the State Department of Agriculture and Public Health are responsible for the safety of food sold in the state. They cooperate with the federal government to ensure the wholesomeness and safety of foods produced and served in the state. State inspectors inspect restaurants, retail food stores, dairies, grain meals, and processing facilities on a regular basis. Some states have authority over inspection of the qualitiy of fish and shellfish taken from state waters. These agencies, if necessary, can embargo illegal food products sold in the state. Federal agencies provide guidelines, when necessary, to the state agencies for the regulations.

C. International Agencies

The Food Standard Commission of the Joint Food and Agricultural Organization and World Health Organization (FAO/WHO), in cooperation with different nations, helps develop international standards in the production, processing, and preservation of foods exported and imported. Codex Alimentarius (CA) is an international food regulatory agency formed by different nations. It helps develop uniform food standards for all countries to ease export and import of foods between countries.

REFERENCES

1. Anonymous, New bacteria in the news: a special symposium, *Food Technol.,* 40(8), 16, 1986.
2. Anonymous, A legislative history of the Federal Food, Drug, and Cosmetic Act, *U.S. Government Printing Office*, Washington, DC, 1979.

INDEX

A

B

C

D

E

F

G

H

J

K

L

M

N

O

P

Q

R

S

T

U

V

W

X

Y